Max Knies

The Relations of the Eye and its Diseases to Diseases of the Body

1894

Max Knies

The Relations of the Eye and its Diseases to Diseases of the Body
1894

ISBN/EAN: 9783337779115

Printed in Europe, USA, Canada, Australia, Japan

Cover: Foto ©berggeist007 / pixelio.de

More available books at **www.hansebooks.com**

THE RELATIONS

OF THE

EYE AND ITS DISEASES

TO THE

DISEASES OF THE BODY

BY

MAX KNIES, M.D.

Professor Extraordinary at the University Freiburg

FORMING A SUPPLEMENTARY VOLUME TO EVERY MANUAL AND TEXT-
BOOK OF PRACTICAL MEDICINE AND OPHTHALMOLOGY

EDITED BY

HENRY D. NOYES, A.M., M.D.

Professor of Ophthalmology and Otology in Bellevue Hospital Medical College ; Executive Surgeon to the
New York Eye and Ear Infirmary ; recently President of the American Ophthalmological Society ;
recently Vice-President of the New York Academy of Medicine ; Permanent Member of the
New York State Medical Society; Member of the American Medical Association, etc., etc.

NEW YORK
WILLIAM WOOD & COMPANY
1894

CONTENTS.

CHAPTER II.

CHAPTER III.

CHAPTER IV.

vi CONTENTS.

THE RELATIONS OF THE EYE AND ITS DISEASES TO THE DISEASES OF THE BODY.

CHAPTER I.

DISEASES OF THE NERVOUS SYSTEM.

THE relations between the eye and nervous system are mutual. It is rare to find eye diseases the starting-point for diseases of the nervous system. Much more frequently diseases of the nervous system, especially of the central organs, cause disorders of the eye. These may be of a functional character or may be objectively visible, such as stasis, inflammation and atrophy of the optic nerve, exophthalmus, spasm and paralysis of the ocular muscles, etc. The functional disturbances of the eye often constitute important local symptoms and are indispensable in the local diagnosis of brain diseases. On the other hand the changes visible with the ophthalmoscope in the optic nerve, retina, vitreous body, etc., are of a more general character, and usually enlighten us rather as to the character than the location of the disease of the brain.

Before entering into details it is necessary to trace, as accurately as possible, the anatomical connection between the eye and the nervous system. This is effected by means of nerves, blood-vessels and lymphatics; in addition, the cerebral meninges pass directly into the sheath of the optic nerve and indirectly into the coverings of the eye. In our description we shall consider the structures in the order named.

1

A. Anatomical Course of the Nerves of the Eye.

We are to consider the optic nerve, the nerves of the ocular muscles, the trigeminus, facial, and sympathetic nerves.   Of these by far the most important is

1. THE OPTIC NERVE.

This collects the centripetal and centrifugal fibres from the nerve-fibre layer of the retina, undergoes partial decussation in the chiasm

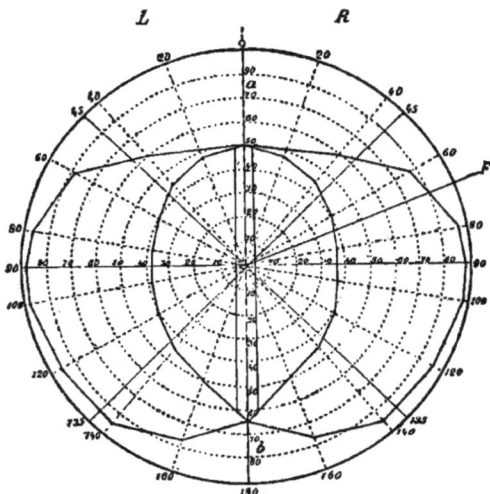

FIG. 1.—Field of Vision of Both Eyes (Schematic). L, Left; R, right half of the field of vision, divided by the vertical line ab which passes through the point of fixation F. The vertical strip is the overlapping portion of the field of vision.

with the nerve of the opposite side, and sends fibres through the optic tracts to both cerebral hemispheres.   The fibres from the inner (nasal) half of the retina (corresponding to the outer [temporal] half of the field of vision) form the larger part of the nerve and undergo decussation.   The fibres from the outer half of the retina, and which answer to the nasal side of the field, pass without decussation to the cerebral hemisphere of the same side.   The optic tract thus contains all the fibres from the halves of the retinæ on the same side, the right tract containing the fibres of the right retinal halves (the temporal half of the right eye and the nasal half of the left eye). Both together supply the left half of the field of vision, so that the right tract corresponds to the binocular left half of the field of vision and *vice versa.* Hence there is complete decussation so far as regards what is seen to the right and left.

But this is only the general course, and the conditions in detail are much more complicated. In the first place it is to be noted that the line of separation between the two halves of the field of vision, which passes vertically through the point of fixation (not through

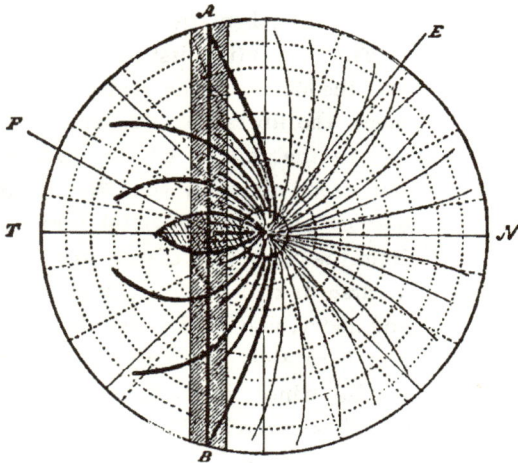

Fig. 2.—Course of the Fibres in the Right Retina, Seen from in Front. *E*, Optic nerve entrance; *F*, fovea centralis; *T*, external (temporal); *N*, internal (nasal) side. From the strip *AB*, corresponding to the overlapping portion of the field of vision, spring crossed (light lines) and uncrossed (heavy lines) fibres.

the entrance of the optic nerve) is usually not a sharp line. As shown in Fig. 1, fibres from both halves of the field of vision pass from one side to the other in a strip which may be 10° in width (overlapping part of the field of vision). Hence, in loss of the region of destination of one optic tract, the boundary of the visual field usually passes to the inside or outside of the point of fixation, and the latter remains intact in both eyes. Fig. 2 gives the probable distribution of the fibres of the optic nerve in the retina of the right eye. E indicates the point of entrance of the optic nerve, F the point of fixation, A B the vertical strip, including the point of fixation, which probably sends fibres to both halves of the brain.

In some cases of hemianopsia the boundary passes exactly through the point of fixation, in others there is merely a slight projection of the intact half of the field of vision at the point of fixation and enclosing the latter. In still other cases the boundary is more or less oblique or not perfectly homonymous, *i.e.*, somewhat incongruent in both eyes. The causes of these differences are unknown, and can only be conjectured. It is evident that we have to deal, at least in part, with individual peculiarities. We will return to this subject again.

The fibres which start from the macula form, as shown in Fig. 2, a triangular bundle on the outside of the optic papilla. As can be seen in the schematic sections of the optic nerve in Fig. 3, *a d* (the first immediately behind the eye, the last from the region of the optic foramen), the fibres from the macula pass gradually to the middle of the optic nerve. The same sections show that the uncrossed fibres (denoted by the horizontal lines) which come from the outer half of the retina are arranged at first in two crescentic bundles on the inside and above, and on the outside and below; farther back they gradually coalesce into a single bundle, which, in the neighborhood of the chiasm, lies to the inside and a little inferiorly.

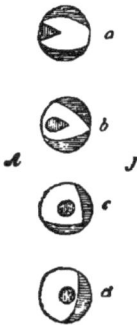

FIG. 3.—Transverse Section of the Right Optic Nerve, Seen from in Front; *a*, in the region of the lamina cribrosa; *b*, in the optic foramen. The macular fibres are shaded vertically; the uncrossed (according to Schmidt-Rimpler), horizontally. *A*, external; *J*, internal.

The course of the macular fibres in the optic nerve was first demonstrated by Samelsohn (*Arch. f. Ophthal.*, XXVII, 1), and subsequently confirmed by Nettleship, Vossius, Uhthoff, Bunge. The course of the fibres of the outer half of the retina is given according to Schmidt-Rimpler (*Arch. f. Augenheilk.*, 1888, p. 296), who was enabled to demonstrate them in a case of "cortical hemianopsia with secondary optic atrophy." These agree with Wernicke's statements ("Lehrb. d. Gehirnkhtn.," I, p. 75). According to Siemerling (*Arch. f. Psych. u. Nervenheilk.*, Vol. XIX) the uncrossed fibres lie within the optic nerve to the outside (laterally). This writer agrees with the majority of other observers, and also with my own observations. It is evident that in this respect there are also individual peculiarities.

A decussation of the fibres coming from the larger inner halves of the retinæ takes place in the chiasm. There is partial decussation of the fibres from the fovea centralis and also of those coming from the previously mentioned (Fig. 2) vertical strip which is common to both halves. The fibres from the outer half of the retina run a direct (uncrossed) course, chiefly in the outer and upper part of the chiasm. In addition there are usually manifold intertwinings of the individual nerve bundles, which are very imperfectly known, and interfere notably with the study of the course of the fibres. These evidently constitute individual peculiarities.

According to Siemerling (*l. c.*) the uncrossed bundle in the chiasm is situated to the side, in the anterior part on the lower (ventral) surface and gradually passing upward (dorsal). According to F. Bernheimer (*Arch. f. Augenheilk.*, XX, 1, p. 133), this bundle passes along the upper half of the chiasm. In one case I was able to ascertain that immediately behind the chiasm, it lay to the inside in the optic tract. As a general rule it is evident that the direct bundle from the outer half of the retina passes through the optic nerve and chiasm in such a way that immediately behind the latter it reaches a position to the outside and above in the optic tract.

The macular fibres decussate only in great part and are situated centrally at the beginning of the optic tract as well as in the deeper part of the optic nerve.

In addition the chiasm contains two systems of fibres which pass from one optic tract to the other and have no connection with the eye, but connect homonymous parts of the right and left halves of the brain. These form the commissure of Gudden, which is about one-third the mass of the entire chiasm, and is situated posteriorly and above, while the much smaller Meynert's commissure is situated in the upper part of the chiasm. The bundles in these commissures, especially in Gudden's, frequently mingle at their borders with adjacent visual fibres, and then a sharp anatomical differentiation cannot be made.

The course mentioned corresponds in the main to the investigations of Gudden (*Arch. f. Ophth.*, XXV) and at the present time probably finds general acceptance. A very few writers, Michel in particular, still support the notion of the total decussation of the fibres of the optic nerve in the chiasm. Michel's views, last presented in detail in his treatise, "Ueber Sehnervendegeneration und Sehnervenkreuzung," Würzburg, 1887, have recently been attacked violently by Darkschewitsch (*Arch. f. Ophth.*, XXXVII, 2, p. 1).

As we shall see later, physiological and clinical data are entirely incompatible with the assumption of a total decussation of the sensory optic nerve fibres in man. The best proof of partial decussation is furnished by Weir Mitchell's case (*Journal of Nervous and Mental Disease*, 1889, p. 44), in which an aneurism had entirely divided the chiasm in the median line. The outer half of the retina of each eye retained the power of vision, while complete blindness would have resulted in the event of total decussation in the chiasm.

In man, Gudden's commissure contains fibres which appear exactly the same as those of the remainder of the chiasm. In the rabbit the former are finer than the others, and in the mole they are the only ones present (Wernicke). The fibres of Meynert's commissure are larger than those of the remainder of the chiasm.

In the optic tract the fibres of the crossed bundles, according to Wernicke, pass below the others; those of the direct bundles pass through the middle, forming, on section, a transverse band through the entire diameter of the tractus. According to Siemerling (*l. c.*), on the other hand, the direct bundle is central and at no place does it reach the periphery. In the vicinity of the chiasm and in the anterior half of the tractus the crossed and uncrossed fibres may run a more or less separate course, but farther posteriorly their complete intermingling is a physiological necessity. Upon entering the cerebral ganglia they are intermingled in such a way that every part of the tractus represents corresponding parts in both retinæ. The macular fibres can only be occasionally demonstrated close to the chiasm as a central distinct bundle (Uhthoff, *Arch. f. Ophth.*, XXXII, 4), farther backward they are not distinguishable. An excellent *résumé* of the views of different investigators on the course of the fibres in the optic nerve, chiasm and optic tract is found in Wilbrand's work, "Die Hemianopischen Gesichtsfeldformen," Wiesbaden, 1890, p. 48 *et seq.*

It has been positively proven that the visual fibres of the optic tract have a threefold termination. The main part enters the external geniculate body; other bundles pass to the pulvinar, thalami optici and to the anterior corpora quadrigemina. These three structures are called the primary optic ganglia. From all three can be traced bands of fibres which pass to the posterior third of the internal capsule (between the lenticular nucleus and optic thalamus) and thence proceed as Gratiolet's optic radiations to the cortex of the occipital lobe, especially to its posterior parts and to the cuneus. Fig. 4 gives a schematic illustration, according to Wernicke, of the principal conditions.

The arm (brachium) of the anterior corpora quadrigemina consists of fibres of the optic tract passing to this ganglion, and of the fibres of the corona radiata which pass from it to the cortex of the occipital lobe. In addition it furnishes a few fibres to the posterior commissure of the brain.

From the anterior corpora quadrigemina pass (or e n t e r) numerous fibres which decussate in the roof of the aqueduct of Sylvius and then enter the so-called fillet which passes in the tegmentum of the cerebral peduncle to the optic thalamus. Meynert has also de-

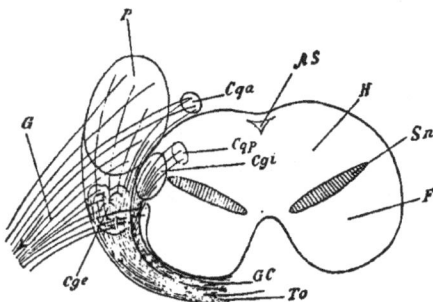

Fig. 4.—Schematic Diagram of the Termination of the Tract, after Wernicke. *To*, Optic tract; *GC*, Gudden's commissure of the same; *Cge*, external geniculate body; *Cgi*, internal geniculate body; *Cqa*, anterior corpus quadrigeminum; *Cqp*, posterior corpus quadrigeminum; *P*, pulvinar of the optic thalamus; *G*, fibres of the corona radiata of the occipital lobe (Gratiolet's visual fibres); *AS*, aqueduct of Sylvius; *H*, tegmentum; *Sn*, substantia nigra; *F*, pes pedunculi.

monstrated association fibres to the nuclei of the ocular muscles (Meynert's fibres).

Gudden's commissure connects both internal geniculate bodies and the posterior corpora quadrigemina and sends a bundle directly into the corona radiata of the occipital lobe on the same side. A free decussation of fibres also takes place between the posterior corpora quadrigemina; these also send a bundle of fibres to the fillet. To judge from its terminations Gudden's commissure seems to be connected with hearing, and in a certain measure to constitute an auditory chiasm. It remains intact after the loss of both eyes. This view is also favored by the fact that the ganglion cells of the internal geniculate body degenerate after extirpation of the cortex in the temporal lobes (Monakow).

 · Meynert's commissure lies only upon the median half of the tractus, enters the cerebral peduncle from below, and is lost in the latter.

It is arranged in loose bundles and consists of larger fibres, with a greater amount of medulla than the others in the chiasm. This commissure is probably connected with the so-called Luys' bodies, which are situated immediately below the optic thalamus.

The roots of the tractus from the internal geniculate bodies, Luys' bodies, etc., mentioned by Hernheimer ("Ueber die Sehnerven-wurzeln des Menschen," Wiesbaden, 1891) may be referred with certainty to these two commissures and hence possess no direct relations to the visual organ. Other tractus roots, which may even be traced into the spinal cord (Stilling) are still very uncertain anatomically, but their existence, as we shall see later, is a physiological necessity. It is not necessary, however, that such fibres, or those which pass to the cerebellar peduncles, etc., must run their course in distinct bundles. The opposite condition appears to hold good, and hence such fibres escape direct anatomical demonstration. The latter would be very desirable in all cases, but is not absolutely necessary to prove the existence of connecting fibres.

Thus far it is possible to follow the visual tracts directly, but their further course to the cerebral cortex must be inferred clinically and experimentally. It is certain that adjacent convolutions of the cerebral cortex are connected with one another by systems of so-called association fibres which lie directly beneath the gray matter, and that between all the cerebral lobes of one side there are other association fibres, whose bundles run more deeply in the medullary substance. The latter include the gyrus fornicatus and uncinate gyrus. The "visual sphere" of the occipital lobe, into which Gratiolet's optic radiations enter, is thus connected by associating fibres with the cortex of the temporal, parietal, and frontal lobes of the same side. Great importance attaches to the connection of the left occipital lobe with Broca's convolution of the left frontal lobe on account of the relations of the latter to speech. For purely topographical reasons this must run its course deep within the inferior parietal convolution. Inasmuch as there are also connections between homonymous parts of the cortex of the two hemispheres by means of the different commissural systems (of which the corpus callosum is the most important), it may be maintained that every part of the cortex stands in direct or indirect mutual connection with every other part, although our knowledge of the precise course of these fibres leaves much to be desired.

The association and commissural systems of fibres are simply connections of the different parts of the cortex with one another. The centripetal and centrifugal connection of the cortex with the rest of the brain is effected solely by the corona radiata, which are collected, from and to the entire cerebral cortex, in the internal capsule between the lenticular nucleus and optic thalamus. Those fibres which pass to and from the occipital cortex (Gratiolet's optic radiations) are situated in the posterior third of the posterior limb of the internal capsule, immediately adjacent to the centripetal sensory tract.

In order to juage correctly the experimental investigations on the course of the fibres in the brain—the results of which have been made to lead to some extent to antagonistic conclusions—the following points must be kept in mind. According to the best recent investigations two ganglion cells are never connected directly by an axis-cylinder process, but the axis cylinders which originate from ganglion cells in one place are resolved, at the other extremity, into a fine network, in which lie other ganglion cells, and these in turn again send out axis cylinders. The other processes (viz., the protoplasmic) of ganglion cells likewise do not pass directly into one another, but are merely intertwined in the most complex manner. Not all ganglion cells give origin to an axis cylinder, but there are some whose processes terminate in a fine network of fibres (Golgi's ganglion cells of the second order).

The connective-tissue neuroglia is found between the nerve fibres of the brain and peripheral nerves. The ganglion cells and the interlacing network of finest fibres (neuropilemma of His) in the cerebral cortex lie in a fine supporting network (neurospongium of His) of originally epithelial origin.

If a nerve is permanently divided or destroyed in the adult, its fibres degenerate as far as the finest terminal network. The degeneration also involves those ganglion cells whose axis-cylinder processes pass through the corresponding nerve. It is only at a late period, and then not constantly, that other groups of ganglion cells and systems of fibres become affected. In my opinion, cases of the latter kind are those in which a more or less inflammatory degenerative process extends along the preformed tracts of the nerve fibres, *i.e.*, we have to deal with a complication. The less the degree of secondary degeneration the more important and correct are the results of the experiment or of the clinical observation.

The conditions are more complicated when the cerebral cortex is destroyed. Apart from the supporting meshwork (neurospongium) this contains:

1. The mesh of nerve fibres (neuropilemma):

    a, of the centripetal fibres of the corona radiata;

    b, of the fibres of the commissures and associative systems which break up at the part in question;

    c, of those processes of the ganglion cells which are not axis cylinders.

    The neuropilemma contains:

2. Ganglion cells:

    a, whose axis cylinders enter the associative fibres and the commissures and break up into a network in other parts of the cortex;

    b, whose axis cylinders run a centrifugal course in the corona radiata; they are generally characterized by their unusually large size;

    c, which only form a network (Golgi's ganglion cells of the second order).

All these elements are found in every part of the cerebral cortex, but in varying proportions in different localities. For example, the large (motor) pyramidal cells, whose axis cylinders conduct centrifugally, are very abundant in the motor parts of the cortex, but are much more scanty in the sensory parts to which the occipital cortex also belongs. From this fact alone it may be maintained that no single part of the cerebral cortex is purely sensory or purely motor, *i.e.*, sends out only centripetal or only centrifugal fibres. Centrifugal fibres are present even in the corona radiata of the occipital lobe, and may even be demonstrated in the optic nerve. It is true that their course is not demonstrable anatomically, but it may be inferred with tolerable certainty.

If a certain region of the cerebral cortex is destroyed, degeneration takes place, first, in the centripetal fibres of the corona radiata, in the associative system of fibres and the commissures whose terminal networks are destroyed, together with the ganglion cells from which they take their origin as axis-cylinder processes; and, second, in all centrifugal (i.e., with reference to the cortical region in question) fibres of these three systems, whose ganglion cells of origin are destroyed, together with their terminal networks in other parts of the brain. In the latter regions those ganglion cells which are merely surrounded by the degenerating terminal networks remain intact.

While the division of a nerve is a comparatively simple experiment, in destruction of a part of the cerebral cortex every other part of the cortex will be implicated, entirely apart from the destruction of the adjacent lower groups of ganglion cells. Other parts are not implicated in a uniform manner, but according to the intimacy of connection with the destroyed region.

In associative and commissural fibres degeneration can commonly be demonstrated only when they are arranged in separate bundles. The degeneration, which affects only certain elements of the neuropilemma, cannot be discovered by the microscope, except in very recent cases.

After division or degeneration of the optic nerve in an adult, atrophy takes place, first, in the ganglion-cell layer of the retina (ganglion optici), whose axis-cylinder processes constitute the main part of the fibres of the optic nerve; and, second, in a group of ganglion cells in the superficial gray matter of the anterior corpora quadrigemina, perhaps of both sides but at all events chiefly on the opposite side (Monakow). Hence, the latter must contain centrifugal axis-cylinder processes destined to the optic nerve.

In the other so-called primary optic ganglia, viz., the external geniculate body, the zonal stratum of the optic thalamus (the pulvinar) and the remaining parts of the anterior corpora quadrigemina, we find merely atrophy of the medullary layer, while the ganglion cells remain intact. Whether single ganglion cells are destroyed in other localities, cannot be determined.

In destruction of *both* optic nerves or of the *chiasm* the lesions are bilateral; in destruction of *one tractus* the above-mentioned ganglion cells of the anterior corpus quadrigeminum on the same side and the ganglion cells of both retinal halves on the same side will undergo atrophy. In addition certain groups of ganglion cells which belong to Meynert's and Gudden's commissures also undergo atrophy after destruction of the chiasm or of one tractus.

What has been stated is true of non-progressive processes and of experiments performed as aseptically as possible. If the degenerative process (perhaps it may be infectious) spreads farther along the atrophic nerve fibres, the ganglion cells of the primary optic ganglia may also be destroyed in part or entirely, and then the fibres of the corona radiata may undergo degeneration. Indeed, the atrophy may even be noticeable as far as the occipital cortex (cases reported by Huguenin, Nothnagel, Kowalewski, Giovanardi, Tomaschewski, etc.). This is undoubtedly very rare in adults. It is possible, however, that after atrophy of the optic-nerve fibres of very long standing, visible atrophy will finally occur in the more central parts, because their function is not sufficiently exercised. But the assumption of a progressive process is more probable even in such rare cases. When, after atrophy of the optic nerve, atrophy has also been found at a comparatively early period in the primary optic ganglia (for example, in tabes and the like), we cannot reject the notion that these ganglia may have been affected primarily and that the optic nerve, whose centripetal neuropilemma was destroyed, has undergone secondary atrophy. Moreover, both may have been attacked at the same time. Compare, for example, Richter, *Archiv für Psychiatrie und Nervenheilkunde*, XX, p. 504.

From the statements made it follows that the optic nerve contains at least two different kinds of fibres, viz., those which are axis-cylinder processes of the ganglion-cell layer of the retina (ganglion optici), and those whose cells of origin are situated in the anterior corpora quadrigemina, possibly also in other parts of the brain. Two kinds of fibres may also be distinguished anatomically in the optic nerve, viz., finer and coarser ones; both are present in approximately equal numbers. The former are the axis-cylinder processes of the ganglion cells of the retina, the latter are derived in great part from the anterior corpora quadrigemina and break up into a fine network, particularly in the internal granular layer (ganglion retinæ). It is an interesting fact that in the eyes of mussels these two parts of the optic nerve pass to the retina as separate bundles (*vide* Rawitz, *Jenaische Zeitschrift f. Naturwiss.*, Bd. XXII and XXIV).

Hence it follows necessarily that either the axis-cylinder processes conduct centripetal and centrifugal stimuli or that conduction in the optic nerve takes place in two opposite directions. The former appears improbable, so that the second assumption must be the correct one, *i.e.*, the optic nerve contains in approximately equal numbers coarser fibres which conduct centrifugally and finer ones which conduct centripetally. It is possible that a nerve fibre may conduct in both directions, but such an assumption with regard to the optic nerve is unnecessary under existing circumstances.

It is to be noted that the optic nerve may not be compared unreservedly with a peripheral sensory nerve. The retina is an outlying part of the cerebral cortex. Hence the optic nerve might, with equal propriety, be interpreted as a system of association fibres between two parts of the brain, which conduct stimuli in both directions and everywhere constitute a mixture of afferent and efferent axis-cylinder processes.

The number of fibres in the optic nerve is variously estimated : by Salzer at a little more than 400,000 ; by Krause at this number of coarse fibres and an equal number of finer fibres ; by Kuhut, at about 40,000 (*i.e.*, only one-twentieth of Krause's figures). At all events the number is considerably less than that of the retinal cones which are estimated by Salzer at about three and one-third millions ; *i.e.*, about eight bacilli to one centripetal fibre of the nerve.

There does not appear to be any dichotomous division of the separate fibres in the optic nerve, but there are numerous intertwinings and anastomoses between the different bundles. According to Michel this is particularly true in the retina and also as they approach the chiasm.

When the occipital cortex is destroyed, its coronal fibres atrophy, together with the majority of the ganglion cells of the external geniculate body, of the stratum zonale of the optic thalamus, and those cells in the anterior corpus quadrigeminum (Monakow) which remain intact after division of the optic nerve. The same ganglion cells also degenerate when the corona radiata of the occipital cortex alone is divided. These groups of cells evidently send their axis-cylinder processes to the occipital cortex, where they end in a fine network. Those fibres of the tractus which pass to the primary optic ganglia also terminate there in a network. In part, however, they merely pass through or alongside the network, and it is impossible to demonstrate their further anatomical course.

Destruction of the occipital cortex destroys a large number of ganglion cells which send axis cylinders into the association fibres and cerebral commissures, and likewise the terminal networks of axis cylinders which are derived from the cells of other parts of the brain. All divided fibres and the corresponding ganglion cells will be destroyed. It is only in the immediate neighborhood of destroyed cortical regions that the association fibres are grouped in imperfect bundles. Monakow observed degeneration of association fibres only in an anterior direction, toward the parietal lobes.

Hence, the distinctly visible atrophy after destruction of the cortex of one occipital lobe is confined to the corresponding part of the corona radiata and to the ganglion cells of the three primary optic ganglia on the same side, with the exception of those in the anterior corpora quadrigemina, which send out centrifugal fibres to the optic nerve. The degeneration cannot be traced farther into the periphery, nor is atrophy of the disc visible with the ophthalmoscope, even after the lapse of many years. It is only in a few cases that visible changes are finally seen in the tractus, chiasm, and optic nerve. Such rare cases are probably due to the extension of a degenerative, inflammatory process which, after destruction of the ganglion cells, also attacks the terminal network of the optic-nerve fibres in the primary optic ganglia, and then leads to degeneration of the fibres of the tractus and optic nerve, and later the ganglion cells of the retina. This is analogous to the previously mentioned cases (page 12), in which destruction of the optic nerve causes visible atrophy in the occipital cortex (vide Monakow, *Corr. Blatt f. Schweizer Aerzte*, 1, VI, 88, p. 346).

All the statements hitherto made hold good only in regard to adults. In the new-born the conditions are entirely different. In them it is evident that many combinations of fibres, which exist in the adult, have not yet developed. The investigations of Bernheimer showed that in the new-born the optic nerve and chiasm contain only a limited number of medullated fibres. According to Flechsig's well-known assumption, they are only partly capable of function. If one eye or both eyes are destroyed at this period the degenerations are much more extensive. The ganglion cells of the primary optic ganglia degenerate almost completely. In the cortex of the occipital lobe much less is noticeable, very probably because, at this time, it is insufficiently connected with the primary optic ganglia and to a certain extent is still "indifferent." In the further development of the brain the occipital cortex appears to assume other functions, probably on account of the more marked development of the system of association fibres, and the entire distribution of the cortical regions becomes different. It is probable that the auditory and particularly the tactile sense acquires a much larger cortical area. But these are merely assumptions which follow in part from the observations of cases of congenital blindness.

To make a brief *résumé*, we find in the optic nerve two kinds of fibres in approximately equal numbers: *a*, narrow centripetal axis cylinders from the cells of the ganglionic layer of the retina, which terminate, in great part, in a fine network in the three primary optic ganglia; and *b*, thicker centrifugal axis-cylinder processes from the ganglion cells of the anterior corpora quadrigemina, which spread out in the internal granular layer of the retina. The remaining ganglion cells of the primary optic ganglia send their axis-cylinder processes (in the posterior third of the posterior limb of the internal capsule, immediately adjacent to the centripetal sensory tracts of the corona radiata) to the cortex of the occipital lobe, especially to the cuneus. The association fibres of the occipital cortex mainly pass forward. [The investigations of Henschen ("Klinische und anatomische Beiträge zur Pathologie des Gehirns," Upsala, 1892, zweiter Theil) have narrowed the region of primary and direct cortical visual impressions to the calcarine fissure which is the inferior boundary of the cuneus.—N.]

2. Course of the Motor Nerves.

The motor nerves connected with the eye are: the motor oculi communis, trochlearis, abducens, facial, and, to a certain extent, the sympathetic. The trochlearis and abducens each supplies a single external ocular muscle, the former the superior oblique, the latter the external rectus. The motor oculi innervates the remaining external ocular muscles and the levator palpebræ superioris. In addition it contains the fibres for the sphincter pupillæ and the ciliary muscle. The facial innervates the muscles belonging to the eye which are situated outside of the orbit, particularly the orbicularis palpebrarum. The portion of the facial nerve which supplies this muscle and the frontal muscle is often called the ocular facial. The

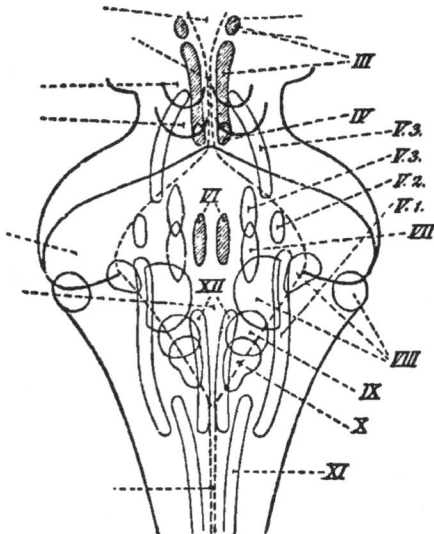

Fig. 5.--Nerve Nuclei of the Motor and Sensory Cerebral Nerves (after Magnus).

sympathetic contains fibres for the dilator pupillæ [if such a muscle exists—N.], the so-called Mueller's muscle, which moderately dilates the palpebral fissure, and for muscular fibres which lie in the fascia that closes the inferior orbital fissure; all of these are smooth muscular fibres. As a matter of course there are also nerve fibres for the muscular coats of the vessels.

The paths of the motor nerves from the eye to the brain lead, apart from the sympathetic, to nuclei in the neighborhood of the aqueduct of Sylvius and in the floor of the fourth ventricle. Fig. 5 gives a schematic representation of their approximate grouping (after Magnus).

a. The motor oculi communis passes from its nucleus through the cerebral peduncle (posterior longitudinal bundle and red nucleus) in a wavy and scattered course, and emerges from the innermost bundles of the peduncle, immediately in front of the pons. It passes between the posterior cerebral and superior cerebellar arteries obliquely outward and forward, and then to the upper and outer wall of the cavernous sinus, where it receives a few filaments from the carotid plexus. Dividing into two branches it then passes through the superior orbital fissure at the outer side of the optic nerve. The upper, smaller branch supplies only the levator palpebræ superioris and the superior rectus, the other supplies the remaining muscles, the internal and inferior recti and the inferior oblique.

The nucleus of the motor oculi lies between the posterior longitudinal bundle (the tegmentum of the peduncle) and the aqueduct of Sylvius, and extends from the posterior extremity of the third ventricle beneath the anterior corpora quadrigemina to a point beneath the posterior corpora quadrigemina. It consists of an accumulation of multipolar ganglion cells, some of which are closely aggregated, others are more scattered, and those from each side come in contact on the median line. Some of the ganglion cells are situated between the fibres of the posterior longitudinal bundle. Shortly after emerging from the lower and outer side of the nucleus the root fibres anastomose freely with one another. The constitution of this nucleus is very complicated, and in the main we follow the description of Perlia, who has published the most recent and exhaustive account of it (*Archiv f. Ophthalmologie*, XXXV, 4, p. 289 *et seq.* Fig. 6 follows his scheme).

We may distinguish an anterior smaller (1–2) and a posterior larger portion (3–8). In the larger nuclear mass is a central group in which the ganglion cells from both sides meet beneath the aqueduct of Sylvius (Perlia's central nucleus 8) and a lateral division (4–7). In the latter Gudden has distinguished an upper (dorsal) portion situated a little to the outside (4, 5), and a lower (ventral) portion, situated to the inside (6, 7). Perlia has rendered it very probable that each of these parts is again subdivided into an anterior and posterior group. To these five groups may be added another (3), first seen by Edinger, and accurately described by Westphal, who distinguished a central and a lateral portion. It has approxi-

2

mately the shape and position shown in the scheme, and its ganglion cells are smaller than those of all other parts of the nucleus.

In the anterior smaller division two groups are recognized on both sides: a median (2) and a lateral (1) group. The latter runs ob-liquely outward and was first described by Darkschewitsch. Ac-cording to him (*Arch. f. Anat. u. Phys.*, 1889, p. 107) this upper nucleus (presumably of the sphincter pupillæ and the muscle of accommodation) pos-sesses smaller cells and sends out finer fibres than the other cell groups (with the ex-ception of the Edinger-Westphal nucleus 3). The fibres of this nu-cleus are said to pass along the posterior cerebral commissure in its more ventral por-

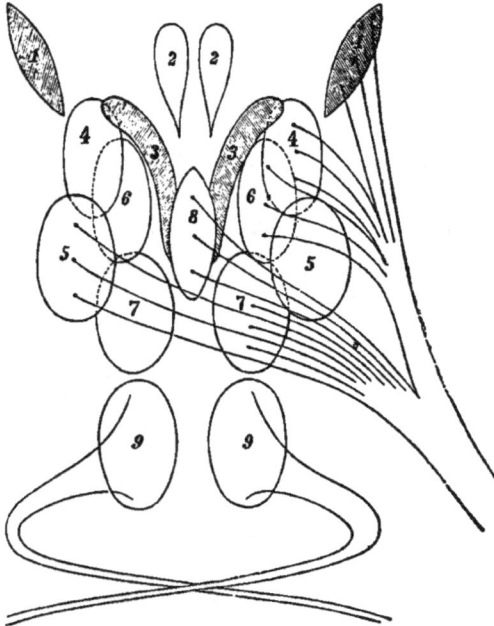

Fig. 6.—Schematic Diagram of the Nuclei of the Oculomotor and Trochlear Nerves.

tions, and become medullated at an earlier period than the dorsal portion. They are said to be derived in part from the pineal gland, whose inferior medullary layer they constitute; fibres from the loops of the lenticular nucleus are doubtful. According to Darksche-witsch the afferent fibres undergo partial decussation. According to Spitzka (*Centralbl. f. Nervenheilk.*, 1889, p. 105) they do not. The anterior lateral nucleus (1) is connected with the main group by bands of fibres; there is also an exchange of fibres with the anterior median double nucleus (2), which lies deeper and more centrally.

The Darkschewitsch nucleus (1) is also said to derive fibres from the region of the brachium of the anterior corpora quadrigemina (tractus fibres).   The complexity of the oculomotor nucleus implies a corresponding richness in the number of its afferent fibres.   We find a thick meshwork of fibres which start from it or go to it, curving upward around the aqueduct through its gray matter.   A prominent addition to the motor oculi nucleus is formed by the fibres of the posterior longitudinal bundle of the tegmentum (sensory fibres from the opposite side of the body) which terminate here in great part (in the neuropilemma).   According to Flechsig and Edinger the fibres of the posterior longitudinal bundle, in a fœtus of nine months, are medullated, and therefore capable of function only in so far as they terminate in the motor oculi nucleus, but not farther forward.   In adults the posterior longitudinal bundle is also said to contain fibres from the loop of the lenticular nucleus.   According to Meynert these represent the projection system from the hemispheres to the gray matter of the central canal (aqueduct).

According to Gudden and Spitzka the motor oculi nerve takes its origin from the nucleus in such a way that its fibres pass out from the anterior division (1, 2) and the inferior part of the posterior division (6, 7) of the same side, and from the superior part of the posterior division (4, 5) of the opposite side.   According to Perlia (*vide* the scheme) the crossed part is derived only from the posterior cell group (5) of the upper part of the principal nucleus, and some of the decussating fibres pass downward in the raphe and then bend into the posterior longitudinal bundle; in my opinion the latter fibres are centripetal.   The anterior division of the superior cell group (4) gives origin to uncrossed fibres.

All the axis cylinders derived from the muscle nucleus become fibres of the motor nerve, so that all the other fibres must pass to the nucleus and—if they do not merely pass through—must terminate there in the network.   From the standpoint of the nucleus they are centripetal.   The fibres which are positively demonstrated to be of this character are: 1, sensory fibres from the opposite half of the body, in the posterior longitudinal bundle of the tegmental region of the peduncle.   In fact they undergo another decussation in the raphe. They are medullated even in the fœtus.   2. Tractus fibres, corresponding to the opposite half of the field of vision, possibly interrupted by ganglion cells. They come from the external geniculate body, passing in great part in the anterior brachium of the corpora quadrigemina and from the anterior corpus quadrigeminum to the oculomotor nucleus (Meynert's fibres). For clinical reasons and from the results of experiments, it is also evident that fibres of the corona radiata must pass to the muscle nuclei. It is very probable that these fibres pass, in great part, along Gratiolet's optic radiations and originate in the cortex of the "visual sphere." We will again return to this point.

Where and how the numerous connecting fibres between the different groups of nuclei begin and end can only be surmised, because we possess no positive data.

The individual cell groups evidently correspond to the individual muscles supplied by the nerve, as was first shown by the experimental investigations of Hensen and Voelkers (*Arch. f. Ophth.*, XIX, 1). There is great difference of opinion regarding the details in man. It appears to be certain, however, that the nuclei of the sphincter pupillæ and the ciliary muscle lie in front of the others, and it is very probable that the nucleus for the levator palpebræ superioris lies in front of those for the other voluntary muscles. Westphal claims that accommodation and movement of the iris are regulated by the nucleus (3) named after him, while Darkschewitsch makes the same claim for the nucleus (1) named after himself. Both nuclei, particularly the former, are characterized by the small size of their ganglion cells.

With regard to the principal nuclear mass we assume that those cell groups (4, 6, 7) which send out nerve fibres on the same side are involved in movement of the eye to the inside, those from which fibres pass to the nerve of the other side (5) take part in movement to the outside. We then have the following distribution: Number 6 and the central nucleus 8 correspond to the internal rectus. We shall hardly go astray in assuming that 8 presides over the movement of convergence and that 6, by means of its connections with the opposite abducens nucleus, presides over the action of the internal rectus in the conjugate movements of the eye. The direct contiguity of the Edinger-Westphal nucleus (3) to the central nucleus (for convergence) testifies to the intimate relations of convergence and accommodation. Both are almost always associated with a movement downward, hence the nucleus for the inferior rectus is to be sought in 7; 4 would correspond to the superior rectus, 5 to the inferior oblique, both of which are concerned in the upward movement of the eye. Experiments and clinical observation make it probable that the nucleus of the superior rectus lies in front of that of the inferior oblique.

Westphal assumes that 3 presides over accommodation and movements of the iris. The case from which he draws this conclusion (*Deutsch. med. Wochen.*, 31, III, 87) was that of a man suffering from chronic progressive paralysis of the external ocular muscles; the pupil reacted to accommodation but not to light. At the autopsy the trunk and nucleus of the motor oculi were found degenerated; groups of ganglion cells were present only in the most anterior portions of the nucleus on both sides of the raphe. This case simply proves that accommodation was still present; the associated movement of the iris occurs in a purely mechanical manner, inasmuch as the blood is forced momentarily into the iris by the contraction of the ciliary muscle. If we regard 3 as the accommodation nucleus, then nucleus 1 would regulate the pupillary movement by the sphincter pupillæ (Darkschewitsch). These assumptions agree very well with our previous anatomical and clinical observations and to a sufficient degree with those derived from experiments on animals. The latter only permit cautious inferences with regard to the conditions in man. Nucleus 2 would then be left for the levator palpebræ superioris. This muscle, in correspondence with the comparatively isolated position of its nucleus, assumes a special position, inasmuch as it is often paralyzed alone or is alone intact in paralysis of the motor oculi nerve. It is not necessary, as Mendel would infer from certain experiments on animals, to locate the nucleus of the "ocular facial" nerve in the hindmost part of the nucleus of the motor oculi. This is disproven by clinical experience in the human subject (*vide Muench. med. Wochen.*, 1887, p. 902).

The scheme adopted does not agree entirely with the well-known one of Kahler and Pick, in which the centre for accommodation is sit-

uated in front of that for the sphincter pupillæ, but agrees very well
with Starr's scheme (*Journ. of Nerv. and Ment. Disease*, May,
1888), which was constructed according to the doctrine of probabili-
ties, from 20 cases of partial motor-oculi paralysis. The nucleus for
the levator palpebræ superioris (2) lies nearer, in my scheme, to the
median line than has been hitherto assumed (viz., corresponding
about to the anterior portion of 4).

Perlia's scheme agrees essentially with that of Siemerling (*Arch. f. Psych.
u. Nerv.*, XXII, Suppl., p. 152). The latter is inclined to place the nucleus for
the levator palpebræ superioris at the posterior extremity of the motor oculi
nucleus, where he found a certain group of cells intact in a case of ophthalmoplegia without affection of this muscle. This would coincide with Mendel's
opinion. According to a more recent publication (*ib.*, XXIII, 3), with regard
to the findings in a case of congenital unilateral ptosis, he again opposes this
view. Positive findings in an acquired unilateral ptosis alone would be convincing. In Boediker's case (*Neurol. Centr.*, 1891, p. 187) the more posterior
part of the motor oculi nucleus was intact, and there was no ptosis. The median
anterior groups (my levator nucleus) were also intact.

Hence the motor oculi nucleus of each side contains the nuclei of
those muscles which take part in the movement of both eyes toward
the opposite side, *i.e.*, the internal, superior and inferior recti of the
same eye and the inferior oblique of the opposite eye. The former
muscles take part in movement of the eye inward, the external rectus, superior and inferior oblique take part in outward movement.

b. The *trochlearis* nucleus is merely the most posterior part of
the motor oculi nucleus; but its ganglion cells are larger than those
of the latter. The fibres originating in it pass backward and inward to the valve of Vieussens at the upper end of the fourth ventricle, where they decussate with those of the other side. The nerve
emerges on the posterior surface of the valve of Vieussens, bends
around the processus cerebri ad corpora quadrigemina and the cerebral peduncle forward and inward, lies immediately beneath the
free border of the tentorium, perforates the dura mater behind the
posterior clinoid process, and passes through the superior orbital fissure into the orbit, where it passes above the origin of the ocular
muscles to the superior oblique muscle.

The root fibres for the inferior oblique also decussate, as we have
seen above (Fig. 6, Nucleus 5).

According to Siemerling (*Neurol. Centralbl.*, 1819, p. 188) Westphal's posterior trochlearis nucleus (*Arch. f. Psych.*, XVIII) has nothing to do with the trochlearis nerve. According to Schnetz it belongs to the "gray matter of the central canal."

 c. The *abducens* nucleus lies about at the middle of the floor of the fourth ventricle beneath its gray lining. The nerve passes forward through the pyramids of the medulla oblongata, emerges at the posterior border of the pons, and then passes to the posterior wall of the cavernous sinus, which it perforates and then extends along the outer side of the internal carotid; here it is said to receive fibres from the carotid plexus. The nerve then passes through the superior orbital fissure into the orbit, perforates the origin of the external rectus and is lost in this muscle.

 According to Spitzka (*l. c.*) the posterior longitudinal bundle contains fibres which connect the abducens nucleus with the nucleus of the internal rectus on the same side. As the roots for the latter muscle do not decussate (*vide* scheme), this is extremely improbable, inasmuch as the roots of the abducens likewise do not cross. According to Duval, fibres from the abducens nucleus pass to the motor, oculi nucleus of the opposite side, and this would correspond better to the actual conditions.

 d. The *facial* nucleus is also situated in the floor of the fourth ventricle, to the outside of and a little behind that of the abducens. Its ganglion cells are larger than those of the adjacent abducens nucleus. The root fibres encircle the latter; after being joined they run for some distance to the outside beneath the gray matter of the central canal and then bend forward (knee of the facial nerve). The nerve makes its appearance at the posterior border of the pons, outside of the olivary bodies, passes into a groove of the acoustic nerve, and then takes its well-known way through the Fallopian canal of the petrous portion of the temporal bone. It emerges from the stylomastoid foramen and then ramifies in the shape of a fan into the muscles of the face.

 In view of the fact that in about ninety per cent of all "central" facial paralyses the orbicularis palpebrarum and frontalis muscles which are supplied by the superior facial escape, while they are affected in peripheral paralysis, Mendel removed these muscles in a rabbit and two guinea pigs. In all three animals atrophy developed in the posterior part of the motor-oculi nucleus on the side operated upon. He is therefore inclined to believe that the nucleus of the ocular facial is separated from the rest of the nerve and is more closely related to the nucleus of the motor oculi. In the discussion on the question Uhthoff and Hirschberg did not accept this suggestion, at least in regard to the human subject (*vide* p. 21). We will hereafter offer another explanation for this fact.

e. The motor *sympathetic* fibres of the eye are derived from the superior cervical ganglion and pass through the carotid plexus. When farther central they are governed by the lowermost part of the cervical cord (about at the level of the sixth and seventh cervical and first dorsal vertebræ). Irritation of this part causes spasm; its destruction causes paralysis of the corresponding muscular fibres (Budge's cilio-spinal centre). The details of the course of these fibres, which cannot be demonstrated anatomically, will be discussed under the heading of muscular disorders.

Apart from connecting fibres and decussations between the muscle nuclei of the same name on both sides (rectus internus and inferior, obliquus superior, sphincter pupillæ, and probably also the ciliary muscle) or different muscle nuclei of the same side (abducens—internal rectus of the other side), the connections with cells of higher cerebral ganglia are known very imperfectly and in the main must be inferred from experiments and clinical observations. There is no doubt of close relationship to the primary optic ganglia (Meynert's fibres between the anterior corpora quadrigemina and the muscle nuclei). Centrifugal fibres of the corona radiata must also terminate in them, as in other muscle nuclei, but we possess no further anatomical knowledge concerning this point. A part of the anterior central convolution, immediately above the speech centres, seems to be connected with the origin of the inferior facial nerve. Injury to a spot in front of and above the one just mentioned, in front of the centre for the upper limb, gives rise to crossed ptosis. This region is accordingly connected with the nucleus of the motor oculi, at least with the nucleus for the levator palpebræ superioris of the other side.

The external ocular muscles may be made to perform co-ordinated movements (chiefly toward the opposite side) from many parts of the cortex, but especially from the "visual sphere."

3. COURSE OF THE SENSORY NERVES.

The eye obtains its sensory nerves from the first and second branches of the trigeminus, the latter supplying only the lower lid. The nucleus of origin (probably the nucleus of termination) for the large sensory root of the fifth nerve extends from the middle of the aqueduct of Sylvius on the outside of the motor oculi and trochlearis nuclei to a point beyond the beginning of the central canal of the spinal cord. It is usually semilunar in shape on transverse section (*vide* scheme, Fig. 5, p. 16). It emerges from a groove in the crus cerebelli ad pontem. Hence we distinguish ascending and descending root fibres. The latter contain an addition of crossed fibres from the cells of the lobus cæruleus. These pass along immediately beneath the floor of the fourth ventricle to the raphe, decussate, are frequently intermingled, and there bend into the common trunk. The descending fibres likewise possess an origin from the raphe itself; these fibres decussate, and, according to Meynert, are derived from the cerebral peduncle. The nucleus of origin at the level of the point of exit of the nerve consists of a clump of gelatinous matter, similar to that in the posterior horn of the spinal cord; into this pass fibres from the cerebellum. The larger number of trigeminus fibres are ascending (Wernicke). If we take into consideration the direction of conduction, which is the decisive element, then the ordinary terminology is incorrect. The "descending" fibres conduct upward, the "ascending" fibres conduct downward toward the spinal cord.

The entire sensory portion of the trigeminus enters the Gasserian ganglion, from which the three branches of the nerve arise. The first and smallest branch runs forward, at first adherent to the outer and upper walls of the sinus cavernosus, and passes through the superior orbital fissure into the orbit. It sends the lachrymal nerve to the conjunctiva and the integument of the outer angle of the eye, the frontal nerve (supratrochlear and supraorbital) to the upper lid

and forehead, and the naso-ciliary nerve which passes with the abducens through the origin of the external rectus.   The branches of
the last-named nerve are the ethmoidal nerve—which passes through
the anterior ethmoidal foramen to the cranial cavity and then
through the lamina cribrosa to the nasal cavity, and supplies the
integument of the ala nasi below the nasal bone—and the infra-
trochlear nerve which supplies the root of the nose, the upper lid,
lachrymal sac, caruncula, and inner half of the conjunctiva.   The
lower lid obtains its sensory fibres from the second branch of the
trigeminus, which passes through the foramen rotundum, then
through the pterygo-palatine fossa, the inferior orbital fissure, the
floor of the orbit, and the infra-orbital canal.   The nerves which
emerge from the latter beneath the middle of the inferior rim of the
orbit, diverge in the shape of a fan and form the pes anserinus
minor.   The motor root of the trigeminus arises in front of the
facial nucleus, passes between the anterior transverse fibres of the
pons, and does not form part of the Gasserian ganglion but enters entirely into the third branch of the trigeminus; it has nothing to do
with the eye.

The destination of the axis cylinders of the ganglion cells of the
trigeminus nucleus is unknown.   It is probable that part, at least,
pass to muscle nuclei in the brain and cord, and that in great part,
together with the sensory track of the entire body, they pass to the
opposite optic thalamus, where the received impressions are meta-
morphosed by ganglion cells and conveyed mainly to the cortex of
the parietal lobe, but undoubtedly also to other regions (cerebellum).

The motor, sensory, and sympathetic nerve fibres for the interior
of the eye and the cornea—with the exception of those which pass
along the optic nerve itself and the central vessels of the retina—pass
through the ciliary ganglion before entering the eye.   This is situated in the posterior part of the orbit between the external rectus and
the optic nerve, and is about 3 mm. in diameter.   It receives posteriorly three roots (which exhibit manifold variations in individual
cases); anteriorly it sends the ciliary nerves to the eye.

The motor root is derived from that branch of the motor oculi
which passes to the inferior oblique (short root), the sensory root

from the naso-ciliary nerve after its passage through the origin of the external rectus (long root); the sympathetic root is derived from the carotid plexus in the cavernous sinus and passes through the superior orbital fissure either to the ganglion itself or to its sensory root. From the ciliary ganglion are derived the long and short ciliary nerves to the choroid, ciliary body, and iris, and particularly to the internal muscles of the eye, the ciliary muscle, sphincter and dilator of the pupil.

The sensory nerves of the cornea (probably also of the sclera) pass along the path of the ciliary nerves. But the sensory tract of the cornea is not sharply defined from the tract of the conjunctiva, which is supplied by twigs from the first branch of the trigeminus (Bouchéron, Compt. Rend. de la Soc. de Biologie, 1809).

The origin of the motor root of the ciliary ganglion from the motor-oculi branch to the inferior oblique makes it possible that the latter muscle, together with the sphincter of the pupil and accommodation, should alone be paralyzed if a peripheral (orbital) lesion affects this branch exclusively.

B. Disorders in the Domain of the Ocular Nerves and their Central Origin.

Here the chief importance attaches to the optic nerve and to those parts of the brain which are connected with its origin; next to the nerves of the muscles, the sympathetic and the sensory nerves. We must distinguish: a, peripheral disorders, extending to the origin or termination in the brain; b, intermediate, i.e., nuclear or ganglionic disorders, when the ganglia of origin are attacked; and c, central disorders proper, whose causes are located more centrally in the brain.

1. OPTIC NERVE.

Affections of the optic nerve and its centre of origin produce visual disorders in the proper sense of the term. The peripheral disorders are located in the eye, in the optic nerve, chiasm, and tractus; the ganglionic, in the three primary optic ganglia; the central, in Gratiolet's optic radiations and the occipital cortex, and also, so

far as visual impressions in the brain can be traced, in the systems
of associative fibres of the occipital cortex and in the other parts of
the cerebral cortex which are connected therewith.

A. Peripheral Visual Disorders.

The peripheral visual disorders really belong to the field of special ophthalmology, and will be discussed here as briefly as possible.

1. *Intra-ocular Visual Disorders.*—These may consist of
opacity of the refracting media and irregular shape (astigmatism) of
the refracting surfaces of the cornea and lens, or of improper focusing on the retina.   An important difference between these two forms
is shown by the fact that, with equal vision, colored (red and blue)
squares are recognized as such at a less distance when the media are
opaque than when we have to deal simply with incorrect focusing
of clear media (uncorrected errors of refraction and astigmatism).
In both cases the visual field is normal.

If the visual disorder is owing to changes in the choroid, retina,
or optic nerve, the color sense is usually disturbed to a greater extent,
especially as regards quality, at least when the retina and optic nerve
are implicated to a notable degree.   We can often demonstrate defects of the field of vision, either concentric or sector-shaped narrowing of the field, or single or more or less numerous scattered blind
or amblyopic spots (scotomata).   These are either seen as mist,
smoke, etc. (positive scotoma, especially when the percipient parts
of the retina, occasionally also the conducting fibres, are not entirely
incapable of function), or they are only demonstrable on testing the
field of vision (negative scotoma).   Total removal or destruction of
an eye also causes a corresponding negative scotoma.

In diseases of the fundus oculi proper the disturbance of the color
sense may be twofold.

a. The same as that which appears in the normal individual
when the illumination is diminished; gradual narrowing of the visible spectrum from both sides, weakness of color with bilaterally narrowed spectrum, because on the one hand red, orange, and yellow, on
the other hand blue and violet can be distinguished less and less distinctly), so that gradually the middle of the spectrum appears color-

less and only a more refractive (cold) color called blue, and a less refractive (warm) color, usually called yellow, rarely red, can be distinguished (green-blindness or two-color vision with bilaterally contracted spectrum). Finally, even these two colors can no longer be distinguished, and only the middle of the spectrum appears colorless (white to gray) (total color-blindness with bilaterally strongly contracted spectrum). This form of color disturbance is observed in diseases of the conducting apparatus, the optic nerve, and probably also the larger of nerve fibres in the retina, and is therefore best called "conduction color disturbance." In these cases it is found not infrequently that, although there is color-blindness in ordinary daylight, the colors are still recognized correctly (Knies, "Heidelberger Berichte," 1889, p. 7) in a very intense light (sunlight).

b. The second form of color disturbance is very closely allied to the condition of one with normal color sense, who looks through a yellow glass, until the yellow sensation is no longer subjectively noticeable, *i.e.*, until yellow and white appear almost identical. It is well known that looking through a yellow glass at first causes a very decided feeling of being dazzled, although objectively less light enters the eye, inasmuch as the violet rays are removed.[1] Apart from the already mentioned confusion of yellow and white, red and green are recognized very well (appearing more vivid), blue to a less extent (with a dark yellow glass it is seen as green; but then the spectrum is also noticeably contracted at the red end). Violet appears black, carmine red (*i.e.*, red and violet) appears dull red. An exactly similar disturbance of color is found in choroidal affections with implication of the external layers of the retina, detachment of the retina, etc. The similarity becomes still greater when, as happens not infrequently, though temporarily, in acute cases, objects are seen colored (usually yellow, more rarely green),

[1] In former years marksmen in shooting for prizes not infrequently used yellow glasses in order to improve the sight. To me, who wear not fully correcting concave glasses, Snellen's test types appear, through a yellow glass, in somewhat sharper outlines than before, evidently because the focus of the rays in yellow light falls a little farther back in the eye than in the case of mixed white daylight. The recommendation of santonin in diseases of the optic nerve probably depended on this subjective increase of distinctness.

analogous to the yellow vision which occurs occasionally in jaundice. By the perimeter the boundaries of blue are found contracted to a greater extent than those of red and green. This is also the case in santonin poisoning.

Evidently we have to deal here with molecular changes or abnormal processes of decomposition[1] in the percipient external retinal layers. Perhaps there is a local hæmatogenous formation of bile pigment. The action of santonin may also be similar. The media are sometimes visibly colored yellow, for example, in detachment of the retina.

If the color disturbance under a is attributed to the conductive apparatus, this second form may be aptly called "perceptive color disturbance." They may occasionally be associated with one another. As both color diseases often appear only in patches in the field of vision, very different fields of color vision are possible. Color scotomata of the second variety are the causes of the very striking fields of color vision with reversal or mutual intersection of the boundaries for the different colors.

After these preliminary considerations certain often-mentioned, more or less peripheral visual disorders will be more easily understood. We refer to day and night blindness, retinal asthenopia, dazzling, anæsthesia and hyperæthesia of the retina. In their description we follow Leber (Graefe-Saemisch's "Handb. d. Augenheilk.," V, p. 980 *et seq.* and p. 1,005 *et seq.*). All these visual disorders are very often the results or concomitants of constitutional diseases.

We must assume that in the act of vision photochemic actions are produced in the external layers of the retina by the entrance of light. The products (decomposition products, carbonic acid, etc.) act injuriously and must be removed by the blood-vessels and lymphatics, while restitution (albuminoids, oxygen, etc.) for the decomposed and used-up material is furnished by the arterial blood-

---

[1] Allied processes (fatty degeneration) in the epithelium of the conjunctiva must also be assumed in so-called xerosis conjunctivæ, which is often associated with evident nutritive disturbances in the external retinal layers (night blindness, torpor retinæ).

vessels. Under normal conditions consumption and supply must stand in a certain harmonious relation. In a like manner the conduction in the optic nerve, by means of which analogous processes are induced in the ganglion cells of the brain, must be carried on normally. As in other localities, the accumulation of products of disintegration causes fatigue.

The nutritive material for the outer layers of the retina is furnished by the chorio-capillaris of the choroid. The products of disintegration may pass in part into the vitreous, but are also partly removed from the eye through the lymphatics of the retina and optic nerve. This affords the possibility of their disturbing action on conduction in the optic nerve.

Retinal asthenopia is abnormally rapid exhaustion of the eye, similar to that found in weakness of accommodation or insufficiency of the internal recti, albeit the latter conditions are not present. There is also absence of conjunctival hyperæmia which may produce similar symptoms. Retinal asthenopia is usually a symptom of a general anæmic-chlorotic or neurasthenic condition. It is the most striking symptom simply because the patients, on account of the inability to follow any occupation, feel the giving way of the eyes with special severity. Recovery will only result from appropriate general treatment which is often very tedious. In severe cases the use of the eyes is directly painful (neuralgia bulbi).

The symptom depends upon imperfect nutritive changes in general, upon insufficient restitution despite normal disassimilation. It is very difficult to determine how much is peripheral, how much central in character, because similar defective conditions of nutrition must also be present in the central organs.

In retinal hyperæsthesia there is excessive sensitiveness of the eyes to ordinary daylight, especially to higher degrees of illumination, while the acuity of vision and the field of vision are normal when the illumination is lessened. It is very often a part of generally increased irritability of the entire nervous system (irritable weakness) and similar symptoms also occur in inflammatory affections of the eye, especially of the cornea and conjunctiva. We must assume that the photo-chemical action of light causes much more considerable, perhaps also more deeply spreading, abnormal disintegrations, whose products themselves act as irritants.

Photophobia may be peripheral in character from causes local to the eye, or it may be the result of brain disease; for example, in meningitis, when the products of inflammation give rise to an irritative condition in the optic nerve or the cerebral cortex. The absence of the excessive sensitiveness to light results in the condition known as day blindness (nyctalopia). This coincides with hyperæsthesia of the retina so far as regards the visual disorder, viz., impairment of vision with ordinary or bright illumination, improved or normal vision with lessened illumination. This is observed when, for any reason, an abnormal amount of light enters the eye (large coloboma of the iris, mydriasis), or in imperfect pigmentation of the fundus oculi (albinism and the like), and also in certain diseases of the retina (Arlt's retinitis nyctalopica) and the optic nerve (intoxication amblyopia). In all these cases the efficient factor appears to be the insufficient restitution of the material employed in the act of vision, this being sufficient only when the action of the light is diminished.

Similar symptoms are found in those who have lived for a long time in the dark, for example, in dark dungeons. The diminished nutritive changes which have become habitual in those parts of the retina which are sensitive to light do not suffice in bright daylight.

V. Graefe originally applied the term retinal anæsthesia to a condition in which central vision was more or less impaired, and the field of vision, in particular, had undergone pronounced concentric narrowing. On pressing upon the insensitive parts of the retina the well-known phosphenes could generally (though not always) be produced, and hence it was inferred that conduction was unimpaired.

Foerster ("Heidelberg. Ber.," 1877, p. 162) has called attention to the fact that often the narrowing of the field of vision can only be demonstrated when the moving object is carried from the point of fixation toward the periphery, but not when it is carried in the opposite direction.

The ophthalmoscopic appearances are normal even after the condition has lasted for years. The condition may appear and disappear suddenly. In the highest grades there is complete blindness

with or without normal reaction of the pupils to light.   Within the
existing field the color sense may be normal.   It may be diminished, even to total color blindness, in the conductive form of disturbed color sense described on page 29.   In rarer cases we find a
disturbance of color which approaches more closely to the disturbance of sensation there described.   Other "nervous" disorders are
often present.

Anæsthesia of the retina is often observed in children at the
period of puberty and in nervous women.   A special form is described as hysterical amblyopia and amaurosis.   It may also be due
to injury (traumatic hysteria).   Both forms will again engage our
attention.

The location and cause of the disease are not at once evident.
The retina, the conducting elements, and the central organs may be
affected, and in different ways in different cases.   Vasomotor disorder is the most probable cause, although not in the distribution of
the central artery of the retina.

If the nutrition of the external layers of the retina by the choroid,
particularly by the chorio-capillaris, were imperfect (as in spasm of
the ciliary arteries), the symptoms of pure retinal anæsthesia would
result.   The periphery of the retina would suffer most; the macula,
which is much more favorably situated as regards nutrition, might
still perform its function in a tolerable degree.   The result would be
concentric narrowing of the field of vision with more or less intact
central vision.   The pigment epithelium would make it impossible
to recognize the condition with the ophthalmoscope.   Moderate
pressure on the eye produces exactly similar phenomena, including
the pressure-phosphenes.

On the other hand, interference with the optic nerve, anywhere
in the region of the optic foramen, also produces concentric narrowing of the field of vision, because the fibres of the macula are most
protected in that locality.   Here the color disturbance, which is
characterisic of interference with conduction, would also be present,
and this is one of the main symptoms of hysterical amblyopia.   Indeed Leber (l. c., p. 985) found even material changes in the peripheral bundles of the optic nerve in a so-called hysterical amblyopia
  3

without ophthalmoscopic findings. Even in this location of the affection the ophthalmoscope would show nothing in the retinal vessels, which only enter the optic nerve immediately behind the eye.

While peripheral and central causes usually act together in the complex of symptoms hitherto discussed, so-called night blindness (torpor retinæ, hemeralopia) must be attributed in the main to a peripheral change in the outer layers of the retina, although the optic nerve may suffer secondarily. Vision is perfectly normal in a good light, but in twilight it fails disproportionately. The ophthalmoscope shows nothing abnormal save more or less hyperæmic signs in the fundus oculi. Adaptation to darkness is also impaired so that the eye suffering from night-blindness requires when the illumination is diminished, four to ten times the time required by a normal eye to attain the corresponding maximum of vision. In addition the subjective light phenomena caused by pressure on the eye (pressure phosphenes) are decidedly diminished, and may even be entirely absent. In pronounced cases the disorder of color perception described on p. 29 is also present, and is characterized particularly by diminished recognition of violet and blue.[1]

The acute forms occur after dazzling, associated with impaired nutrition; for example, during the strict Lenten fasting in Russia, where at the same time the snow reflects a glaring sunlight; in besieged fortresses, on shipboard, during scurvy or other severe general disturbances, after infectious diseases, etc. In these cases the fundus oculi is found normal. At the same time a characteristic xerosis of the conjunctiva is often present. It is to be assumed that deeply spreading changes in the external layers of the retina impair the perception of light quantitatively and qualitatively. This is analogous to the fatty degeneration of the epithelium in conjunctival xerosis within the inter-palpebral fissure, and which results from abnormal irritants (reflection from the snow, etc.) combined with insufficient nourishment.

In other cases night-blindness is a sign of more or less acute cho-

---

[1] In night-blindness the violet sensation is very often diminished at the periphery. This will escape discovery if, as ordinarily happens, only the tests for red, green, and blue are made.

roidal diseases— perhaps preferably chorio-retinitides and retinal affections with decided implication of the external layers of the retina and intact conduction. Corresponding ophthalmoscopic appearances are then found. In such cases the color disturbance is also more pronounced. Its analogy with vision through a yellow glass is still further increased by the fact that in very acute cases subjective colored sight occurs (usually yellow vision, more rarely green vision).

I may here mention that every healthy individual by looking through a yellow glass, can not only imitate the corresponding disturbance of color, but also exhibits the symptoms of night-blindness, viz., considerable slowing of adaptation and marked impairment of vision in the dark.

Day-blindness also occurs as a congenital affection without findings. It is best known as a symptom of a definite form of chorioretinitis, which is called retinitis pigmentosa. In this disease the degeneration of the pigment epithelium, the atrophy of the choroid and of the outer layers of the retina in the later stages are also visible with the ophthalmoscope. On account of the very chronic course a color disturbance is usually not demonstrable during the degeneration of the light-perceiving layers of the retina (*vide* the note on the preceding page). In the places which still perform their function the color sense is normal, on account of the intact conduction. The marked impairment of vision in diminished illumination and in the dark alone shows the beginning degeneration of the percipient layers of the retina.

When there is simply dazzling of an otherwise healthy eye, *i.e.*, excessive photo-chemical decomposition without corresponding restitution, the subjective symptoms are always the same as if less and less light were entering the eye, *i.e.*, obscuration of the field of vision extending to complete blindness (for example in snow-blindness proper). The normal condition is soon restored after rest and darkness. On the other hand, it is easily understood that, in individual cases, dazzling may give rise to the opposite condition so that hyperæsthesia as well as anæsthesia of the retina, day-blindness as well as night-blindness, may be brought about. We must not overlook the action of heat on the pigmented parts of the fundus, which may be intensified into the burning in of a solar image into the pigment epithelium (in observing an eclipse of the sun without a smoked glass). The nutritive disturbances in the choroid as the result of radiating heat must also be regarded as the probable cause of the development of cataract in glass-blowers.

To sum up briefly: asthenopia of the retina is its easy exhaustion; hyperæsthesia of the retina is excessive irritability of the entire peripheral and central visual apparatus, although certain parts may be implicated to a more marked degree. The essential element of day-blindness, which is, in the main, peripheral, is imperfect restitution of the material used up in the process of vision. In retinal anæsthesia there is diminished and often qualitatively changed action of light upon the percipient retinal elements, or diminished conducting power of the optic nerve, or both; in night blindness and torpor of the retina there is, in the main, only the former condition. All of these terms are not very happily chosen. There are also considerable differences of opinion with regard to details, but we cannot enter upon them here (compare, for example, Treitel, *Arch. f. Ophth.*, XXXIII, 1 and 2; XXXV, 1; XXXVI, 3, and XXXVII, 2). To mention two technical expressions which are often used, initial stimulus (Reizschwelle) means the faintest objective light stimulus which can be perceived, and the differential stimulus (Unterschiedsschwelle) is the amount of difference between two impressions of light which can just be distinguished as possessing different degrees of brightness.

Although the visual disorders under consideration are, in great part, of a peripheral character, we have discussed them somewhat in detail because we will again come in contact with them under the heading of the most varied general diseases. It is also easily understood that to a certain degree there may be combinations of the forms mentioned, for example, of retinal anæsthesia with hyperæsthesia (photophobia).

On account of the specific energy of the nerves of special sense, neither the retina nor optic nerve can be the seat of real pain. But pains may result from the entrance of light on account of the implication of other parts of the eye or of the central organs.

2. *Diseases of the Optic Nerve.*—If conduction is entirely interrupted in one optic nerve, there is unilateral blindness. The

pupils of both eyes will not react on the entrance of light into the blind eye, but both react to an equal degree on the entrance of light into the healthy eye (consensual pupillary reaction), unless there are mechanical hindrances, such as synechiæ. Hence, in unilateral blindness whose cause is situated in the optic nerve (or in the retina), the pupils of both eyes are equal in width, unless complications render this impossible.

If an optic nerve is only partly destroyed, there is loss of function of all those parts of the retina to which fibres are sent by the destroyed section. Hence, according to the site of the lesion, there will be concentric or sector-shaped narrowing of the field of vision, central or peripheral scotoma, or various combinations (*vide* Fig. 3, p. 4).

If the optic-nerve fibres degenerate slowly, as often happens, the impairment of vision for the corresponding parts of the retina occurs in exactly the same way as if these received a much more feeble illumination (amblyopic spots in a monocular field of vision). The scotomata are usually negative.

At the same time there occurs, as a very characteristic feature, the color disturbance which we have called conduction disorders of color and have described on page 29. It is only absent in very chronic cases. The field of vision when taken with the perimeter shows narrowing of the boundaries of the colors, especially of green, red, and violet; subsequently these disappear completely, until finally even blue and yellow can no longer be distinguished. If the color boundaries, as compared with the boundaries of the field of vision in general, are disproportionately narrowed, this indicates a rapidly progressing process. If the narrowing for white and for colors is approximately uniform the process is slowly progressive or stationary, although there are frequent exceptions to this rule. In partial disease of the optic nerve the color disturbance is confined to the corresponding part of the visual field. If the point of fixation or its immediate vicinity is not affected the subjective disturbance will be very slight, and the physician may readily overlook it, unless direct search is made for "peripheral color scotomata."

Despite the impairment of vision a sort of day-blindness or hyperæsthesia to light in the amblyopic spots may be present, and is shown by improved vision with poorer illumination. This is especially true in inflammatory processes in the connective tissue of the optic nerve, for example, in the central scotoma due to central (axial) neuritis in so-called intoxication amblyopia (day-blindness, nyctalopia).

Sooner or later (at the latest, in six weeks) the ophthalmoscope will reveal corresponding findings, when there are material changes in the nerve fibres, viz., partial or total atrophy of the optic nerve. This will only be wanting when the nerve fibres themselves remain intact despite the interference with or even abolition of conduction.

3. *Disorders of the Chiasm.*—Complete destruction or frontal division of the chiasm results in bilateral total blindness with abolition of the pupillary reaction to light, because all the optic-nerve fibres are divided.   But if the chiasm is divided in a vertical direction from before backward, both external halves of the retinæ and therefore both internal halves of the fields of vision remain intact.   Hence there is bilateral temporal hemianopsia.   This was present in Weir Mitchell's case mentioned on page 6.   The hemianopsia was not pure because it was not due to a clean cut exactly in the median line but to a progressive, destructive process (aneurism) which also caused degeneration of adjacent fibres.

The visual disorders will vary according to the part of the chiasm which is destroyed, inasmuch as sometimes the crossed, sometimes the uncrossed, bundles are more affected.   The defects in the field of vision are almost always bilateral and often more or less homonymous, *i.e.*, they occur in similar positions on both sides: to the right, above, below, etc.

If the lesion is situated in the anterior or posterior angle of the chiasm, the crossed bundles are usually mostly affected and more or less complete bilateral, temporal hemianopsia, *i.e.*, a bilateral loss of the outer halves of the fields of vision, may be the result.   A quite considerable number of such cases have already been published. Hemianopsia is rare in diseases of the chiasm and then is only sharply defined for a time, because the affection is rarely situated exactly in the median line and affects both sides uniformly.   Usually we have to deal with progressive processes, tumors, syphilitic and tubercular proliferations, etc. The anatomical conditions in the chiasm explain the fact that hemianopic symmetrical defects in both nasal halves of the fields of vision are so rare. A single lesion could produce them only under very peculiar conditions, but they might result from bilateral symmetrical lesions.

Foerster (*l. c.*, p. 113) even denies the occurrence of binasal hemianopsia, but since then a few cases have been reported, for example, by Herschel (*Jahr. f. Aug.*, 1883, p. 111). But if we as-sume that the course of the fibres in the optic nerve is like that claimed by Wernicke and Schmidt-Rimpler (*vide* p. 4), and illustrated in Fig. 7—this course appears to be the exception, not the rule— then a single lesion in the anterior angle of the chiasm, between the two optic nerves, may give rise to a condition somewhat like binasal hemi-anopsia.

From the blind parts of the field of vision the light reaction of both pupils cannot be obtained; from those parts

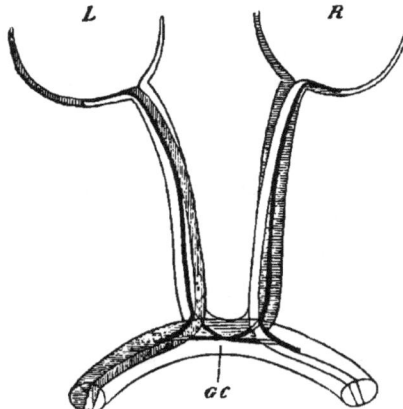

Fig. 7.—Schematic Diagram of the Chiasm, after Wernicke, but including the course of the macular fibres (black); the bundles corresponding to the right half of the field of vision are shaded. *L*, Left; *R*, right eye; *GC*, Gudden's commissure.

of the field which are still present—unless there are complications— both irides are uniformly innervated. The pupillary reaction to con-vergence, accommodation, and cutaneous irritation is present and equal on both sides.

As we generally have to deal with progressive processes other cerebral nerves in the vicinity of the chiasm are very often attacked, for example the olfactory, the motor nerves of the eye, the trigemi-nus, etc. Not infrequently they are attacked in succession in a typi-cal manner.

It is preferable not to employ the term hemianopsia for such disorders of the field of vision, as the half-blindness is almost always only an approximate one. It would be better to speak of symmetrical defects. Hemianopsia and half-blindness would then be employed only for bilateral homonymous forms (right-sided or left-sided). In incomplete hemianopsia we would speak of homonymous defects in contrast to symmetrical ones.

4. *Disorders of the Optic Tract.*—In the optic tract at a little distance from the chiasm, the fibres from both sides are intermingled in such a way that destruction of any part causes homonymous defects in the visual fields of both eyes. Slight differences in the position and extent of the defects do occur, because the combination of the fibres from both sides is not yet completely uniform in the beginning of the optic-nerve tract. The very fact that the fibres from both optic nerves, or at least the majority of them, do not run in bundles, might give rise to the mistaken opinion that an ascending degeneration of the optic nerve stops at the chiasm. On account of the intimate commingling of fibres, atrophies of considerable extent cannot be demonstrated, but there is merely more or less uniform diminution in the volume of the entire tract (or both tracts). Even this may be compensated in a measure by a corresponding increase in the amount of interstitial tissue.

Destruction of an entire tract causes typical homonymous hemianopsia, the retinal halves on the same side or the halves of the visual fields on the opposite side being lost. At the same time there is loss of the light reflex of the pupils, which could be excited from the blind halves of the retinæ. There is hemianopic pupillary reaction (also called Wernicke's symptom), *i.e.*, the light reaction of both pupils is intact on illumination of both halves of the retinæ which are sensitive to light, but does not follow when light falls upon the insensitive halves of the retinæ. "Hemianopic light rigidity" is a better term for this condition.

Hemianopic light rigidity of the pupil is an extremely rare symptom, because isolated disease of one tract or degeneration confined to one optic tract (this is most apt to result from a hemorrhage) is equally rare. Its occurrence has been doubted by a number of writers, but a few typical cases have again been published recently, so that the possibility and hence the diagnostic value of this symptom have been assumed (*vide* Martius' *Charité Annalen*, XIII, 1888; Seguin, *Journ. of Nerv. and Ment. Dis.*, Nov., 1887).

In incomplete destruction of one optic tract the hemianopsia is also incomplete and the defects are more or less accurately homonymous. The defects are probably so much more distinctly homonymous the nearer the lesion is situated to the primary optic ganglia. This cannot be proven by examples, however, on account of the small number of diseases of the tractus which have been thoroughly observed clinically and anatomically.

When conduction in the chiasm and optic tract is merely impeded but not abolished, the same symptoms appear with regard to the recognition of colors as in interference with conduction in the optic nerve, except that they are more or less accurately homonymous in both eyes.

It is noteworthy that, in very pronounced interference with conduction in one tract, the hemianopic or homonymous defect may be seen as a positive scotoma, as a mist or smoke. This was true, for example, of Spierer's case (*Mon. f. Aug.*, 1891, p. 218), in which a hemorrhage probably took place into one tract as the result of fright following an earthquake. In such cases a positive scotoma is prognostically favorable, because complete interference with conduction cannot have occurred. As a rule positive scotomata are observed only in diseases of the outer layers of the retina, so long as the latter are not entirely destroyed. They also occur occasionally in diseases of the optic nerve. I have observed this a number of times in so-called toxic amblyopia.

When actual destruction of nerve fibres takes place in the peripheral visual disorders hitherto discussed, total or partial atrophy of the optic nerve becomes visible sooner or later with the ophthalmoscope. Not infrequently, however, there is a certain lack of harmony—sometimes quite pronounced—between the visible findings in the optic nerve and the disturbance of vision. In the first place, a certain time elapses before the abolition of conduction in the deeper parts of the nerve, the chiasm and the tract, becomes noticeable at the entrance of the optic nerves. In addition, the visual disturbance is relatively marked and the ophthalmoscopic appearances are relatively slight when there is only interference with conduction but the optic nerve fibres are still intact, as, for example, in so-called toxic amblyopia, in the optic-nerve affections of multiple sclerosis, etc. On the other hand, the atrophic discoloration of the papilla may be very pronounced, and the disturbance of vision may be very slight or absent. The latter condition is observed particularly after the recovery from optic neuritis. If the disturbance of vision is located in the eye or at the entrance of the optic nerve the vessels of the retina are usually more or less narrowed. If it is located behind the entrance of the central vessels into the optic-nerve, the calibre of the retinal vessels may be normal for a long time or permanently, even if there is complete blindness.

In symmetrical or homonymous defects of the field of vision from partial disease of the chiasm and tract, the atrophy visible with the ophthalmoscope is often confined to the corresponding sector of the papilla, in the same way that, in axial degeneration of the optic nerve, the atrophic discoloration is often confined to the outer half of the nerve disc.

If there is simply interference with or abolition of conduction without destruction of the nerve fibres, and the cause is located behind the entrance of the central vessels into the nerve, all ophthalmoscopic abnormalities may be permanently absent despite long duration of the process and serious disturbance of vision. Other reasons must then determine the diagnosis of a peripheral affection (for example, its unilateral character, because unilateral disturbances of vision cannot be located in or beyond the chiasm). This is also true of typical hysterical disorder of vision, unilateral anæsthesia of the retina (*vide* p. 33), which will be considered in detail at a later period.

It is a common feature of all peripheral visual disorders that no light reaction of the pupils can be obtained from those parts of the monocular or binocular field of vision which are insensitive to light, although both pupils react uniformly to the entrance of light upon sensitive parts. This is demonstrated most easily in homonymous hemianopsia from loss of function of an entire optic tract, for the recognition of which the hemianopic light rigidity of the pupil is decisive.

B. Intermediate Visual Disorders.

In the so-called primary optic ganglia (anterior corpora quadrigemina, pulvinar of the optic thalamus and external geniculate body), the visual fibres for the first time enter into relations with other systems of fibres.

In the primary optic ganglia, as in all other ganglia, are found: 1, a fine network of the entering nerve fibres; 2, a fine network of the protoplasmic processes of the ganglion cells; 3, ganglion cells which, in part, send out axis-cylinder processes, in part do not; 4, nerve fibres which may simply pass through; 5, the neuroglia and the vessels.

In any ganglion there may be either an increase or a diminution of the nerve fibres, because more axis cylinders may emerge than are lost in the fibrillary network; and, on the contrary, the opposite condition may hold good. Fibres of different origins usually meet in one ganglion, and the efferent axis-cylinder processes have different destinations; but certain lines of direction usually predominate. Fibres may also run in both directions between two ganglia.

Experiments on animals with regard to the primary optic ganglia may only be utilized in man with great caution, because their function differs markedly in different classes of animals.

Positively convincing cases of lesions of the primary optic ganglia are very few if we regard only their clinical symptoms. It is known that destruction of one pulvinar causes crossed hemianopsia, and that the *anterior* corpora quadrigemina may be destroyed without notable disturbance of vision.

Gowers observed choked disc, but without complete blindness, after considerable destruction of the anterior corpora quadrigemina.

Nieden observed complete destruction of the corpora quadrigemina without pronounced disorders of vision. The slighter the symptoms, the purer and more important is the case and the experiment. Destruction of the parts with almost intact vision is convincing. It is evident that the anterior corpora quadrigemina have nothing to do with conscious vision.

According to Nothnagel there are found, in disease of the corpora quadrigemina, cerebellar ataxia and disorder of the ocular movements, but not of the pupillary movements and of vision. According to Eisenlohr (*Muench. med. Wochen.*, 20, p. 90), on the other hand, the movements of the pupils are controlled by the anterior corpora quadrigemina. After a shot from a revolver the bullet was found encapsulated in the right anterior corpus quadrigeminum; in addition to the other symptoms, the right pupil was nearly twice as large as the left, and the associated movements of the eyes upward and downward were interfered with.

Knoll found the pupillary reaction intact after experimental destruction of the corpora quadrigemina.

According to Adamuek the corpora quadrigemina are the centre for associated movements of the eyes. The right anterior corpus controls the movements toward the right, the left corpus those toward the left. Irritation in the median line between the two causes movement upward with marked mydriasis; irritation of the most posterior parts causes pronounced convergence downward and myosis.

According to Bechterew the anterior corpora quadrigemina perceive light, and exhibit relations to sensibility, to blood pressure, and to the vasomotor nerves. He denies their connection with pupillary reflexes.

Knoll obtained dilatation of the pupils on feeble irritation of one anterior corpus quadrigeminum; Ferrier saw the same results in addition to dilatation of the palpebral fissure and rotation of the head and eye toward the opposite side.

In Ziehrer's experiments (*Arch. f. Psych.*, XXI, 3) every deep irritation of the anterior corpora gave rise to motor phenomena: acceleration of respiration, manifestation of anger, violent movements, rapid running movements, and often nystagmus and dilatation of the pupil on the same side. Traction on the brachia of the corpora quadrigemina gave rise to rapid running movements, but division of the brachia caused no symptoms. Strong irritation of the posterior corpora quadrigemina caused tetanic rigidity which outlasted the irritation; the eye was often closed by an almost tonic spasm. The motor effects on the same side of the body predominated.

In the main, all these experiments show the relations of the anterior corpora quadrigemina to the external ocular muscles, the nuclei of which are situated, in part, directly beneath them. These muscles control the associated movements.   The fibres to the nucleus of the sphincter of the pupil very probably pass in close proximity. Relations to conscious vision in man are wanting or very slight. On the other hand there is a good deal in favor of the view that the anterior corpora quadrigemina convey to the occipital cortex the knowledge of involuntary ocular movements which follow upon light stimuli; and these movements are either directed or inhibited by the occipital cortex.

The pulvinar is simply the most posterior portion of the optic thalamus.   The larger part of the fibres of the optic tract end in it, and terminate in a network.   Its ganglion cells are known to send their axis-cylinder processes to the occipital cortex and degenerate after destruction of the latter (Monakow).

Hughlings Jackson (*Lancet*, 1874) and Pflueger ("Ber. d. Berner Augenkl.," 1878) observed crossed homonymous hemianopsia in destruction of the pulvinar by hemorrhage.   In a tumor of the thalamus and pulvinar Dercum (*Journ. of Nerv. and Ment. Dis.*, 1890, No. 8) observed, in addition to hemianopsia, hemianopic light rigidity of the pupil, although not in a perfectly pure form.   Hence the visual fibres to the nucleus of the sphincter must either pass through the pulvinar or near by it.   According to Schiff, Magendie, and others, irritation of one pulvinar has no influence on the ocular movements.   Bechterew, on the other hand, often observed nystagmus or, rather, nystagmus-like, slow movements.   It is to be remembered, however, that, in the experiment of exposing the thalamus, the cerebral cortex was removed.   The results in cases of tumor are less useful on account of the effect on neighboring parts. In one case of tumor Mills (*Journ. of Nerv. and Ment. Dis.*, 1887, p. 707) observed drawing of the face toward the same side (crossed paralysis) and bilateral lagophthalmus (spasm of the levator?).

The external geniculate body is an accumulation of ganglion cells within the large band of fibres of the optic tract, entering the pulvinar.   Its cells send their axis cylinders to the occipital cortex (**Huguenin, Meynert**) in the region of the sulcus hippocampi.   Only a part of the fibres terminates here in a network, another part passes alongside and through the external geniculate body to the pulvinar.

The only cases of isolated destruction of the external geniculate which are at all useful were observed by Tuerk (*Zeitschr. d. G. d. Wien. Aerzte*, XI, 1, quoted by Wilbrand ("Hemianopsia," p. 103). In both cases there was a cicatrix in and upon the ganglion of one side; during life there was notable (*sic!*) disturbance of vision. Tuerk (Wilbrand, p. 104) also reports three cases in which numerous granular corpuscles were present in the tractus from the chiasm to the external geniculate body, while they were absent in the corpora quadrigemina and optic thalamus.

After destruction of the optic nerves, tractus, and chiasm in the adult the ganglion cells of the primary optic ganglia remain intact, with the exception of a group in the anterior corpora quadrigemina. Only its outer layer of tractus fibres degenerates, and in this way the external geniculate body may appear to be notably flattened, although its ganglion cells are intact. Exceptions are very rare, and must be attributed to the fact that we have to deal either with an ascending degenerative process, or that the primary optic ganglia were first diseased, as is very apt to happen in tabes, multiple sclerosis, etc. In congenital blindness these ganglia are not developed; the same result is produced by experiments on newborn animals (Hadlich, *Ctlbl. f. d. med. Wiss.*, 1880, p. 539). After destruction of the occipital cortex or its medullary fibres the ganglion cells of all three primary optic ganglia undergo degeneration (Monakow).

The pulvinar and external geniculate body are unquestionably of great importance in conscious vision.

Inasmuch as clinical experience, pathological findings, and experiments furnish very incomplete data, we shall gain the best understanding of the functions of the parts under consideration by careful study of the anatomical course of the fibres. Into the anterior corpora quadrigemina pass a certain number of fibres of the optic tract, and also a number of fibres from the so-called fillet. The latter conduct sensory fibres from the entire opposite half of the body to their terminal ganglia (corpora quadrigemina, but particularly the optic thalamus), not directly, but after a double interruption by ganglion cells, first in the spinal cord and then in the medulla oblongata. Sensory stimuli from the entire cutaneous surface of the opposite half of the body and light stimuli from the opposite half of the field of vision meet in the anterior corpora quadrigemina and optic thalamus. Both anterior corpora quadrigemina are united by numerous fibres which decussate in the median line, hence the stimuli probably act on both.

We must probably also attribute to the pulvinar, in addition to the afferent fibres of the optic tract and the efferent corona radiata, as in the optic thalamus itself, a connection with the nucleus of the facial nerve (mimicry).

The external geniculate body, in the main, receives optic-tract fibres and sends out medullary fibres to the occipital cortex. Hence it follows that, among other functions, the primary optic ganglia must convey or permit to pass through them involuntary reflexes from the organ of vision not alone to the ocular muscles but also to the muscles of the head and neck and of the entire body, first on one side, then on the other. The ganglia to which the fibres from the fillet pass preside over the involuntary reflexes from the cutaneous sensibility to the muscles of the eyes and face.

What the primary optic ganglia are to the optic tract, the internal geniculate body appears to be to the acoustic nerve; it presides over the involuntary reflexes from the organ of hearing to the muscles of the eyes, head, face, and then to the entire muscular system.

This must be one of the functions of the primary cerebral ganglion, in particular of the pulvinar and anterior corpora quadrigemina, because it is hardly possible in any other locality.

For example, if a strong sensory stimulus unexpectedly acts upon the dorsal surface of the left hand, then, apart from the local reflexes, the muscles of the neck will turn the head in that direction and the external ocular muscles will focus the eye upon the irritated spot. The same thing happens when a noise falls upon the ear, etc. If a violent sensory irritant acts upon the conjunctiva, then in addition to local reflexes (closure of the lids, etc.) there are others in the arm and hand, which are not even always useful, as, for example, rubbing the eyes. Or if a light falls upon the retina, not alone are the head and eyes turned in its direction, but in the degree that the stimulus is intense and unexpected, motor effects follow in the

entire muscular system : a terrified start, outcries, running, etc. All
these movements are involuntary. They may be inhibited by the
higher centres and can also be replaced by others, but this is so much
more difficult, the more intense, unexpected, and unusual the stimulus has been.

It is still uncertain whether these reflexes are determinately
acted upon by the ganglion cells in the primary ganglia of the brain
or are simply conveyed through them.   The latter is not improbable,
because, as experiments show, the axis cylinders of the cells of these
ganglia radiate almost exclusively into the corona radiata and to the
cerebral cortex; they degenerate after destruction of the latter.
The afferent sensory nerves are in their behavior similar to the posterior roots of the spinal nerves, which directly connect with all
regions of the spinal cord by ascending and descending fibres.   The
visual fibres in an analogous way resolve into a network in all
parts of the posterior horns (primary optic ganglia), and also pass
into the anterior horns (nuclei of the ocular muscles) without demonstrable anastomoses (Koelliker, *Muench. med. Woch.*, March 18th,
1890).   In the spinal cord we find also abundant decussations between the anterior and posterior horns on both sides (in the anterior
and posterior commissures), as we also find between the corpora
quadrigemina and the nuclei of the ocular muscles.

According to this view, the larger part of the fibres of the optic
tract would be resolved, in the primary optic ganglia, into fine networks amid which lie the ganglion cells.   The latter send their axis
cylinders centrally through the corona radiata to the cortex.   To a
less degree the fibres of the optic tract pass through the optic ganglia and are only resolved into a network in the muscle nuclei,
whose ganglion cells preside directly over the reflex involuntary
movements.   This is also true of those fibres of the fillet which are
connected with the optic ganglia and the muscle nuclei.   The fillet
also sends its fibres into the neuropilemma of the optic thalamus,
whence they are switched indirectly to the cerebral cortex as well as
directly to the nuclei of the motor cerebral nerves.   By means of the
former the sensations are carried to consciousness, by means of the
latter the direct involuntary reflexes are executed.

The peripheral stimuli which are constantly entering from every part of the body are distributed among the different muscle nuclei and thus effect a certain tension (tonus) of the entire muscular system. If stronger stimuli enter through any organ of special sense, then they will produce, in the nuclei, a condition of increased irritation which corresponds to the point of entrance of the stimulus and diminishes as the distance from the latter increases (involuntary reflex).

The more delicately graded and complicated are the muscular movements, the more numerous are the axis cylinders and the ganglion cells which belong to them in comparison with the size of the muscle. The same thing is true of efferent sensory nerves and nerves of special sense in relation to the corresponding ganglia, the more finely graded the perceptive qualities are. If we assume that an afferent fibre, which is split up into a network of fine fibres, conducts a stimulus to a ganglion, then the ganglion cell which lies nearest to the point of entrance of the fibre in question will be irritated most intensely, the remoter ones less intensely, in a definite arithmetical or, more probably, geometrical proportion to the distance. All stimulated ganglion cells conduct, with corresponding intensity, the stimulus to the muscle nucleus or to the cerebral cortex, and here again the same thing is true with regard to the ganglion cells, viz., that they are all innervated, but that each one is stimulated with different intensity. Hence, if there is a considerable number of cells, great multiplicity of conscious or purely reflex movements is possible.

This happens in the case of the eye. The number of visual fibres from the optic nerve and tractus, in proportion to the size of the surface of the retina, is enormously large when compared with other localities. This is also true of the nuclei of the ocular muscles in comparison with the size of the muscles innervated by them. According to the location of the light impression on the retina, and according to its quality and quantity, the involuntary reflexes will be very finely graded. For this very reason it will possess a high degree of responsiveness to the character of the light stimulus, especially in regard to frequently recurring stimuli of the same or

4

similar quality. Hence they will closely resemble intentional, voluntary manifestations, and only when the stimulus is very unexpected, unusual, and intense, will they possess the character of involuntary reflexes and extend to the general muscular system of the body.

An exceptional position is occupied by the movements of the internal smooth muscles of the eye, which are influenced by the will to a very slight degree or not at all. According to Mendel's latest investigations (*Deutsch. med. Woch.*, 1889, 471), the fibres which convey the reflex reaction of the pupil to light run directly from the optic tract to the ganglion habenulæ, which consequently effects the further conduction to the muscle nucleus of the sphincter pupillæ. He found in new-born cats, rabbits and dogs, from which the iris had been totally removed, that the corpora quadrigemina and external geniculate body remained intact, but that the ganglion habenulæ of the operated side was atrophic. According to Darschewitsch the pupillary fibres of the optic nerve pass uncrossed to the ganglion habenulæ of the same side (and to the pineal gland), at least in animals. Both ganglia habenulæ are freely connected with one another through the posterior commissure of the brain, and this readily explains the consensual reaction of the pupils. In accordance with these views Knoll found the movements of the pupils retained after destruction of the corpora quadrigemina.

These observations are interesting on account of the relations of the ganglion habenulæ to the pineal gland, which, as is well known, forms in part the rudiments of the so-called parietal eye. A certain support for this view is afforded by the relation of the pineal gland to the termination of the optic tract and to the nucleus of the motor oculi nerve.

From these statements we can recognize how difficulties beset experimental investigation of the primary cerebral ganglia, how complicated the effects of irritation and destruction, and how coarse the experimental results must be. The individual elements of such a ganglion cannot be destroyed directly. We merely know that, after destruction of the occipital cortex or of the medullary fibres leading to it, the ganglion cells of the primary optic ganglia undergo degeneration. We assume that, after destruction of the optic tract, its terminal network of fibres in these ganglia and the fibres which merely pass through are also destroyed. This can only be demonstrated in part, however, by anatomical means.

The principal function of the primary optic ganglia is undoubtedly the transfer to the occipital cortex of the stimuli which have been conveyed to them from the periphery. This takes place through the axis cylinders of their ganglion cells. The fibres passing through the optic tract thus experience a very considerable increase in numbers, and this, as a matter of course, renders possible a much finer differentiation of the visual impressions received in the cerebral cortex.

It is probable that the union of the binocular visual impressions of the corresponding halves of the field of vision into a single sensation takes place in the primary optic ganglia.

The anterior corpora quadrigemina are traversed by the fibres of the optic tract, which pass to the nuclei of the extrinsic ocular muscles. If they were destroyed separately, the involuntary reflex movements of these muscles would be abolished, but only in so far as they are aroused from the opposite half of the field of vision.[1] But inasmuch as the impressions of light are conveyed by the other primary optic ganglia to the occipital cortex, corresponding or similar movements of the extrinsic ocular muscles (especially movement of the eyes toward the luminous object and fixation of the latter) will there be provoked as conscious movements. The loss of reflex motion could only be determined by accurate measurement of the interval between the stimulus and the resulting movement, which in this case is prolonged by a fraction of a second (on account of the longer distance).

The pulvinar really constitutes only the posterior extremity of the optic thalamus. The fibres of the optic tract which pass to the former and the fibres of the fillet which pass to the latter probably form in great part a network in which lie ganglion cells whose axis cylinders enter the corona radiata. By means of these, visual impressions from the opposite half of the field of vision are conveyed from

---

[1] Probably also from the opposite half of the entire body.

the pulvinar to the occipital cortex, and sensory impressions from
the opposite half of the body are conveyed from the optic thalamus
to the parietal cortex.   Some of the fibres, however, merely pass
through and terminate in the nuclei of the facial muscles.   It is also
certain that descending fibres of the optic tract pass down to the
spinal cord, and subserve the involuntary reflexes of the general
muscular system due to visual impressions from the opposite half of
the field of vision.

Inasmuch as auditory fibres also pass to the optic thalamus (to
the internal geniculate body), and probably also fibres from the other
sense organs, we must regard the thalamus and its appendages as an
organ through which the tracts for the involuntary reflexes of all the
senses pass to the facial muscles, perhaps also those of the higher
senses to the general muscular system of the body.   The relations of
the thalamus to involuntary mimicry (Bechterew, Nothnagel) are
thus easily understood.

On the whole, the involuntary reflex movements which are produced will be simple: rotation of the head and eyes toward the
location of a sensory stimulus, distortion of the face (mimicry), de-
fensive movements, inarticulate cries, movements of flight, perhaps
manifestations of anger (preparing for resistance), starting of the
entire body, when the irritant is very unexpected or intense, etc.
The reflexes after sensory stimuli on the part of the eye (closure of
the lids, rubbing, etc.) also probably pass through the thalamus.

Inasmuch as the sensory fibres which pass to the individual
nuclei (as far as the lower end of the spinal cord) come from all
parts of the body, it will be impossible to cause degeneration of com-
pact bundles of fibres from any sensory organ.

The destruction of an optic thalamus, which naturally involves
ganglion cells, network and fibres which pass through and near the
organ, would, therefore, include abolition of the involuntary facial
movements as reflexes from the opposite half of the body, i.e., the
mimetic movements in particular.   The proof of this fact is difficult
because, on account of the intact centrifugal connection with the
muscle nuclei the same movements can still be produced voluntarily.
Special attention must therefore be paid to this point, as Nothnagel

has recently emphasized. But since, in destruction of the thalamus, a large number of ganglion cells are destroyed which send axis cylinders to the sensori-motor cortex, disturbances of sensation on the opposite side of the body, probably in the form of paræsthesiæ, might also be expected. Involuntary reflex movements of the trunk and limbs after cutaneous irritation of these parts are conveyed through the spinal cord and would remain unimpaired.

The general remarks here made concerning the relation of the thalamus to the sensibility of the opposite half of the body are also true of the pulvinar and the visual impressions received from the opposite half of the field of vision. On account of the hemianopsia after destruction of the pulvinar, the loss of reflex movements from the side of the blind half of the field of vision will not appear as a special symptom, while their retention will so appear in central disorders of vision. Perhaps the fibres of the corona radiata which emerge from the anterior corpora quadrigemina and a part of those from the pulvinar serve solely to convey to the cerebral cortex a knowledge of the occurrence of the involuntary reflexes at the time of their production. If these do not occur on destruction of these structures, this function is also abolished without producing a noticeable symptom. Such an assumption would enable us to understand how reflex inhibition in general takes place.

The external geniculate body is one of the principal terminations of the fibres of the optic tract, but in the cases reported (Tuerk, p. 46) its lesion produced no notable disorder of vision. This is very surprising because the ganglion lies directly among the fibres of the tractus which are evidently concerned in vision, and it sends its axis cylinders to the visual cortex. I know only a single possibility in which the destruction of a large number of visual fibres of the tractus could take place without notable visual disturbance, viz., when macular fibres are affected, because these, as a rule, pass into both hemispheres.

At first this view seems paradoxical. On more careful consideration, however, there is much in support of the opinion that it is the macular fibres which terminate, at least in great part, in the external geniculate body. The mid-position which they occupy among the fibres to the pulvinar, which pass everywhere past and through the ganglion, would correspond to the position of the macular fibres in the optic tract, where they in all probability also run centrally. A fraction of the macular fibres probably passes through the external geniculate body to the pulvinar and subserves the involuntary reflexes; the largest part, on the other hand, is resolved in this ganglion into a network of fibres. On account of the numerous ganglion cells which send their processes to the visual cortex, there is made possible a very fine differentiation of the sensory stimuli from the macula lutea, the site of most distinct vision in the retina. As regards the conscious voluntary motor manifestations called forth with the aid of the occipital cortex, preference would be given to the fibres coming from the macula lutea, but this would be true to a much less extent with regard to the involuntary reflexes.

Tuerk (*vide* p. 46) reports three cases in which very numerous granular corpuscles were present in the optic tract from the chiasm to the external geniculate body, but were absent in the thalamus and corpora quadrigemina. In another case he found merely fatty degeneration of the medullary layer upon the external geniculate body, while the latter itself was intact. This would show that, as is true of the optic nerve, so also in the optic tract, there is a certain degree of antagonism, as regards disease, between the macular fibres and the peripheral fibres.

I am well aware that the assumption that the macular fibres terminate in the external geniculate body is by no means proven, but it is worthy of discussion and is stimulating to special investigation which, unfortunately, I am unable to pursue at the present time. In destruction of the macular zone of the visual cortex the ganglion cells of the external geniculate body would necessarily be the principal site of degeneration.

With regard to those fibres of the optic tract which pass to the nucleus of the sphincter pupillæ, it is still doubtful whether they pass through the anterior corpora quadrigemina or the external geniculate body; they probably pass, close to both, directly to the ganglion habenulæ (Mendel). If they alone were destroyed on one side, or at the same time with the anterior corpora quadrigemina (whose elimination causes no visual disturbance), a peculiar symptom, which has not been observed hitherto, would develop, viz., hemianopic pupillary reaction without hemianopsia. Since this symptom must be specially looked for, it may very readily escape observation. If the site of disease is suspected in the region of the corpora quadrigemina and optic thalamus, the hemianopic pupillary reaction without hemianopsia would be a very important focal symptom, which could only be produced in the narrow space between the termination of the optic tract and the nucleus of the sphincter on the same side. We will soon report a case of this kind, although it is not very sharply defined.

In regard to the diagnosis of a lesion of the primary cerebral ganglia, the following points may be emphasized:

Destruction of one pulvinar causes hemianopsia of the side opposite to the lesion; the external geniculate body is usually destroyed at the same time. After a shorter or longer period bilateral hemiatrophy of the optic nerve without diminution of the calibre of the vessels will become visible with the ophthalmoscope. At the same time there will be abolition of the involuntary reflexes of the face and the entire muscular system (therefore, also, of mimetic movements), but only in so far as they can be excited by visual impressions from the opposite half of the field of vision. As the latter is blind, this symptom will not be especially striking, but it must be referred to, because, in affections of the thalamus alone without implication of the pulvinar, abolition of the involuntary mimetic movements occurs without visual disturbance. The involuntary associated movements of the eyes (coarse binocular adjustment toward the site of the stimulus) as well as the reaction of the pupil to light (if the adjacent pupillary fibres are not also destroyed), could also be excited from the blind half of the field of vision. A special examination must be made with the aid of sudden and intense light stimuli.

In destruction of one anterior corpus quadrigeminum the involuntary conjugate and associated movements of the eyes would be abolished, but only in so far as they are excited by light stimuli from the opposite half of the field of vision. They might develop after cutaneous stimuli, because the latter are conducted in the so-called posterior longitudinal bundles directly to the nuclei of the ocular muscles.

As the involuntary associated movements of the eyes produce approximate binocular adjustment toward or away from the situation of the stimulus, *i.e.*, as they are adapted to an end, they may be replaced, inasmuch as vision itself is not disturbed, by the voluntary performance of the same or similar movements, and may thus escape diagnosis.

Hence the condition of the pupil is important because the fibres of the optic tract which pass to the nucleus of the sphincter pupillæ pass close to the pulvinar as well as the anterior corpora quadrigemina, and because the reaction of the pupils to light cannot be influenced by the will. If these "pupillary fibres" are also destroyed, heminanopic pupillary reaction will develop, as in destruction of an entire optic tract; hemianopsia is not, however, necessarily present and will be absent so long as the primary optic ganglia are in the main preserved. According to Mendel's investigations (*l. c.*) the same thing would happen if one ganglion habenulæ were destroyed. Hence there is always sufficient reason for searching for the hemianopic pupillary reaction without the corresponding restriction of the field of vision, in suspected cases of disease in the region of the primary optic ganglia.

Isolated disease of one external geniculate body does not appear to be capable of diagnosis. That it does occur is shown by Tuerk's (*l. c.*) cases, in which notable disturbance of vision was absent. If my assumption that the macular fibres of one optic tract terminate in the external geniculate body is correct, then the destruction of the latter would give rise to paracentral homonymous amblyopic spots in both opposite halves of the fields of vision, while central vision might be entirely normal as in hemianopsia. With the ophthalmoscope slight atrophic discoloration of both outer halves of the papillæ might be expected after a certain lapse of time. If the lesion were situated on the left side the visual disturbance would be especially noticeable in reading, as in right homonymous hemianopsia.

Such cases are not extremely rare. I have under treatment at the present time a man, aged sixty-four years, with symptoms of atheroma of the cerebral arteries, in whom a disturbance of vision suddenly developed five days ago and was especially noticeable in reading. On both sides $V = \frac{2}{3}$ and the boundaries of the field of vision were normal; but reading was peculiarly slow and syllabic. The cause was found, not in a scotoma proper, but in a spot to the right of the point of fixation on each side, in which everything appeared much "duller." The reaction of the pupils to light was retained but was not very free. It was much more pronounced when the illumination came from the left side than when the light entered from the right. On the whole, the disturbance of sight was similar to that in so-called dyslexia (which see later). The ophthalmoscope showed on both sides a somewhat cloudy, grayish-red optic nerve, whose vessels were enclosed in white streaks, though only upon the papilla and its immediate vicinity. I can explain this case only as the result of a small hemorrhage either in the central bundles of the left optic tract or in the terminal ganglion of the central fibres of the tract, the external geniculate body. In the former case the dimness of sight would be greater. The latter view appears to me the more probable one. As in Tuerk's cases, the expected post-mortem findings may consist of a cicatrix in the external geniculate body, and the reduction of sight ($V = \frac{2}{3}$) might also have been regarded as not "notable," had not a special search for scotomata or amblyopic spots in the field of vision been made with the perimeter. A cortical disturbance of vision is excluded by the hemianopic diminution of the reaction of the pupils to light.

C. Cortical Disorders of Vision.

Under the term central disorders of vision we include those which are situated on the central aspect of the primary optic ganglia, *i.e.*, in Gratiolet's optic radiations and in the cortex of the occipital lobe. Those disorders which are located in associative and commissural fibres will be discussed later under the heading of transcortical disorders, for reasons which will soon be explained.

The previous considerations and the anatomical course of the fibres of sight have proven conclusively that: 1, central disorders of vision must be homonymous; 2, they furnish no ophthalmoscopic findings; 3, the involuntary reflex movements due to light impressions remain intact even if the perception of light does not awaken consciousness.

1. Experiment and clinical experience in man show that when the entire occipital cortex (visual sphere) is removed or destroyed, permanent hemianopsia sets in, with loss of both opposite halves of the fields of vision. Degeneration takes place, as we have already said, in the ganglion cells of the pulvinar and external geniculate body and in a great part of those in the anterior corpora quadrigemina (*i.e.*, with the exception of those which send centrifugal fibres into the optic tract). If both visual spheres are destroyed, there is total blindness (cortical blindness), with slight exception as seen by Foerster. Subjective sensations of light, such as photopsias and visual hallucinations, are no longer possible. In cortical hemianopsia they are possible only in the opposite half of the field of vision. Their presence proves that the occipital cortex is still capable of function.

The visual disorder in hemianopsia is not infrequently regarded by the patient as unilateral and as affecting the eye in which the more important external (temporal) half of the field of vision is wanting. Right hemianopsia is regarded as a disturbance of sight in the right eye. Such mistakes are also made by physicians, and only a careful taking of the field of vision will prevent this error.

In cortical hemianopsia the boundary of the field of vision may pass vertically through the point of fixation or may leave this free on both sides, or the intact fields of vision may have a vertical band in common (residual part of the field of vision). The boundary may also run somewhat obliquely on both sides, and the defect is not even necessarily exactly uniform in both eyes. In short, there are subjective differences within certain limits as in hemianopsia from peripheral causes. The larger the number of cases which were carefully studied, clinically and anatomically, the more assured became the fact that in the distribution of the visual fibres there were no inconsiderable individual differences even in the occipital cortex, against which all theoretical considerations are useless. Hence it is impossible to construct a scheme which will hold in all cases for the finer localization in the occipital cortex. It is certain, however, that the macular region of the field of vision corresponds to the cuneus, and perhaps the first occipital convolution (Nothnagel's perceptive centre for visual impressions), and that, of the remainder of the occipital cortex (Nothnagel's memory centre for visual impressions), the anterior parts correspond to the inferior part of the opposite field of vision, the lateral parts to the outer, and the posterior parts to the upper part of the field.

In partial destruction of one occipital cortex, accordingly, the visual disturbance will vary greatly according to the location. Destruction of the cuneus and its vicinity produces approximately the same effect, at least in man, as that of an entire occipital cortex, · viz., homonymous hemianopsia, or at least marked hemiamblyopia in the opposite half of the field of vision. This was the reason that Nothnagel applied the term perceptive centre for visual impressions to this part of the cortex. The part of the cerebral cortex corresponding to the macula lutea is evidently of much greater importance to conscious vision than the macula lutea itself, the destruction of which merely produces a central scotoma. If the lesion is situated in another part of the occipital cortex, the visual disturbance is much slighter, and is less noticeable the closer it is to the outer boundary of the visual sphere. According to the summary scheme furnished above, there would be present homonymous, more or less extensive (negative!) peripherally situated scotomata, or merely homonymous amblyopic spots, both of which produce practically no subjective symptoms and would not be discovered, on examination with the perimeter, unless attention was specially directed to them. On the other hand, lesions of the cortical periphery may produce all sorts of disturbances in visual perceptions and memories and with the impressions of the other senses. These are mainly temporary, and, strictly speaking, included among the transcortical visual disorders which will be discussed later.

Hence, it follows that only Nothnagel's perceptive centre exhibits a tolerably pure optical character. The peripheral parts of the so-called visual sphere are of a mixed character sensorially, the optical portion predominating, but diminishing progressively toward the periphery.

According to theoretical considerations, partial destruction of one perceptive centre would give rise to a strictly homonymous loss of the field of vision (negative scotoma) of a corresponding size in the vicinity of the point of fixation of each eye. Wilbrand ("Die hemi-anopischen Gesichtsfeldformen," Wiesbaden, 1890, p. 5) is also inclined to assume a circumscribed hemorrhage into the perceptive centre of one side in a case of this kind under clinical observation. It might also be due to a small hemorrhage at the entrance of the tractus into the primary optic ganglia, perhaps even into the latter. Post-mortem examinations alone can decide such points. Hun's case (*Amer. Journ. of Med. Sc.*, Jan., 1887) is the only one known to me. After an attack of apoplexy a man lost the left lower quadrants of the fields of vision and the peripheral parts of the left upper quadrants. A lesion was found in the lower (?) half of the right cuneus. [For similar cases see Henschen.]

Experiments on animals (the last ones by Obregia, Schaefer, Munk, etc.), with which clinical experience in man agrees, reveal another function of the occipital cortex. By stimulating the occipital cortex of one side with feeble currents conjugate movements of both eyes toward the opposite side are produced, in a somewhat upward direction when the posterior part is irritated, in a downward direction when the anterior part is irritated. Hence the occipital cortex also exercises motor activity and, as we shall see later, brings about voluntary movements of adjustment of both eyes upon an object which is appearing in the opposite half of the field of vision. Further details will be furnished under the heading of the central disorders of the ocular muscles.

Experiments on animals can be utilized for man only after the exercise of caution. The most reliable are those on monkeys, although even they present considerable anatomical differences from man. The different condition in many animals as regards decussation in the chiasm in itself enjoins caution, and still more the fact, for example, that birds are probably able to see after removal of the cerebral cortex, while the mammalia are not. In such cases parallels cannot be established even between larger parts of the brain which apparently are perfectly homologous.

Among clinical observations cases of softening are the most valuable, because they are most sharply defined ; hemorrhages are less valuable, tumors least of all. Every rapidly or suddenly developing lesion gives rise, apart from the symptoms which correspond to its situation, to other more or less remote symptoms, at first in the hemisphere of the same side, then in the opposite one (remote effects). Only the symptoms which remain after a certain lapse of time are the real signs of the loss of tissue which can be utilized in localization. For example, if an embolism causes softening at the periphery of the occipital cortex, complete hemianopsia or even total cortical blindness may be present immediately afterward. In the latter event the opposite hemisphere first resumes its function and the cortical blindness passes into hemianopsia : then the latter is contracted into an homonymous scotoma, later still into an homonymous peripheral amblyopic spot in the opposite half of the field of vision, which produces no subjective symptoms and is recognized objectively with difficulty or not at all. Only accurate clinical observation of the manner in which, in such a case, the hemianopsia recovers and what part of the field of vision is finally obliterated, will render a local diagnosis possible. Otherwise we have only the clinical diagnosis of no hemianopsia at the time of examination, and the anatomical finding of a small spot of softening at the periphery of the visual sphere (vide, for example, Henschen, *l. c.*, Cases 27 and 28). No case has been reported of actual destruction of the "visual sphere" in man, without the corresponding homonymous disorder of vision.

Bilateral destruction of the visual sphere produces total cortical blindness—bilateral hemianopsia. But hemianopsia on both sides does not lead necessarily to complete blindness. In two very similar cases [1] (Foerster, *Arch. f. Ophth.*, XXXVI, 1, and Schweigger, (*Arch. f. Augenheilk.*, Bd. XXII) typical hemianopsia with retention of a small zone around the point of fixation occurred first upon one side, probably as the result of embolic softening. Subsequently this also developed on the other side. Instead of complete blindness there was left a small field of vision (only a few degrees in diameter) which enclosed the point of fixation, and in which there was comparatively good central vision. Foerster draws the conclusion that it is not necessary to assume a connection of the macular fibres with both hemispheres. This is true; the above assumption is not necessary to explain the so-called residual part of the field of vision. It is sufficient that the macular portion of the cortex, the perceptive centre, obtains its blood supply from two arteries, so that on closure of one the supply necessary to the nutrition and function of that part of the cortex may be furnished collaterally. Nevertheless there may be a distribution of the macular fibres of both eyes to both visual spheres, and this appears probable to me for other reasons. Individual differences must occur in both assumptions. In one they affect the

---

[1] This category probably includes Berger, Bresl. aerztl. Zsch., 1885, No. 1, and Groenouw, Arch. f. Psych., 1891, p. 339.

distribution of the blood-vessels, in the other the distribution of the fibres (or both) to the cerebral cortex. The residuum of the field of vision may be color-blind (Foerster) or sensitive to color (Groenouw); in the latter case a larger part of the field of vision had been spared. As a matter of course the entire visual cortex is not destroyed in such cases.

Diseases of the brain exhibit a tendency to occur symmetrically in both hemispheres, although not always at the same time on both sides, and often with quantitative differences. For example, Edinger (*Deutsch. Zsch. f. Nerv.*, 1, p. 265) reports a case of bilateral softening of the occipital lobes in which suddenly a dazzling light was perceived and then permanent complete blindness set in. Apart from diffuse processes which involve the surfaces of the occipital lobes (for example, in so-called uræmic and diabetic cortical blindness), blindness as the result of the successive or simultaneous development of hemianopsia on both sides may be due to embolism, hemorrhages, even tumors. The cortical disease on both sides may also be partial and may give rise, for example, to loss of both upper halves of the field of vision on account of destruction of the posterior halves of both visual spheres of the occipital cortex. Hoche's very interesting case (*Arch. f. Psych.*, XXIII) of bilateral hemianopsia inferior in an insane patient, with hallucinations in the intact field and like phenomena in the blind field, was only observed clinically. The occurrence of visual hallucinations points to an affection of the corona radiata.

We have already considered those cases in which only an homonymous section of each visual field has been lost. There are also some which are confined apparently to color perception (so-called color hemianopsia or heminchromatopsia). When this is central in character it must also be homonymous, *i.e.*, it must have the same position in the visual field of both eyes.

Cases of this category are quite rare. In pure cases the boundaries are normal, or almost normal, and so likewise is central vision. The color limits either pass vertically through the point of fixation or they leave a spot around it free, as happens so often in complete hemianopsia. It is an astonishing fact that in such cases very little attention has been paid to the "overlapping" color field. Bjerrum (*Centralbl. f. Augenheilk.*, 1891, p. 120) has shown that whenever there was disturbance of color in the field of vision, a disorder of vision could also be demonstrated if the objects examined were sufficiently small. This was even true of a case which, with the ordinary method of examination, had appeared to be pure hemiachromatopsia. A relative amblyopia in the peripheral parts will be disclosed most readily, as in cases of complete hemianopsia, when situated on the right side on account of the interference with reading to which it gives rise. Hence such cases offer the first incentive to an examination in regard to the diminution of peripheral vision. In fact, Epéron (*Arch. d'Ophth.*, 1884, p. 356) and Verrey (*Arch. d'Ophth.*, 1888, p. 289), who mention hemiamblyopia in connection with the disturbance of color, had to deal with cases of right hemiachromatopsia.

Diminished function of the visual sphere thus seems to produce the same disturbance of color sense as interference with conduction in the optic nerve. Central vision is good, on account of the better supply of vessels to the macular part of the cortex (*vide* Foerster's and Schweigger's above-mentioned cases on p. 61), and on account of the presence of a concurrent part of the field of vision. At the periphery vision is diminished to correspond with the perception of colors; there is relative amblyopia, with normal or nearly normal external boundaries. Thus there is no color disturbance without a corresponding disturbance of sight.

If this is true, we do not need to assume with Epéron that, in the cerebral cortex, the elements for the light, space, and color sense lie in juxtaposition like the squares of a chess-board, nor with Wilbrand that they are distributed above one another, in the above-mentioned order, in the occipital cortex, so that those for the color sense are situated to the outside. According to the latter writer hemiachromatopsia is due to superficial disturbances of the cortex, but anatomical proof of this statement is entirely wanting. We need merely assume that the function within the cortex is not abolished, but merely disturbed or rendered difficult,—interference with conduction in the neuropilemma,—or that the ganglion cells exhibit diminished function and capacity for reaction, which are temporary and may also remain stationary.

The occurrence of total hemianopsia with color-blindness of the intact half of the field of vision (Schoeler, in Michel's "Jahresber." 1884, p. 386, twice among fifteen cases), or of hemiachromatopsia, together with loss of an homonymous quadrant of the field of vision (Swanzy, *Lancet*, 1883, II, p. 103), is thus explained much more simply and by a single lesion which abolishes the function of one visual sphere and merely impairs that of the other.

If no overlapping area of the fields is present, then the disturbance of central vision in hemiachromatopsia will also be much more noticeable, as, for example, in Verrey's case (*l.c.*) of hemorrhagic cyst in the lower part of the occipital lobe.

Hemiachromatopsia may also be peripheral in character—for example, in neuritis of one optic tract. If it is binasal, *i.e.*, confined to both inner halves of the field of vision, as in Galezowski's case (*Gaz. Méd.*, 1880, p. 163), it is certainly peripheral (chiasm), and also when it is unilateral.

A temporary disturbance of nutrition must be assumed in a typical hemianopic disorder, the so-called scintillating scotoma (teichoscopia, amaurosis partialis fugax, etc., preferably called hemianopsia completa or incompleta scintillans). It is probably due to spasm of the artery supplying the occipital lobe (posterior cerebral artery). This furnishes a certain connection with migraine, to which scintillating scotoma exhibits great resemblance. French writers often term it "migraine ophthalmique."

After lively scintillation the field of vision contracts to complete or incomplete hemianopsia, or the point of fixation is chiefly affected. The scotoma is always strictly homonymous and negative, *i.e.*, it does not appear dark or black. This disorder of vision may last from a few minutes to half a day, either alone or in combination with headache (migraine), nausea, etc. It is very rarely present on both sides at the same time, and then causes temporary cortical blindness, which is easily recognized, however, on account of the characteristic scintillation. This was observed on several occasions by my former teacher, Professor Horner. *Per se* the affection is harmless. On account of its great frequency it is often said to be connected with material diseases of the eye, for example, with glaucoma. This is decidedly incorrect, because glaucoma, which is almost always bilateral, is surely not related to a unilateral disorder of the cerebral cortex. It would be more likely to be associated with brain disease, but this is also very rare. It is often merely a symptom of insufficient nourishment, for example, in individuals who do not eat anything until noon, and its development is also favored by mental strain.

Irritation of the cortex of the visual sphere causes visual hallucinations, *i.e.*, subjective visual perceptions varying from simple to quite complicated processes. The visual hallucinations are usually present on both sides, but they may also be confined to one-half of the field of vision (homonymous hemiopic hallucinations) and then develop in the opposite visual sphere.

Half-sided hallucinations are observed most frequently in migraine, epilepsy, and hemiplegia, more rarely in other brain diseases, including insanity. Usually—especially in the first-named diseases —there is also hemianopsia and the hallucinations are seen upon the blind half of the field of vision. They also occur when the functions of the central visual apparatus are, in other respects, entirely normal.

In irritation of one visual sphere conjugate deviation of the eyes to the opposite side is also observed; this will be discussed later. In unilateral visual hallucinations we often find conjugate deviation of the eyes toward the supposed site of the hallucinations.

Whether central disorders of vision are ever unilateral is still an open question. For example, Charcot regards the majority of the unilateral hysterical disorders of vision as central, and explains them by decussation of the uncrossed visual fibres on the central or cortical side of the peripheral optic ganglia. Such an assumption has no anatomical justification. This does not combat the view that unilateral visual disorders may be produced by unilateral irritations and destructions of the brain (Lannegrace). Here we have to deal probably with disorders of circulation which, provoked by a lesion of the brain, exercise their effect upon peripheral parts of the visual fibres. A disease of the frontal lobe, which implicates the optic nerve on the

5

same side, causes merely a peripheral disorder of vision, despite the fact that the cause is located in the central organ of the nervous system. We will enter more fully into this question in the consideration of hysterical disorders of vision.

2. In central disorders of vision the ophthalmoscopic appearances are normal. The ganglion cells which perish in the primary optic centres do not send axis cylinders into the optic nerves. Atrophic degeneration of the optic nerve is therefore not seen with the ophthalmoscope, especially as the function of the nerve, as we shall soon see, is not entirely abolished in destruction of the visual cortex. On page 14, we have considered the exceptions in which finally the degeneration does become noticeable at the periphery.

Cases of ophthalmoscopically visible bilateral atrophy of the optic nerves and total blindness with intact pupillary reaction to light, as in Jessop's case (*Lancet*, July 11th, 1891, p. 73), may be explained on the assumption that the ophthalmoscopic picture of atrophy of the optic nerve also occurs when vision is good, and that the symptoms in question must follow, if a central disorder of vision is superadded.

3. In all central disturbances of vision the involuntary and unconscious reflexes on illumination of the insensitive parts of the field of vision remain intact, particularly the movements of the pupil to light. These are most easily observed and tested. The direct and consensual pupillary reactions to light are retained, despite the fact that the latter produces no sensation of light from the corresponding part of the field. Hence, in central hemianopsia the pupillary reaction will be the same whether the light enters the eye from the seeing or the blind half of the field of vision, while in the much rarer hemianopsia from disease of an optic tract no pupillary reaction to light follows on illumination within the blind half of the visual field (hemianopic pupillary reaction). This symptom must be specially looked for, particularly when we have to deal, not with hemianopsia, but with smaller homonymous defects in the field of vision. In complete bilateral cortical blindness the reaction of the pupils to light, despite the complete absence of all perception of light, is extremely striking and has long been known (Graefe).

Other involuntary light reflexes are observed with much more difficulty. We should notice whether, in the field of an homonymous scotoma, very sudden and intense light stimuli are capable of producing involuntary movements of the éyes and head toward or from the source of light, closure of the lids, etc. This can be shown most clearly when the central disorder of vision is bilateral, *i.e.*, in cortical blindness.

I have observed one case of bilateral central blindness (uræmic blindness in a boy of about ten years) in which, apart from the intact pupillary reaction to light, movement of the eyes and head toward the source of light occurred on the entrance of direct sunlight from the side; on the entrance of light from the front, closure of the lids followed, although the patient had not the faintest perception of light.

But the involuntary reflexes from the visual organ (and also probably from other higher organs of sense, for example, the ear) evidently play no great part in man, particularly in adults. Apart from the pupillary movements, which are independent of the will, they are replaced mainly by conscious and voluntary movements which are intended, in general, to effect the same objects as the involuntary reflexes; viz., adjustment of the eyes upon an object which appears in the field of vision, looking away from the object, closure of the lids in a bright light, etc. It is not impossible, especially in adults, that these involuntary reflexes from the visual organ which are subject to outside influence are almost entirely suppressed by practice and experience and that they only take place after very unexpected, sudden, and intense light stimuli (fright). This readily explains the fact that the involuntary light reflexes were especially striking in my youthful patient. Further investigations in this direction would be very interesting.

Reflexes to more minute visual stimuli, such as the approximation of a needle or a finger to the eye, do not belong to this category, because they require actual seeing, differentiation and judgment of objects. Such movements are conscious and performed with the aid of the occipital cortex. We will again refer to these conditions.

Central disorders of vision from lesions in the corona radiata of the occipital lobe (Gratiolet's optic radiations) will be very similar to those caused by lesions of the occipital cortex. The only possibility of differentiation—unless other local symptoms are present—is furnished by the presence of visual hallucinations. These presuppose a visual cortex which is still capable of function. Hence, if visual hallucinations on the blind side are present in hemianopsia (with intact pupillary reaction on illumination of the blind half of the field of vision), the lesion is situated within the corona radiata, and the cortex of the visual sphere is still exercising its function to a greater or less extent. If this symptom is absent, a differential diagnosis is impossible unless the accompanying conditions furnish some data.

It would be advisable to employ special technical expressions for the different kinds of central and peripheral visual disorders. I would recommend that amblyopia and amaurosis be used solely for peripheral (and the rare intermediate) disorders, and the words terminating in "opsia" for central disorders of vision. In the former the light reflexes of the pupil are also disturbed, and atrophic changes in the nerve, which are visible with the ophthalmoscope, usually occur sooner or later; in the other forms, this is not true. The visual disorders of the tractus and those which are located in the chiasm would then be called hemiamaurosis and hemiamblyopia (homonymous, binasal, bitemporal, incomplete, symmetrical). Hemianopsia would be reserved for central half-blindness and would always be homonymous. Anopsia would mean central blindness (cortical blindness), and the term miopsia (= seeing less) could be introduced for central diminution of vision. Hemimiopsia would then mean those cortical homonymous disorders of vision which do not amount to complete hemianopsia.

Before proceeding further in the discussion of the central disorders of vision, we must trace, if possible to the cortex, the motor and sensory disorders of the eye which, as we have seen, are never purely motor, purely sensory, or purely sensorial. A critical sifting of the different theories with regard to localization in the cerebral cortex is only possible after the fullest possible recognition of the relations of the latter to the centripetal and centrifugal tracts.

2. Disorders of the Voluntary Ocular Muscles.

The muscular disorders consist of paralyses and spasms, in part of a very peculiar and characteristic kind. In general spasms play a subordinate part; they are due to irritation of those parts whose destruction would cause paralysis. Hence, spasms and paralyses furnish the same local diagnosis. In many cases the spasms are merely the first stage of a subsequent paralysis of the muscle.

The paralyses, like the disorders of vision, may be peripheral, intermediate (nuclear), and central. We speak of paralysis when the affection is unilateral and the paralyzed muscle or muscles are supplied by a single nerve; all other cases are known as ophthalmoplegia. This distinction, however, is not always carried out strictly.

a. The peripheral paralyses are characterized by the fact that the voluntary as well as the involuntary movements of the muscle are abolished. The so-called degeneration reaction appears very early and the nerve-fibres degenerate, together with the ganglion cells of the nuclei from which they originate. Finally, the muscular fibres also undergo fatty or connective-tissue degeneration (really only two stages of the same process), as the result of which secondary changes (secondary contractions, etc.) usually set in. Recovery will take place only when conduction has been abolished temporarily, not when it has been interrupted or destroyed for too long a period.

The peripheral paralyses of the ocular muscles may be thus located : 1. Intra-ocular (iris and ciliary muscle), for example, in injuries. 2. Within the orbit; the muscles themselves or the nerve trunks and twigs may be affected. 3. Intracranial, within the skull to the entrance of the nerve trunks into the brain. 4. Within the brain itself (fascicular, Mauthner and Dufour; preferably called root paralyses), between their entrance and their termination in the nuclei whose position has already been described. We will devote only a brief space to the peripheral paralyses. Their location will be inferred mainly from the accompanying circumstances, the coexistence of other disorders, etc.

An orbital cause is probable if the inferior oblique and the internal ocular muscles alone are affected, because the latter receive their nervous supply from the branch of the motor oculi which passes to the inferior oblique (*vide* p. 27). Otherwise the conclusion depends mainly on the complications, such as exophthalmus, unilateral or bilateral character, number and kind of the muscles. Primary affections of the muscles are rare and usually diagnosed with difficulty unless œdema of the insertion of the muscle into the eye, pain on attempting movement, etc., point to such a condition. In orbital inflammation the eye is more or less immovable, either as a whole or chiefly toward one side. Exophthalmus is present or the eye is pushed forward toward the side opposite to that in which movement is abolished. A similar condition is observed in orbital tumors. There is often coincident disorder of sight and even complete blindness. After a certain lapse of time atrophic conditions may be found at the entrance of the optic nerves. Pressure on the eye is almost always painful. The internal ocular muscles may escape when the cause of paralysis is situated within the orbit.

In congenital paralyses the muscle in question is very often absent (usually the levator palpebræ superioris [congenital ptosis], more rarely the superior rectus or some other muscle).

Intracranial or basilar paralyses are not easily diagnosed if only a single nerve is involved. In many cases several adjacent nerves at the base of the brain are implicated, particularly those on the same side, or disorders are found in the distribution of the olfactory, optic tract, chiasm, etc. (meningitis, tumors, aneurism). Paralysis of the motor oculi with exemption of the internal ocular muscles can hardly be basilar. On the other hand post-mortems have shown that the majority of so-called "periodical" paralyses of the ocular muscles are basilar in character (multiple neuritis of the nerve roots and origins).

Fascicular location of a paralysis (root paralysis) is assumed, for example, when an ophthalmoplegia (*vide* above) is complicated with crossed hemiplegia, because in the cerebral peduncle the roots of the nerves of the ocular muscles intermingle with the still uncrossed motor portion of the former. The decussation of the motor fibres in the pes pedunculi only takes place in the medulla (decussation of the pyramids). Such paralyses are extremely rare without coincident lesion of the nucleus.

Peripheral paralyses which furnish no data whatever concerning their cause or location are not very frequent. They are then usually called " rheumatic."

b. The so-called nuclear paralyses present, like peripheral paralysis, total suspension of the voluntary and involuntary movements; in them, likewise, the nerve and muscle degenerate at an earlier or later period. The degeneration cannot be followed farther centrally, although the cerebral cortical cells, whose terminal network of fibres in the nerve nuclei is destroyed, are undoubtedly also destroyed. It is evident that the fibres which pass from the cortex to the nuclei of the ocular muscles, and indeed all the centrifugal fibres which pass to the latter, do not run in close bundles. Nor do they originate in a circumscribed part of the cortex, because otherwise degenerations would be found in the corona radiata and the cerebral cortex.

In the diagnosis of nuclear paralysis as distinguished from the peripheral forms, the anatomical conditions of the nerve nuclei are especially decisive (*vide* Figs. 5 and 6). For example, if the sphincter pupillæ of only one eye is paralyzed and all the other muscles of that motor oculi nerve act normally, this can hardly be explained by anything but a nuclear affection (apart from local drugs). The diagnosis is still more positive when the sphincters of both pupils, whose nuclei lie close together, are alone paralyzed. When the cause is in the nuclei, only single muscles of one or both motor oculi nerves will be affected, for example, isolated ptosis. It not infrequently happens that only the internal or the external muscles of one or both eyes (internal or external ophthalmoplegia) will be involved or that similar muscles are paralyzed in both eyes. Isolated paralysis of the internal ocular muscles of one eye may also be due to ocular (atropine, trauma) or intraorbital causes; for example, destruction of the ciliary ganglion, which would also produce anæsthesia of the cornea. In such cases, however, sufficient diagnostic and anamnestic data can usually be found. The diagnosis is often more difficult in so-called multiple neuritis, especially root neuritis, when only individual bundles of the motor oculi degenerate, and when, as sometimes happens, this is almost symmetrical on the two sides. In fact, several cases which were diagnosed during life as nuclear paralysis were found, on autopsy, to be basilar in character, or the findings were entirely negative (Eisenlohr, *Neur. Centralbl.*, 1887, p. 337).

The diagnosis is positive when a slowly progressive process attacks the ocular muscles in the order in which their nuclei are arranged anatomically alongside of one another. This may occur in an ascending or descending direction, so-called progressive paralysis of the ocular muscles or superior polio-encephalitis (Wernicke) as opposed to inferior polio-encephalitis or bulbar paralysis which is a closely allied process. Nevertheless, these two forms of disease do not often pass into one another. An exhaustive monograph on chronic progressive paralysis of the ocular muscles has been furnished by Siemerling (*Arch. f. Psych. u. Nerv.*, Bd. XXII, Supplement) from which it appears that the disease is not a unit. It is generally a nuclear affection, but the anatomical appearances may vary greatly.

From an analysis of two hundred and twenty cases Dufour distinguishes acute and chronic forms of nuclear paralysis. The chronic cases may remain stationary for ten years or they may be slowly progressive. In the latter event they may be confined to the principal nucleus or, beginning at other muscle nuclei, they may extend to those of the eye, or they may follow the opposite direction. Nuclear paralysis may complicate bulbar paralysis, progressive muscular atrophy, tabes, disseminated sclerosis, or it may constitute the initial symptom of tabes. Of the hitherto known cases of periodical paralysis of the ocular muscles (*vide* the cases collected by Moebius, Schmidt's *Jahrb.*, Bd. 207, p. 244) Dufour regards only four as surely of nuclear origin (Pflueger, *Tagebl. d. Naturforsch.-Versamml. z. Strassburg*, 1885, p. 491; Vissering, *Muench. med. Woch.*, 1889, p. 699; Camuset, *Gaz. des Hôp.*, 1875, p. 259, and Dubois). All others are probably due to a basilar cause which was found in every case in which an autopsy was obtained (exudation around the motor oculi, Gubler; eruption of tubercles in this region, Weiss; fibrochondroma of the nerve, Richter).

Among the acute nuclear paralyses there is one severe form which usually proves rapidly fatal. This is due generally to an acute hemorrhagic polio-encephalitis at the floor of the fourth ventricle (in drinkers), but solitary tubercles or a cyst (Bull) have also been recognized as causes. Sometimes the post-mortem findings were entirely negative (Eisenlohr, *l. c.*, etc.). The benign acute form includes almost the half of all known nuclear paralyses. This variety may be due to injuries, chemical poisons, constitutional anomalies, acute and chronic infectious diseases (for example, paralysis of accommodation after diphtheria).

In those cases of nuclear paralysis in which autopsies were obtained there were usually found hemorrhages or hemorrhagic inflammations of the nuclei, much more rarely acute or chronic inflammation or simple atrophic degeneration. In one-sixth of the cases the findings were negative. The complete bibliography will be found in the two principal works on nuclear disorders of the ocular muscles, by Mauthner, Wiesbaden, 1885, and Dufour, *Annal. d'Oculist.*, 1890, p. 97.

The congenital absence of muscles and congenital paralyses are probably always due to intra-uterine peripheral or nuclear causes. Moebius (*Muench. med. Woch.*, 1892, Nos. 3 and 4) states that in so-called "infantile atrophy of the nuclei," the facial nerve is also often affected, but not the internal ocular muscles. After the process has run its course, it can hardly be decided whether the nucleus, nerve, or muscle was primarily affected.

The very rare symptom that one eye is deflected downward and outward, the other inward and upward, points to the nuclear region or at least to its immediate vicinity. According to Nothnagel it is a focal symptom of disease of the middle crura cerebelli.

c. Central Disorders of the Voluntary Ocular Muscles.—On the central side of the nuclear region the motor fibres no longer run in bundles. Hence the anatomical demonstration of their course is practically impossible, and this must be inferred, in the main, from non-anatomical premises.

We have intentionally used the term "disorders" because, as in all central paralyses, we have to deal, not with complete paralysis, but merely with disorders of motion, *i.e.*, with the loss of certain qualities of motion.

In a motor nerve nucleus we find, apart from the connective-tissue constituents and the vessels:

1. Motor ganglion cells, whose axis-cylinder processes form the motor nerves.

2. Networks of nerve fibres (neuropilemma). The latter undoubtedly has different origins. For the nuclei of the ocular muscles the following three sources of origin can be demonstrated:

a. Fillet fibres which convey sensory impressions from the opposite half of the body to the nuclei of the ocular muscles. This connection evidently subserves the involuntary ocular movements after sensory impressions from the opposite half of the body; associated deviation of the eyes (and head) toward the approximate position of a tactile impression, etc. The great mass of the fillet fibres passes to the optic thalamus.

b. Tractus fibres from the anterior corpora quadrigemina (Meynert's fibres), corresponding to the opposite half of the field of vision. These subserve the involuntary ocular movements after impressions of light from the opposite half of the field of vision; these are also conjugate movements of adjustment; *a* and *b* are centripetal tracts.

c. The third place from which fibres pass to the nuclei of the ocular muscles is the cerebral cortex, particularly the visual sphere in the occipital lobes, in whose corona radiata they run. These are centrifugal tracts.

It is an astonishing fact that definite relations have not been discovered between the ocular movements and those parts of the cortex which are called "motor." A definite relation exists only between a spot in front of the upper extremity of the anterior central convolution (viz., Fig. 8, 1) and the levator palpebræ superioris and orbicularis palpebrarum, *i.e.*, with movements of the eyelids, not of the eye itself.

Ocular movements toward the opposite side (conjugate deviation) can be obtained occasionally on irritation of almost all parts of the cortex of a hemisphere, but this can be done most certainly from the cortex of the occipital lobe, *i.e.*, from the visual sphere. The investigations of Munk and Schaefer in particular have shown that irritation of one visual sphere with a feeble induced current causes associated movements of both eyes toward the side opposite to the irritation. The eyes are also directed downward when the irritation affects the anterior zone of the visual sphere, and upward when in the posterior zone. Under certain conditions, dependent upon the site of irritation and the position of the eyes, there are (apparently) no associated movements, but movements of adduction of one or both eyes (convergence!) or a continuance of the eyes in the same position.

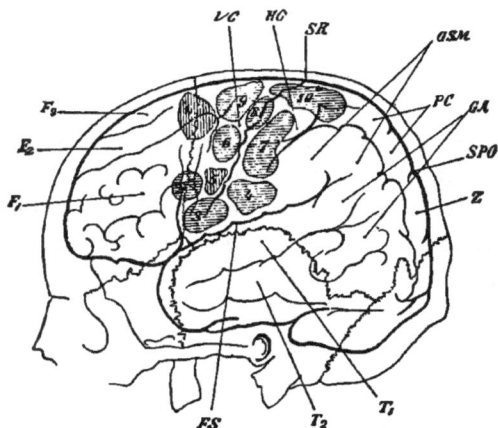

Fig. 8.—The Left Motor Cortical Centres in Man, after Bergmann. *SR*, Sulcus of Rolando; *FS*, fissure of Sylvius; *SPO*, parieto-occipital sulcus; F_1, F_2, F_3, first, second, third, frontal convolution; *VC* and *HC*, anterior and posterior central convolutions; T_1, T_2, first, second, temporal convolutions (T_2, auditory centre; T_1, Wernicke's convolution, so-called); *GA* and *GSM*, angular gyrus and supra-marginal gyrus, both together form the inferior parietal lobe; *Z*, cuneus; *PC*, præcuneus; 1, oculomotor (levator palpebræ superioris); 2, hypoglossus; 3, motor speech centre; 4, oral part of facial; 5, facial part of same; 6, 7, 8, 9, upper extremity (6, abductors; 7, flexion, supination, and pronation-motor writing centre; 8, extensors); 10, lower extremity.

Obregia (*Arch. f. Anat. u. Phys.*, 1890, p. 260) arrives at essentially the same results. He also noted the important fact that irritation of the macular region of the visual sphere hardly excites any movements, but that the conjugate movements of the eyes toward the opposite side become much more extensive, the more the periphery of the visual sphere is approached. These are evidently conjugate movements of adjustment of the eyes. The visual sphere contains a motor projection field for movements of adjustment of the eyes, so that, on irritation of a definite part, that conjugated movement is produced which turns the fixation point of both eyes toward the

corresponding point in the opposite half of the field of vision. If
the irritation affects the macular region of the visual sphere, no
movement follows; the irritation appears to come from the point
of fixation, and this can only happen when the eyes are already
adjusted.

These ocular movements also take place when the associative fibres
to the motor cortex are divided, and the occipital lobe, in a measure,
is isolated (Munk, *Centralb. f. Augenheilk.*, 1890, p. 149). The
muscle nuclei are accordingly stimulated along the direct path of the
corona radiata, and both eyes are always impelled to perform conjugate movements.

These experiments agree with clinical observations, in which irritation of one hemisphere causes conjugate deviation toward the opposite side; paralysis causes abolition of the voluntary conjugate
movements toward the opposite side. If the other hemisphere is
in a condition of irritation at the same time, it will cause conjugate
deviation toward the site of disease. Newman (*Berl. kl. Woch.*,
1890, p. 403) published an apparent exception to this rule. In his
case there was a hemorrhage into the frontal lobe. But this does
not necessarily cause paralysis of the entire hemisphere. Indeed,
those parts of the cortex which are remote from the lesion—in this
instance the occipital cortex—will be more apt to be in a condition
of irritation. Hence it is not astonishing that the eyes were deflected to the side opposite to the lesion. Such exceptions merely
prove the rule.

The influence of the inferior parietal lobe on the conjugate movements of the eye (Wernicke) is easily explained by its close proximity to the periphery of the visual sphere, which has a special
motor efficiency and whose destruction causes only slight disturbance
of vision (*vide* p. 59).

The ocular movements which Munk obtained in his experiments
by stimulating the visual sphere he regards as the lowest visual reflexes, which do not presuppose visual perceptions but merely sensations of light. In fact the movements obtained in such an experiment are involuntary. In my opinion, however, the performance of
these movements takes place along those tracts which, in the normal

condition, convey the voluntary movements of the eyes after conscious visual impressions.

In the visual sphere the afferent optic and the efferent motor fibres are distributed in such a way that irritation of any part of a visual sphere adjusts both eyes in the same way as if an object, appearing at the corresponding spot in the opposite half of the field of vision, were fixed binocularly. Hence, the optical cerebral cortex is at the same time the motor centre for the voluntary ocular movements, in so far as the latter are provoked by conscious light impressions. Each visual sphere controls these movements chiefly in the domain of its half of the field of vision on the opposite side.

In Obregia's experiments the conjugate movements of the eyes were so much more extensive, the more peripheral the site of irritation. A peripheral position in the visual sphere corresponds to a peripheral part of the opposite half of the field of vision. The farther the image of an object is from the point of fixation, the more extensive, as a matter of course, must be the conjugate movement of the eyes necessary for adjustment. If the image of an object falls upon the fovea centralis of the retina and is conducted thence to the macular portion of the visual sphere, no conjugate movement will result because the eye is already adjusted. Now convergence and accommodation for the object are alone necessary, and for these two movements—viz., the fine adjustments of the eyes which are already turned in the general direction of an object—which take place under the control of conscious vision, we must regard the macular portion of the visual sphere as the cortical centre; from its periphery are produced voluntary associated movements after conscious light impressions and also under the control of conscious vision.

As the nucleus of the motor oculi of one side sends its fibres to all the muscles which are innervated in conjugate movements of both eyes toward the opposite side (*vide* p. 22), as the roots of the trochlearis decussate while those of the abducens do not, each visual sphere stands in centrifugal connection mainly with the nuclei of the motor oculi and trochlearis on the same side, and with the opposite abducens nucleus. But if it is subsequently proven that the root fibres of the abducens nucleus do decussate, as is positively maintained by some writers—this is denied with equal positiveness by the majority of writers, but it is certain that numerous fibres between the two nuclei undergo decussation—then the visual sphere would be in centrifugal connection essentially with the nuclei of the ocular muscles on the same side. In the main the macular locality of the visual spheres would send its motor fibres to nuclei 8 and 3 of Fig. 6 (p. 18), which are employed in convergence and accommodation. From this part contraction of the pupil as an "associated movement" must also be provoked, and this, as a matter of course, is bilateral. The cortical impulse for accommodation and convergence, likewise, is always equally strong for both eyes.

All movements which are excited by the cortex affect both eyes and are associated and conjugate; this was also found to be true of the involuntary reflex movements.

If a visual sphere is destroyed, then, apart from the consequent disturbance of vision, the following phenomena will be produced:

1. The voluntary ocular movements after light impressions within the lost half of the field of vision are abolished. It goes without saying that when an object is not seen, adjustment upon it will not be made, and that the disturbance of motion will be entirely overlooked on account of the disorder of vision. If the object were visible there is no doubt that voluntary adjustment of both eyes would take place. In addition it is to be noted that not all the conjugate ocular movements toward the side of the lost half of the field of vision are abolished, but only in so far as they are induced by conscious visual impressions. This disorder of motion is capable of direct demonstration. If I attempt to imitate the visual disorder in right hemianopsia by attempting to read immediately behind a screen, this is indeed made somewhat difficult, but not by any manner of means to the same extent as in cortical right hemianopsia. As the visual disturbance in both cases is the same, this difference can only be explained by a disorder of motion; the aid of the constant movements of adjustment which are excited by peripheral visual impressions, and which are necessary for fluent reading, are lacking. The voluntary ocular movements toward the opposite side, which are excited from other parts of the cortex, are only ap-

proximate and only replace imperfectly the movements performed with the aid of the occipital cortex; hence also the rapid exhaustion.

In the same way that the visual sphere is active optically and oculomotorially, the auditory sphere is active acoustically and otomotorially. According to Baginsky irritation of the auditory sphere with feeble electrical currents also excites movements of the ears, even dilatation of the palpebral fissure, which I can only compare with those observed in listening. The so-called motor region also contains an abundance of sensory fibres, in great part from the optic thalamus (whose ganglion cells perish after destruction of the motor cortex (Monakow). Hence the optic thalamus constitutes the intermediate ganglion of the motor cortex, analogous to the relation of the three primary optic ganglia to the visual sphere and (probably) of the internal geniculate body and posterior corpora quadrigemina to the auditory sphere.

In circumscribed destruction of one visual sphere it is difficult to demonstrate the homonymous disorder of vision; this is equally difficult with regard to the consequent conjugate disorder of movement. This could be done more easily if the macular locality of one visual sphere were alone affected. The visual disorder is then slight because the fixation point is intact in both eyes. On the other hand, the unequal motor innervation of the centre for convergence will also be made evident by the fact that the closely approximated images of both eyes can be made to unite with difficulty or not at all, a condition usually known as imperfect fusion of the double images. I have observed this symptom as an entirely isolated temporary phenomenon which was substituted for an attack of scintillating scotoma and was attended with the same subsidiary symptoms as the latter. In such cases the cortical location is undoubted. In others the lesions seem to be subcortical (corona radiata or vicinity of the muscle nuclei) as, for example, in the imperfect fusion which is an initial symptom of locomotor ataxia. The principal point is that the cortical innervation of the nuclear centre for convergence in the motor oculi nucleus does not take place uniformly from both macular localities of the visual sphere.

2. In destruction of a visual sphere the following functions remain intact:

a. The involuntary reflexes after unconscious light stimuli, particularly the reaction of the pupils to light on illumination of the blind half of the field of vision. Conjugate movements of the eyes and corresponding movements of the head, closure of the lids, etc., may also take place, despite the abolition of the perception of light, as is shown by my case mentioned on p. 67. *b.* Involuntary movements of the eyes after other than light stimuli (cutaneous irritants, noises, etc.). *c.* Voluntary movements of the eyes after other than light stimuli. The retention of the latter conceals, to the greatest extent, the disorder of motion in cortical hemianopsia, because they affect the same object as voluntary movements of the eyes after conscious light perception (adjustment of the eyes toward the locality from which the stimulus starts). But the conjugate movements of the eyes, without control of vision, are much more awkward—for example, those produced at the word of command. Under normal conditions it is sufficient that the movement of the eyes is an approximate one, that it brings the object in question into the field of vision. Then follows the fine adjustment with the aid of the sense of sight. The voluntary conjugate movements of the eyes (for example, those at the word of command) in the blind are remarkable from the fact that they again pass gradually and unconsciously into the position of rest (Raehlmann).

As this voluntary innervation of the nuclei of the ocular muscles also takes place when the visual sphere is destroyed, their tracts cannot pass by associative fibres to the visual cortex and from thence through the corona radiata to the muscle nuclei, but they must pass directly to the latter. A few nerve fibres through which the eyes are incited to perform conjugate movements upward, downward, to the right or left, suffice for this purpose. The rest is done with the aid of the visual cortex. It follows again that the cortex is nowhere purely motor, purely sensorial, or purely sensory, and that there is nowhere a localization in the strictest sense of the term, although in general certain parts of the cortex chiefly exercise certain functions, and the individual parts are nowhere absolutely co-ordinate.

Binocular adjustment is learned by the child at the same time with the judgment of fixed objects, with the development of conscious vision and the recognition of objects and processes of motion. The necessary fibrous connections between the retina and cerebral cortex, between the fovea centralis and the perceptive centre, only develop fully after birth and gradually become medullated and thus capable of function. If this development is disturbed for any reason, for example, by early acquired opacity of the refracting media, by the impossibility of perfect vision on account of congenital anomalies, etc., then a peculiar disturbance of movement, viz., nystagmus, is constantly noticed. By its conjugate occurrence this points to a cause which is situated centrally from the muscle nuclei. If visual disturbance or blindness begins in later life, the development of nystagmus is much rarer. Hence, nystagmus may be defined as imperfect cortical innervation of the voluntary muscles of the eye (as a peculiar form of cortical paralysis agitans). Its real cause may be peripheral, central, or both.

If insufficient vision is obtained by an eye from optical causes (opacity of the media, albinism, marked astigmatism), or because, for example, the fovea centralis cannot be utilized in fixation (early convergent squint, etc.), then the visual impressions from the cerebral cortex do not suffice to excite, on the part of the visual cortex, an impulse to movement which is sufficiently intense to cause proper adjustment of the eyes. The motor cortical innervation is too feeble and assumes the character of paralysis agitans. Under such circumstance, it is also understood that the ocular movement does not correspond to an apparent motion of objects because the individual, so far as he remembers, has never seen otherwise and knows that the visible objects are not moving. The apparent rest of the objects despite the movement of the eyes is a psychical process.

If the visual disorder is bilateral, the nystagmus is constant; if it is unilateral, nystagmus is usually not present when an object is fixed with the good eye. But binocular nystagmus will develop forthwith when the poor eye is employed in fixation (strabismus convergens concomitans).

6

The visual disturbance in early strabismus convergens—unless this is alternating—is due to the fact that, from the squinting eye, the fibres from the macula lutea and fovea centralis are not connected in the normal manner with the macular locality in the visual cortex, but that this takes place from a peripheral part of the retina, upon which the image of the object looked at falls on fixation with the other eye. This peripheral part of the retina does not possess the organization of a fovea centralis, and hence the acuteness of vision is more or less diminished (vicarious macula). Nevertheless it occasionally happens that crossed double vision appears in the correct position of the eyes after the operation for strabismus, because relative external squint now obtains for this acquired macular locality. Such cases, however, are quite rare.

The disturbance of movement in the ordinary amblyopia of strabismus differs in no essential respect from the congenital or early acquired disorder of vision due to anatomo-pathological changes. In the latter (nystagmus proper) there is a constant pendulum movement around the position of equilibrium; in the former there are twitching movements to one side. This form is preferably described as nystagmus-like twitchings. In both forms there is deficient cortical innervation. In true nystagmus there is a steady, constant, uniform disorder of innervation; nystagmus-like twitchings appear in unequal and changing disturbance of innervation, as, for example, in rapid cortical exhaustion after unusual or forced movements. The number of oscillations varies from two to ten in a second; the average is four to five. The higher numbers correspond to the "tremor" of the eyes in alcoholics and cases of paretic dementia, the lower numbers to the "explosive" twitchings of the nystagmus-like disturbance of movement.

In nystagmus proper and also in the lesser twitchings we can distinguish a centripetal or optical form, a centrifugal or motor form, and a cortical form. As a matter of course, these cannot always be sharply separated, because there are mixed forms, even of all three, as in the nystagmus of miners. As the disturbance in the optical forms is usually constant and permanent, and often variable in the motor forms, this explains in a very simple manner the fact that true nystagmus is so often observed in the former, and usually nystagmus-like twitching in the latter.

Nystagmus rarely begins in adults after unilateral or bilateral disturbance of vision from peripheral disease of the eye, but it may develop in other ways. The best known is the so-called nystagmus of miners, especially of coal-miners after protracted spells of work. Disturbance of vision is often present at the same time, such as diminution of central vision, night-blindness, etc. The cause is the strain of the eyes in imperfect illumination, and constrained movements as in looking upward. Nutritive disturbances, chronic poisoning with miasmatic gases, etc., are predisposing factors. It is easy to understand that, under such circumstances, an optical as well as a motor exhaustion of the cortex must set in. One or the other will predominate, according to circumstances. Hence, disturbance of vision may be absent or slight, or the nystagmus may appear only when looking in certain directions, especially upward. In this nystagmus which is acquired in later life, a corresponding apparent movement of the objects—in a direction opposite to that of the ocular movement—is the rule. We may apply to this form the term motor weakness of the occipital cortex. The wagging of the head, which is occasionally present, is explained in like manner as cortical motor exhaustion of the centres for the muscles of the neck.

The nystagmus of multiple sclerosis, tabes, etc., probably develops in another way; in the former disease it is a very characteristic symptom. Sclerotic foci in the vicinity of the muscle nuclei (hence the frequent nuclear paralysis) probably give rise to the disturbance of motion. In such foci the nerve fibres are not divided—as happens in this disease, according to Uhthoff, in the optic nerve—but merely deprived of their medullary layer; hence conduction is not entirely abolished, but merely interfered with. In consequence of this the motor cortical impulse to the nuclei of the ocular muscles is more or less weakened; there is no complete cortical paralysis but merely "paralysis agitans."

If the theory that nystagmus is due to imperfect innervation is correct, it is easily understood that it may occasionally be checked by forced conjugate movements (for example, by forced convergence,

v. Graefe). The notion of a peripheral cause (myopathy) is opposed by the circumstance that the movements are conjugate, but peripheral processes may act as predisposing factors.

Priestley Smith maintained that ocular movements in general have an intermittent character, and bases this view on the fact that the after-image of the sun, during ocular movements, does not form a band but is composed of individual little images of the sun, standing alongside of one another. This is not entirely correct. If we look, in the ordinary way, from one point to another along the shortest line, then the after-image of the sun is really a band. It is only when the eye intentionally changes its direction slowly, that the movement becomes discontinuous. This movement is jerky but not oscillating, and leads to very rapid optical and motor exhaustion of the eye (*vide* Landolt, "Beitrag z. Physiologie d. Augenbewegungen," 1892).

Nystagmus may also be the result of other cerebral diseases (subdural hemorrhages, pachymeningitis hæmorrhagica, cerebral hemorrhage, cysticercus, etc.), and is often complicated with a corresponding disorder of vision. In such cases we may also assume an incomplete motor paralysis of the occipital cortex. Magelsen (Michel's "Jahresber.," 1883, p. 611) reports an interesting case of nystagmus, attended with pains in the back of the head, which developed in an overworked seamstress.

In rare cases rhythmical changes in the size of the pupils (hippus) are observed (Michel's "Jahresber.," 1887, p. 511, etc.).

Cortical spasms of the eye muscles are likewise always conjugate and associated. They include tonic and clonic movements to the right and left, rolling of the eyes, spasm of convergence, etc.

Optic and motor fibres are freely intermingled in the occipital cortex and its corona radiata, but in the vicinity of the nuclei of the ocular muscles the motor fibres are evidently separated from the rest. Hence it follows that lesions of the fibres of the corona radiata, either in the immediate vicinity of the nuclei or between the individual groups of ganglion cells, may produce purely motor symptoms. These will exhibit a somewhat cortical character and will not be attended by degeneration of the nerves and muscles (perinuclear and internuclear disorders of motion). In this way we may account for many a so-called paralysis of convergence or imperfect fusion of the double images, imperfect conjugate rotation to the right or left, etc. This is particularly true of diseases which often result in circumscribed lesious of the nuclear region, such as multiple sclerosis, tabes, etc.

As we have already mentioned on p. 75, a portion of the cortex, which is situated in front of the upper extremity of the anterior central convolution (Fig. 8, 1), is definitely related to the levator palpebræ superioris of the opposite side.[1] Destruction of this centre causes crossed paralysis of voluntary raising of the lid. In this centre and in that of the facial nerve must also be located the cortical termination of the trigeminus nerve. Irritation of this region may cause conjugate movements of the eyes to the opposite side. These are evidently the voluntary movements of the eyes and lids after conscious sensory stimulation from the lids, conjunctiva, and cornea.

This explains the well-known fact that ptosis is so often absent in central paralysis of the ocular muscles, or that central ptosis may be isolated (often associated with central paralysis of the facial nerve). The cortical centres are widely separated; the muscle nuclei, on the other hand, may be grouped in an entirely different manner.

Experiments on monkeys have shown that mild electrical irritation of a definite, circumscribed part of the cortex will produce movement of the thumb alone, but more pronounced irritation of the same part will produce movements of the other fingers, the hand, arm, etc.

The remarkable cases of associated movements of facial muscles when the eyelids are widely stretched open may be explained in a similar manner. The first two cases of this kind were reported by Helfreich (v. Reuss, *Wien. kl. Woch.*, 1889, No. 41). Congenital incomplete ptosis was present in almost all the cases. If by an extremely vigorous cortical innervation the attempt is made to raise the lid, the impulse will be communicated to adjacent parts of the cortex, and in this way give rise to both voluntary and involuntary movements of other parts.

Numerous examples of similar processes in other parts of the

---

[1] Some locate the region in question in the vicinity of the angular gyrus (Lemoine, Rev. de Méd., 1887, No. 7). Others deny, for the present, the possibility of determining a definite cortical region for "central ptosis."

body could be adduced. [See case of congenital ptosis with synchronous movements of the affected lid and lower jaw, by Hubbell,
Archives of Ophth., Knapp and Schweigger, Am. ed., January, 1893,
p. 63. For other cases see Michel, "Jahresbericht. f. Oph.," 1892,
p. 354.—*N.*]

3. TRANSCORTICAL DISORDERS OF VISION.

It follows, from our previous considerations, that an optical
as well as a motor projection field must be assumed to exist in the
visual cortex of the occipital lobe.

According to Munk the visual cortex of each hemisphere constitutes a projection field of the opposite half of the field of vision.
Each part of the field corresponds to a definite part of the cortex,
and Nothnagel's perceptive centre corresponds to the macula lutea.

Wilbrand ("Die hemianopischen Gesichtsfeldformen," Wiesbaden, 1890) has formulated a similar scheme, but describes the
course of the fibres somewhat more in detail.

Schaefer holds a different view. According to him the visual
cortical surfaces of both hemispheres form a single visual field upon
which both retinal surfaces are projected in approximately the same
way as if the visual fields of both eyes were drawn upon one field
(Fig. 1, p. 2).

On the whole the projection takes place in such a way that right,
left, above and below in the visual field correspond to left, right,
posteriorly and anteriorly in the occipital cortex. At the same time
every part of the cortex may induce a conjugate movement of the
eyes in such a way that it corresponds to an adjustment of the fixation point of both eyes upon the corresponding part of the field of
vision.

We cannot, however, represent the matter as if the retina and
the cortex were simply connected by wires between corresponding
localities; the conduction along the wires is transmitted in various
ways.

As a result, every point in the visual field, if stimulated with
sufficient intensity, will stimulate the entire occipital cortex of the
opposite side, although a certain part of the cortex is stimulated

most strongly. Furthermore, every part of the occipital cortex will innervate all the muscles which rotate the eyes toward the opposite side, but the intensity of the stimulus to the different muscles will vary with the different parts of the cortex. The stimuli which come from all parts of the body keep the muscles in a state of tension (tonus). If a stimulus from any part of the field of vision predominates, this will furnish an excess of stimulation to the corresponding part of the cortex, and this in turn induces a corresponding conjugate movement. Stimuli which pass to the macular locality give rise to convergence and accommodation upon the part from which the stimulus takes its origin.

Let us assume a stimulus is applied to a single retinal cone. This will be conveyed through the two granular layers until it finally stimulates a number of cells in the ganglion-cell layer of the retina, one cell most intensely, the others less intensely in proportion to their remoteness. The centripetal fibres of the optic nerve conduct

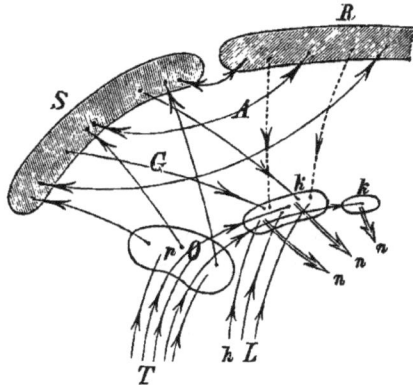

Fig. 9.—T, Optic tract; pO, primary centres of the optic nerve; kk, nuclei of the eye muscles; hL, posterior longitudinal bundle; nnn, motor nerves of the eye muscles; G, Gratiolet's visual fibres; S, visual sphere; R, remaining cerebral cortex; A, association fibres.

these stimuli mainly to the three primary optic ganglia, where the fibres, in great part, break up into the neuropilemma. Some of the tractus fibres (Fig. 9, T) pass, with or without interruption by ganglion cells, to the nuclei of the ocular muscles (k k)—probably also to other muscles—and convey the involuntary reflexes which follow unconscious impressions of light. This path, viz., optic tract—muscle nuclei—motor nerves, n n, we will call the first or lowest motor reflex arc. It exists for all senses and groups of muscles. In adults it plays a subordinate part as regards the eye muscles, apart from the movements of the pupils.

In the primary optic ganglia the stimuli which enter are distributed in the neuropilemma according to the number of conducting fibres. The ganglion cells within the neuropilemma will be stimulated with varying degrees of intensity, according as they are more or less remote from the conducting fibres. The stimulation of these ganglion cells is conveyed to the cortex of the occipital lobe, in which the visual concepts are produced and come to the notice of consciousness. As a result voluntary conjugate movements of adjustment are produced in such a way that the image of the object in the field of vision falls upon the foveæ centrales of both retinæ, and thence can be conveyed to the perceptive centre with the finest possible differentiation. It is only along the tract from the fovea centralis to the perceptive centre that visual perceptions can be conveyed of sufficient delicacy to produce optical memory pictures which can be utilized.

These paths—optic tract, primary optic ganglia, corona radiata, occipital cortex, corona radiata, muscle nuclei, motor nerves—are known as the second or middle motor reflex arc, within which voluntary movements follow conscious sensations of light.

This does not complete the process. By means of associative fibres the visual perceptions are: a, combined within the cortex into optic concepts and memory pictures; b, they are also combined with those pictures of the other senses which refer to the same and similar objects. These combinations take place particularly with the auditory concepts in the auditory centre of the temporal lobe, with the tactile and sensory concepts in the so-called motor cortex, with the so-called motor sound pictures in the speech centre (Broca's convolution) and with any existing olfactory and gustatory concepts of the object in question. The two latter concepts play a subordinate part and may be disregarded. By means of these combinations of the various memory pictures the sum of all our knowledge of an object, in other words, the concept or idea of the object, is produced.

c, The visual impressions are also conveyed, by means of associative fibres, to the frontal cortex. The latter is connected with all other parts of the brain by centripetal and centrifugal fibres. For this reason the frontal cortex is known to dominate the rest of the cerebral cortex. In it all conscious percepts are united and from it all conscious movements may be incited, modified, and inhibited. As is well known this very inhibition of lower reflexes is one of the chief physiological functions of higher nervous centres.

This combination of fibres, viz., a sensory nerve, intermediate ganglion (optic ganglia for the optic nerve, optic thalamus for the tactile nerves, internal geniculate body and posterior corpora quadrigemina, probably, for the auditory nerve), corona radiata, sensory cortical apparatus, associative fibres, frontal cortex, associative fibres, motor cortex, corona radiata, muscle nuclei, motor nerves—forms our third or highest motor reflex arc, in which perceptions that have been thought out are manifested by well-considered movements or by the inhibition of such movements. Hence the close relationship between the cortical centre of the most carefully elaborated movements and the motor speech ·centre, which is innervated (inhibited) from the frontal brain, either spontaneously or as the result of sense-perceptions. Speech without the act of speaking is equivalent to thought, and in this sense we may apply the term thought-centre to the frontal cortex, especially on the left side. As a matter of course this function can only be exercised in connection with the remainder of the cerebral cortex. Hence diseases of the frontal cortex do not permit of localization because every portion takes part, in a more or less uniform manner, in the functions of all the organs of sense and motion. It does not necessarily follow, however, that all parts of the frontal cortex are strictly co-ordinate. Indeed, certain observations compel us to assume the opposite. But a localization, for example, as regards numbers, musical notes, etc., cannot be carried out because such details depend evidently on the characteristics of development. Diseases of the frontal brain are manifested by symptoms which vary from mere qualities of character to insanity and complete dementia. The local symptoms are either general or entirely absent. A sort of local symptom is furnished when an ataxic disorder of speech follows a change of character in progressive disease of the frontal brain (general paresis).

In diffuse disease of the cortex hallucinations (sight, hearing, feeling, taste, smell), central paralysis and spasms may indicate the implication of its sensorial and sensori-motor parts. Insanity proper is produced only when the frontal cortex and its medullary fibres are affected.

It is evident that the purely anatomical course of the fibres in the brain, so far as known at the present time, permits wide inferences, which agree entirely with the results of clinical experience. But we will return to the consideration of the transcortical disorders of vision.

If the visual sphere, with its peripheral connections, is intact and capable of function, but is cut off from the rest of the cortex, a very peculiar condition is produced, which is known as psychic blindness. Vision is good and the finest objects are seen and fixed, but they are not recognized. Recognition takes place at once when the object is perceived by another sense, for example, touch. In these cases we have the visual sphere sharply isolated from all its connections—while sight and the power of fixation are undisturbed.

It appears, however, that vision may not be perfect in psychic blindness; a disorder of the color sense is also found occasionally (Charcot).

Psychic blindness is only possible under two conditions: 1. The causal lesion must not be too extensive. Otherwise local symptoms will be present either in the visual sphere itself (homonymous defects of its field of vision, conjugate spasm or paralysis) or in neighboring parts of the cortex (auditory sphere, motor cortex). Furthermore the lesion must not extend too deeply into the brain. For this reason psychic blindness is almost always a temporary condition which either recovers or, if the process is progressive, passes into cortical blindness.

2. The functional isolation of the occipital cortex must either be bilateral and hence be due to a symmetrical disturbance (which is comparatively rare), or one visual sphere is completely eliminated. This gives rise to cortical hemianopsia, and the condition of psychic blindness is then found in the intact half of the field of vision. This is the more frequent condition.

Psychic blindness, therefore, exhibits no constant anatomical appearances. It is due generally to the remote effects of a lesion situated in another part of the brain. If, for example, a strip of superficial softening occurs at the anterior border of one visual sphere while the function of the other is abolished (hemianopsia), the associative fibres of the occipital cortex may be interrupted during the acute initial stage and psychic blindness develops. At a later stage the remote action ceases, the psychic blindness disappears, and the narrow peripheral band of cortical softening causes no noticeable local symptoms or, at the most, a few degrees narrowing of the visual field inferiorly.

Psychic blindness is a local symptom, only in so far as it indicates that the visual sphere and its peripheral connections are still capable of function. It occurs very rarely as an isolated symptom.

A similar condition in those upon whom the operation for congenital cataract has been performed, in imperfect adjustment of the eyes, particularly in monochromatic light, etc., is sharply distinguished from true psychic blindness by the fact that vision in them is either decidedly impaired or that objects do not present the same appearance to such persons as they do to others.

The remaining transcortical disorders of vision are really of an aphasic or paraphasic character. Indeed, psychic blindness is a speech disturbance, in so far as the distinct sight of an object does not call forth its being named because its name cannot be given.

The motor speech centre is dominated from the first frontal convolution of the left hemisphere (Broca's convolution) and its vicinity. Its destruction produces motor aphasia, i.e., the impossibility of spontaneous innervation of the larynx and the walls of the buccal cavity in the manner necessary to articulate speech. On the other hand, the involuntary movements are intact, as in all cortical paralysis. The power of mechanically repeating words after another may also remain more or less intact.

The motor speech cortex is situated, in the main, in the anterior and posterior central convolutions and their vicinity. Fig. 10 shows the approximate distribution of those cortical regions which are chiefly involved in "speech" in the widest sense: *MSp*, is the cortical origin of the muscles employed in speech.

The cortical centre of the right hand, which is especially important in writing, is situated about the middle of the posterior central

convolution (Fig. 10, *MSch*).   The perceptive centre for auditory
impressions is situated about the middle of the second temporal convolution.   The cortical centres of taste and smell, whose location

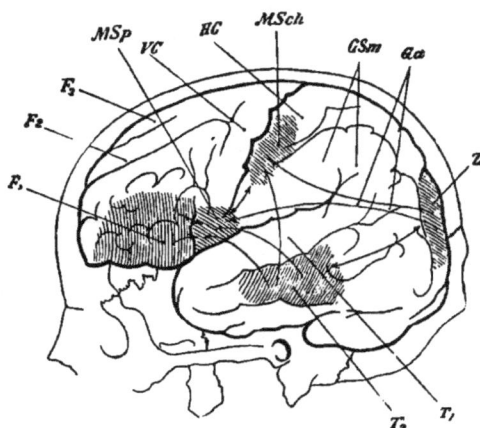

FIG. 10.—F_1, $_2$, $_3$, Frontal convolutions (F_1, Broca's convolution); T_1, $_2$, first, second, temporal convolutions; (T_1, Wernicke's convolution, so-called); *VC*, *HC*, anterior and posterior central convolutions; *GSM* and *Ga*, supramarginal and angular gyrus, together—inferior parietal lobe; *Z*, cuneus; *MSch*, motor writing centre; *MSp*, motor speech centre.

is still unsettled, may be disregarded on account of their slight importance in man.

The acquisition of ideas and concepts begins long before that of
speech.   At first the new-born babe is a simple reflex machine.
Gradually the central nerve fibres become medullated and thus capable of function.   At about the age of five months this has taken
place in the entire white matter of the brain.   The ocular movements, which were at first irregular, are then converted into conjugate and associated movements.   Objects are fixed and their movements are followed.   This can usually be done at the age of three
weeks.   Optical impressions of the objects are then obtained and
are retained as memory pictures.   These are combined, with tactile
and gustatory impressions and pictures of the object, into ideas and
concepts.   Objects are recognized long before the first attempts at
speech, and this is manifested by voluntary imitative movements.
Every sensory impression calls up, by means of association, other
sense-impressions of the object, and then excites movement.

At first the impressions of sight, feeling, taste, and smell play the principal part; auditory impressions are subordinate and rarely characteristic. This changes with the acquisition of speech.

At first conscious auditory impressions are imitated voluntarily in a purely mechanical manner. The auditory impressions conveyed from the ear (O, Fig. 11) to the auditory sphere (H) are conducted by associative fibres to the speech centre (Sp) and are imitated by means of movements, which are innervated from that centre under the control of the auditory sphere. If an object, which is already known to the individual, is often exhibited or is felt by the hand at the same time that the name of the object is pronounced, then the already existing concept will be associated with the corresponding sound-picture. The object, when seen or felt,

recalls the sound-picture and the latter recalls the object. This constitutes the first step in conscious speech, which becomes more perfect and fluent with time. Objects, which are seen, heard, felt, etc., call forth in the cerebral cortex, by means of association, the memory pictures of all other senses and, at the same time, the motor sound-picture in the speech centre, which undergoes innervation. The object is then recognized and called by name. At the same

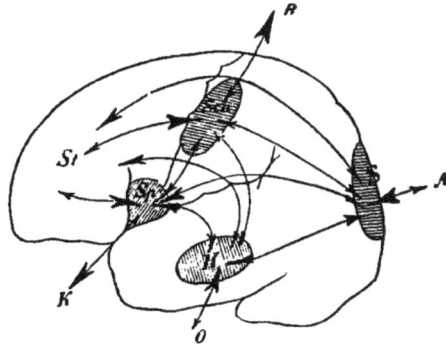

Fig. 11.—Schematic Diagram of the Disturbances of Speech. O, ear; H, auditory sphere; A, eye; S, visual sphere; R, right hand; Sch, writing centre; K, muscles of larynx and mouth; Sp, motor speech centre; St. frontal brain. The arrow-heads are made heavier in proportion to the preponderance of the conduction in the directions they indicate.

time all these parts of the cortex innervate the frontal brain and there the perception is brought into relation with the sum of all other present and former perceptions. From the frontal brain the speech centre is inhibited or innervated in another manner. The latter connections are absolutely necessary to rational speech and also to spontaneous speech. Hence the great importance of the first left frontal convolution, in and beneath which pass all the associative fibres from the frontal brain to the "motor speech cortex" proper.

Speech is under the constant control of hearing. If this is abolished in the adult (interruption of $O\ H$, peripheral deafness), a disturbance of speech gradually develops. The voice becomes peculiarly hoarse, monotonous, finally almost unintelligible. In congenital deafness speech is not acquired without special instruction. If deafness takes place in the first few years of life, the words already learned are forgotten. In both instances deaf-mutism develops.

Deaf-mutes learn to speak by seeing and touching the mouth as it takes the required positions for speech, and by seeing the movements of the larynx, and these impressions take the place of auditory impressions. Under good instruction the individual may learn to speak tolerably well. As a matter of course, however, all his notions lack auditory concepts and memory pictures (acoustic sound pictures).

The paths of rational speech (not the mere repetition of words spoken by another) pass through the frontal brain ($H—St—Sp$), where the sensory impressions and memory pictures of all the senses meet, and under whose influence speech is uttered or repressed (inhibited). As a matter of course, auditory impressions play the most important part.

If any combination of words is frequently repeated (learned by rote), direct connections evidently develop between the auditory sphere and the speech centre ($H—Sp$), and make it possible to repeat the combination "without thinking" and without passing through the frontal brain. This category includes the multiplication table, the names of the days and months, certain songs, prayers, etc., which may be repeated in a perfectly mechanical manner.

The part played in speech by these direct connections is subject to great individual variations. They greatly facilitate fluent speech. When they are lost, speech becomes labored. Even the most trivial and familiar things must then be reconstructed anew, through the medium of the frontal brain and the remainder of the cerebral cortex. To this condition the term dysphasia may be applied, i.e., speech is possible, but is attended by such mental strain that it is soon abandoned. The condition is similar to that which would be experienced in solving a complicated mathematical problem if the formulæ were not known by heart and it became necessary to work them out anew, or as if we were compelled to calculate the necessary logarithms instead of employing the table of logarithms.

But if the influence of the frontal lobes and the associated sensory cortex upon the speech centre is abolished, the individual must rely solely upon the direct connections between the auditory sphere and the speech centre. Much may still be spoken in a mechanical manner, words and short sentences may be repeated, and objects may be correctly named; but the control over the sense of the language is lost, as soon as it goes beyond the simplest things. Letters, syllables, and words are apt to be omitted or mistaken for similar sounds—in brief, paraphasia develops. This is observed with the greatest distinctness when diseases of the frontal lobe extend gradually toward the motor speech centre, particularly in paralytic dementia.

The conditions become more complicated when reading and writing are learned.

Speech is usually developed to a considerable degree when the individual begins to learn to read. At first a series of images of the letters is committed to memory and each is combined with a definite sound picture. At the beginning the reading is done aloud, *i.e.*, the visible "writing-image" (which does not exhibit the slightest resemblance to the object which it designates) calls forth the corresponding sound in the auditory sphere and in the motor centre.

If a word, composed of several letters, is read, its meaning is recognized more rapidly and easily when, by successive enunciation of the letters, the sound-image of the entire word is produced, because the latter has been made familiar while learning to talk. But soon this is no longer necessary. Even without this assistance the writing-image of an object directly calls forth the acoustic and motor sound-image. This shows the importance of the connection between the visual and auditory spheres in learning to read. The predominant influence of the auditory sphere sometimes continues through life. But if reading is practised very extensively, direct connections of the visual sphere gradually develop and render unnecessary the

deflection on the one side to the motor cortex, and on the other side
to the frontal cortex.   Reading thus becomes more and more independent of the auditory sphere and the acoustic sound-images.   It
depends upon circumstances whether one path or another is mainly
utilized in adults.   Hence, as clinical experience proves, the symptoms will vary greatly in any individual according as one path or
another is obstructed.

When reading aloud is practised extensively, direct paths develop
in particular between the visual sphere and the motor speech centre,
so that finally the act of reading becomes purely mechanical.   If the
individual reads much to himself, special direct connections with
the frontal lobes are gradually developed.   Finally, the concepts of
objects and processes may be called forth as readily or even more
readily, by writing-images
as by the sound-pictures of
the auditory sphere.   (A
written article can then be
read and understood by the
individual much more rap-
idly than when the article
is read to him.)   The manner of education, etc., are
the main factors in the production of these individual
differences.

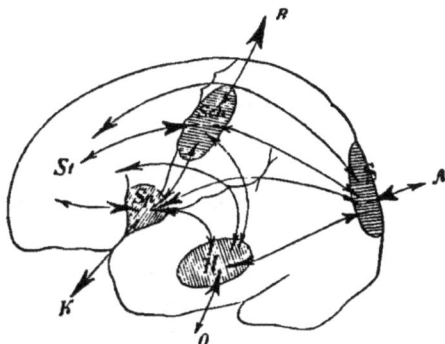

Fig. 12.—Comp. Fig. 11, p. 93.

Similar processes take place in learning to write.   Originally the
script-images are imitated mechanically, and this is evidently done
by connections between the left visual sphere and the cortical centre
for the right· hand, situated in the left central convolutions (S—
Sch, Fig. 12).   In learning to write, speech is first employed to
pave the way for a comprehension of what has been written; S—H
are innervated synchronously with S—Sch, and the sound-pictures
are aroused in the auditory sphere.   In fluent writing the former
path is abandoned more and more, and the direct tracts S—Sch, are
perfected, especially if copying is extensively practised.   But if the
individual writes mainly from dictation, the direct connections

H—Sch, will be chiefly developed. In adults, accordingly, the different tracts may vary greatly in importance. If the writing is done with thought, the frontal brain is innervated at the same time and controls the act of writing. The oftener the writing is done from dictation, the oftener the frontal lobes are avoided, and the more mechanical does the process become.

On account of learning to write with the right hand the predominance of the left hemisphere in all forms of speech becomes permanent. If the left hand is used in writing (in left-handed individuals), this predominance is transferred to the right hemisphere. We will not discuss the question whether the ordinary right-handedness is the result of a congenital predominance of the left side or whether the latter is the effect of education. At all events speech, with all its complicated relations, appears to be so difficult that it can be learned thoroughly only by a single hemisphere. In adults the loss of the speech centre cannot be completely repaired, but as late as the age of seventeen years, the loss of the right arm may induce a transfer of the speech centre to the right hemisphere.

The following brief *résumé* indicates the tracts in which the different speech functions run their course (*vide* Fig. 12):

1. *Speaking:*

   a, is originally learned in a purely mechanical manner:
   $$O—H—Sp—K,$$
   b, is spoken understandingly in conversation:
   $$O—H—St'—Sp—K,$$
   c, is spoken spontaneously:
   $$St—Sp—K.$$

2. *Reading:*

   a, reading aloud in a purely mechanical manner:
   $$A—S—H—Sp—K,$$
   b, reading aloud, understandingly, at first:
   $$A—S—H—St—Sp—K,$$
   later:
   $$A—S—St—Sp—K,$$

---

[1] St indicates not alone the frontal brain but also the associated sensorial and sensory cortical regions.

7

c, reading softly, understandingly, at first:

A—S—H—St—Sp!,¹

   later:

A—S—St—Sp!,

d, frequently practised reading:

A—S—Sp—K.

3. *Writing :*

   a, purely mechanical writing of the alphabet:

A—S—Sch—R,

   then the letters pronounced at the same time;

$$A—S \begin{cases} \text{Sch—R,} \\ \text{H—Sp—K,} \end{cases}$$

   b, copying understandingly, at first:

A—S—H—St—Sch—R, .

   later:

A—S—St—Sch—R,

   finally in a more mechanical manner:

A—S—Sch—R,

   but more perfectly than in the beginning,

   c, writing from dictation, at first:

O—H—St—Sch—R,

   later, more mechanically:

O—H—Sch—R,

   d, writing voluntarily:

St—Sch—R.

As a matter of fact, however, the conditions are much more complicated (*vide* Malachowski, Volkmann's "Klin. Vortr.," No. 224; Wysmann, *Deutsch. Arch. f. kl. Med.*, 47, p. 27) because it has been found clinically that the ability to speak, write, and read may be lost only in part. For example, the knowledge of numbers and the ability to count may be lost; indeed the individual may even be able to reckon with numbers up to three, while the higher numbers are lost. The knowledge of musical notes may be lost or the ability to sing, and this may even be restricted to certain notes and tones. Or an acquired language alone is lost while the speech faculty other-

---

¹ ! indicates an inhibitory action.

wise remains intact; or certain groups of words, such as nouns, or even individual words and letters, may alone be lost. In short, the clinical symptoms may exhibit the greatest possible variations.

In the central disorders of speech (*vide* Naunyn, Congress f. innere Medicin, Wiesbaden, 1887) there are two principal groups, viz., motor and sensory aphasia (Wernicke). In pure motor aphasia (also called ataxic in contrast to anarthric aphasia, which is due to paralysis and weakness of the speech muscles) it is impossible for the patient to utter the word, which is on the tip of his tongue, although he recognizes it when spoken and read and also the corresponding object. This affection is located in the left inferior frontal convolution, the destruction of which interferes with the connection between the left frontal lobe and the motor centre of the speech muscles in the lower third of the anterior central convolution. This variety contrasts somewhat with amnesic aphasia (Ogle, Bastian) in which the "internal" word is forgotten, *i.e.*, the motor tone-picture.

A mild grade of ataxic aphasia is known as paraphasia (also called conduction aphasia) in which letters, syllables, and words are omitted or are confused with others which have a similar sound. Here conduction is not interrupted, but is merely interfered with or there is irregular innervation.

As in all other central muscular disorders the involuntary movements (deglutition, respiration, etc.) are entirely intact.

In sensory aphasia we may distinguish, in general, two groups, according as the relations of hearing or sight to speech are disturbed. On account of the greater importance of the sense of hearing in speech, acoustic aphasia or aphasia with word-deafness is the more frequent. Despite the patient's ability to hear, the comprehension of spoken words and sounds is disturbed in this affection. It is located in the upper part of the first temporal convolution (Wernicke's convolution), beneath which the connections of the auditory sphere (Fig. 10) are situated anteriorly.

There have been very few accurate reports of cases of optic aphasia (aphasia with word-blindness) in which, despite the ability to see, there is lack of comprehension of visible objects and words (*vide*

Freund, *Arch. f. Psych.*, XX, p. 387). In Sigaud's case (*Prog. Méd.*, 1887, No. 36) there was found, for example, a spot of softening, as large as a walnut, in the left inferior parietal convolution (Fig. 10, *Ga* and *Gsm*), *i.e.*, in a place where the association fibres of the visual sphere must run forward. Other cortical disorders of vision, particularly hemianopsia, are very often combined with word-blindness. It has often been found that the disease was situated in the right hemisphere, even in right-handed individuals. We have already emphasized the fact that such interruptions of conduction are necessarily bilateral, and are therefore more apt to occur when the function of one visual sphere has already been abolished, *i.e.*, when hemianopsia is present. A lesion on the right side which has caused left homonymous hemianopsia will, therefore, be more apt to produce, by "remote effect," a condition of word-blindness or interruption of conduction in the associative fibres of the other visual sphere which run anteriorly.

There are also a number of "undefined" forms of aphasia which are located in the vicinity of the three important cortical speech centres. According to Grashey aphasia may also be produced by a simple diminution of the duration of sensory impressions, although the tracts or centres in the brain are not destroyed.

It is certain that the sensory disorders of speech develop when the connection of the auditory or visual sphere, or both, with the anterior parts of the brain is abolished or interfered with. As Fig. 10 shows, the disorder will be purely or chiefly acoustic when the disturbance of conduction is located beneath the first temporal convolution, and chiefly optical when it is located in the inferior parietal convolution (supramarginal and angular gyri). The same figure also shows that tracts whose functions differ widely may be affected in places which are closely approximated.

We have seen (p. 59) that unilateral destruction of Nothnagel's perceptive centre produces crossed homonymous hemianopsia, and bilateral destruction results in cortical blindness. In this locality optical impressions mainly undergo associative connection with one another. The more they are situated toward the periphery of the visual sphere, the more the optical impressions are combined with those of the other senses. To all parts of the visual sphere which are situated outside of its perceptive centre Nothnagel applies the term "memory centre for visual impressions." If a considerable part of this is destroyed, a part of the optical images already known lose their association with the memory pictures of the other senses. The optical image does not recall to memory the object and its name (partial psychic blindness). If the perceptive centre has remained intact, new memory pictures may be acquired, and so much more easily the younger the individual. This locality alone receives from the macular portion of the retina visual impressions which are sufficiently detailed to be employed in the production of optical memory pictures. It is also easy to understand the fact that in optical aphasia concrete nouns are especially concerned. These alone possess real optical memory pictures. The optical memory picture of a verb or of an abstract noun is merely a writing-image.

After these preliminary observations we will briefly mention the more important clinical forms of optical disorders of speech. In the nature of the case these are chiefly disorders of reading and writing.

Word-Blindness.—Despite perfect vision the names of objects cannot be recalled, although they are recognized. This fact distinguishes the condition from psychic blindness in which the objects are not recognized. In word-blindness proper (*cécité verbale*) printed or written words cannot be spoken; in other cases the writing-image of a word recalls its tone-image to memory and it can then be correctly spoken (optical word amnesia, *amnésie verbale visuelle*). In the main such cases result from considerable loss of the peripheral parts of the visual sphere, together with abolition of the association between the visual and auditory spheres, each of which remains intact. In word-blindness proper there is also loss of association between the visual spheres and the motor speech centre.

Optical aphasia can only be distinguished from word-blindness by regarding the former as motor aphasia plus word-blindness.

Alexia is the inability to understand writing or print, or both, although the characters are seen and can be reproduced, and the words themselves are recognized and can be uttered (optical alexia). Or written and printed matter is understood but cannot be read aloud (motor alexia), although the words, when spoken, can be repeated by the patient. In optical alexia the memory pictures of the signs of writing and words are lost, probably in the peripheral parts of the visual sphere. In motor alexia they are retained, but their association with the motor speech centre (probably also with the auditory sphere) is interrupted.

Dyslexia is the term applied by Berlin ("Ueb. eine besondere Art d. Wortblindheit," Wiesbaden, 1887) to a condition in which the patient could read but, after uttering a few words, a peculiar uncomfortable feeling was produced and compelled him to stop. It is a notable fact that this condition was temporary in all cases. The disorder of reading usually ran a favorable course, but severe cerebral symptoms developed at a later period. Definite localization was impossible because the disorder must always be a remote effect. It is due to impeded conduction between the speech centre and the centres of special sense. Later publications on dyslexia coincide in the main with Berlin's statements. It is possible that the principal part in this disorder is played by obstructed cortical "oculomotor" innervation of the left occipital cortex on moving the eyes toward the right side (*vide* page 78).

Paralexia is a term (analogous to paraphasia) applied to a condition in which single letters, syllables, or words are omitted in reading, or are confused with others, especially those which have a similar sound.

Agraphia is the inability to write, although the movements of the hand and arm in other respects are approximately normal. We may distinguish a motor form (probably due to destruction or complete isolation of those parts of the cortex from which the right hand is innervated in writing) and a sensory form (in which conduction is interrupted between the last-named centre and the visual and auditory centres). The latter form may be subdivided into two varieties. In one variety, the patient cannot copy but can write on dictation or when the words are at the same time read aloud (optical agraphia). In acoustic agraphia the individual can copy but cannot write on dictation. The loss of both these functions at the same time would be tantamount to sensory agraphia.

Paragraphia is the confusion or omission of letters, syllables, and words in writing. This may only be present or predominate in spontaneous writing, in copying, or in writing from dictation.

All these disorders of speech may easily be made schematic (*vide* Wysmann, *l.c.*). In reality, however, there are numerous transitions and combinations; pure forms are very rare and are almost always temporary, either ending in recovery or passing into more severe forms.

Interesting as it is to analyze the process of speech, all these different forms of disease permit only a very limited local diagnosis with regard to interference with or interruption of conduction between different parts of the cerebral cortex. If the motor factor predominates in a speech disorder, the interruption is to be looked for near the motor centres; if the acoustic or optic factor predominates, it is to be looked for nearer to the corresponding part of the cortex.

Together with the disturbance of conduction there is more or less disturbance or loss of the optical, acoustic, and motor memory-pictures of speech, which are situated at the periphery of the corresponding motor or sensory parts of the cortex. If the latter disturbance exceeds that of conduction, the speech disorder assumes an amnesic character.

In order to make a thorough examination of a central disorder of speech we must note:

1. Whether the individual can speak and write spontaneously and rationally, *i.e.*, whether speech is normally influenced by the frontal brain. If the latter is extensively diseased, thought, speech, and writing are irrational (insanity). Paraphasia and paragraphia are apt to develop when the frontal brain functionates but does not influence speaking and writing in a normal manner, either on account of irregularities of conduction (progressive paresis) or even in the normal individual who reads and writes mechanically, while absent-minded. If Broca's convolution is destroyed, spontaneous and rational speech is impossible, but mechanical repetition is still possible. Even this is no longer possible when the cortical origin of "the nerves of speech" (Fig. 10, *MSp*) is destroyed (typical motor aphasia).

2. Whether acoustic disorders are present, *i.e.*, whether the individual can repeat or write dictated words. In pure motor aphasia spoken words and sentences are understood; they are not understood if word-deafness is also present. If the latter exists alone, the individual can speak voluntarily and can read writing aloud, but he cannot repeat or write what he hears. If, at the same time, letters, syllables, or words are omitted or confused with others, especially with those of similar sound, the condition may be called acoustic paraphasia.

3. Whether optical disorders are present, *i.e.*, how objects shown to the individual are called or their names written, and how written and printed matter is read aloud or copied. In word-blindness objects are not recognized; in writing-blindness written or printed objects are not recognized, although they are seen with perfect distinctness and their names can be repeated. We should examine whether and how objects shown are named, and written and printed matter is read or copied. At the same time it should be noted whether letters, syllables, and words are omitted or mistaken, as this merely indicates irregularity of conduction.

Educated patients are to be examined in the same way, and also in regard to musical ability, foreign languages, mathematical knowledge, etc. This may render the examination very complicated, but will furnish no further diagnostic data with regard to more accurate localization.

In conclusion, a few words with regard to

4. THE FRONTAL BRAIN.

In regarding this part of the brain as superior to the remainder of the cerebral cortex we are opposed to the majority of writers, who consider such a higher "centre" unnecessary and superfluous. We are forced to this conclusion, however, by the teachings of anatomy, inasmuch as the frontal cortex receives fibres from and sends fibres to all other parts of the cortex. The chief functions of the frontal cortex are assumed to be:

1. Inhibition of voluntary movements following conscious sensory impressions, or the execution of deliberative movements with coin-

cident consideration of all sensory impressions, including those which are remembered as well as present impressions.

2. The execution of spontaneous movements, likewise upon the basis of all past and present sensory impressions of all the senses.

As a matter of course the frontal lobes do not perform these functions "detached from all other parts of the cortex," but in association with them.

In view of the great importance of language, which distinguishes man from the lower animals, the most important function of the

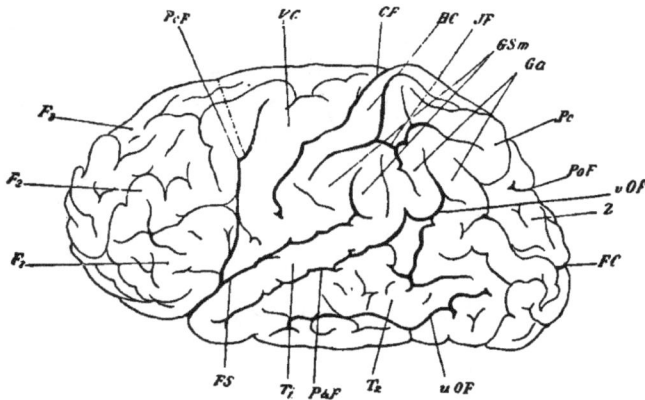

Fig. 13.—Normal Human Brain, after a figure from Wernicke, the frontal brain limited by a dotted line. Lettering as in Fig. 14; in addition: *CF*, calcarine fossa; *uOF*, inferior occipital sulcus; *Z*, cuneus; *Pc*, præcuneus; *Ga*, angular gyrus; *Gsm*, supramarginal gyrus; *VC* and *HC*, anterior and posterior central convolutions.

frontal lobes consists in their relations to speech. This is manifested by the close proximity of the corresponding parts of the cortex. The more the frontal brain approaches the cortical centre for the speech muscles (Fig. 10, *MSp*) the more it assumes the rôle of a motor organ superior to the latter. The chief difference between the brain of monkeys and of man consists in the much more pronounced development of the frontal lobes in the latter (*vide* Figs. 13 and 14).

We have seen that no part of the cortex is purely motor or purely sensory, but that it exercises both functions, *i.e.*, it receives and sends out stimuli, although these vary qualitatively and quantitatively in different places. Nevertheless, the individual regions of the cortex possess characteristic functions so that localization is possible. This is not true of the frontal lobes. Here every part evidently receives stimuli from all the sensory organs and can send stimuli to all motor cortical regions. Although from a purely theoretical standpoint not all parts of the frontal cortex can be co-ordinate in the strictest sense of the word, yet a local diagnosis is not possible. The greatest possible subjective differences will be present in different individuals, as a result of the character of the education.

Hence, affections of the frontal cortex, which are not too slight

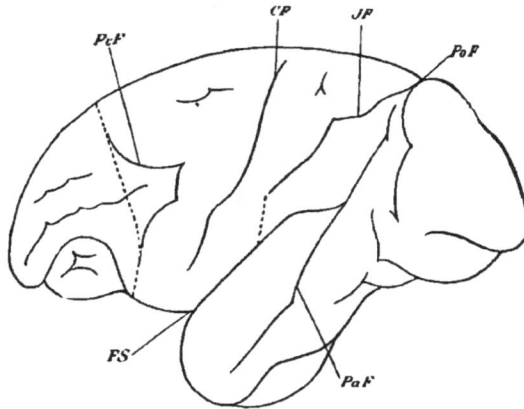

Fig. 14.—Brain of Monkey, after a figure from Horsley and Schäfer; the dotted line separates the motor region from the frontal brain. *FS*, Fossa of Sylvius; *PcF*, precentral fissure; *CF*, central fissure; *JF*, interparietal fissure; *PoF*, parieto-occipital fissure; *PaF*, parallel fissure.

in extent, will not be manifested by disturbances of individual sensory organs or groups of muscles, but by changes in the general impression produced by the outer world, by changes in thought and voluntary action, *i.e.*, by changes of character.

Diseases of the frontal cortex, especially those of a diffuse character, may be called mental diseases, although not every circumscribed and slowly progressive lesion (tumor or abscess) in this region necessarily produces a typical form of insanity. Such diseases may run their course without noticeable or characteristic symptoms, on account of the relatively co-ordinate value of all parts of the frontal brain and the possibility that the different parts may act vicariously for one another.

Extensive surface disease of the frontal cortex will always be manifest as a mental affection (in the broadest sense of the word) and irritative symptoms (hallucinations) will be present or absent according to the implication of sensorial and sensory regions of the cortex. To this extent the presence and absence of hallucinations may be properly employed as a principle of classification in insanity.

It is plausible to attribute maniacal conditions to diffuse arterial hyperæmia and increased irritability, and melancholic conditions to passive hyperæmia, but for the present such assumptions are purely hypothetical.

I would suggest, however, that when anæmia, hyperæmia, or both are present in patches in the frontal cortex, increased and diminished or even normal excitability will be found coexisting. The result must be a complete confusion both as regards centripetal sensory impressions and motor manifestations.

5. DISORDERS IN THE DOMAIN OF THE SENSORY NERVES.

Peripheral irritation of the sensory nerves will result in neuralgias; peripheral paralysis will result in anæsthesia of the corresponding branches, and contact or irritation will then fail to produce reflex movements. Peripheral anæsthesia of the cornea—whether the trigeminus is affected in the orbit, the base of the skull, in the Gasserian ganglion, or between the latter and its origin in the brain —often leads to so-called neuro-paralytic keratitis. This may recover at any stage, but is usually followed by loss of the eye. Without entering further into the clinical relations of this disease or into the question of the existence of trophic nerves, it may be remarked that we evidently have to deal with a traumatic loss of substance (this is very apt to occur on account of the insensibility of the cornea) which is infected from the conjunctival sac. The further course of this condition is modified by the fact that the vascular reflexes in the remote conjunctival, scleral, and episcleral vessels, which are necessary to the healing of an infected ulcer, remain absent. Hence there is no reaction of the tissues and the process makes a characteristic, steady progress.   The assumption of special trophic nerves is superfluous.

In hysterical anæsthesia of the cornea neuro-paralytic keratitis is never observed, because the involuntary vascular reflexes are intact, and are often exaggerated.   The nerve fibres are not interrupted or destroyed, although the sensory irritants do not enter consciousness.

We can only entertain surmises concerning the central course of sensory trigeminal stimuli.   With the exception of those roots which pass to the cerebellum and the spinal cord, the remainder—such as the ascending sensory spinal fibres of the tegmentum—probably pass, in the main, to the great centre of involuntary mimicry, the optic thalamus, and there terminate in the network of nerve fibres. The ganglion cells of the thalamus send their axis cylinders to the cortex of the parietal lobe, the so-called motor, preferably motor-sensory cortex (Monakow).   We will hardly go astray in assuming that the cortical termination of the trigeminal fibres of the eye, which are interrupted in the thalamus, is that part of the "motor" cortex which presides over the movements of the ocular muscles (Fig. 8, 1).

A lesion of this part in man appears to produce central paralysis of the opposite levator palpebræ superioris, the only positively dem-onstrated central paralysis of a single ocular muscle.   This part of the cortex may be regarded as the locality from which voluntary movements of the eyelids are executed after conscious stimuli on the part of the sensory nerves of the eye.

Movements of the eyes—conjugate movements toward the opposite side—may also be produced by irritation of this locality (Ferrier, Beevor and Horsley).   Mott and Schaefer (*Brain*, February, 1890) succeeded in making a still more minute differentiation.   According to them irritation of the uppermost part caused conjugate deviation outward and downward; in irritation of the middle portion purely lateral deflection followed, and irritation of the lower portion caused conjugate deviation outward and upward.   The irritation evidently passes along those tracts in which voluntary (always conjugate) movements of the eyes follow conscious sensory stimulation of the cornea, conjunctiva and eyelids.

These movements probably do not pass, by associative fibres, to the visual sphere on the same side, because they remain intact after destruction of the occipital cortex. They pass directly through the corona radiata to the nuclei of the ocular muscles. The remarks made on page 80, concerning cortical conjugate movements of the eyes, also hold good with regard to these movements. They are always incomplete, and occur approximately in a certain direction. Delicate voluntary movements of adjustment can only be innervated from the visual sphere under the control of vision.

The long-known fact that, in cerebral paralyses of the ocular muscles, the levator palpebræ superioris often is not affected, or that it is paralyzed alone, perhaps together with the ocular facial, is easily explained. It is unnecessary to assume that the nuclei of the levator palpebræ and the ocular facial nerve are situated in close proximity (Mendel, Siemerling). It is sufficient that their cortical origins are closely related, and that the chief paths of innervation from the motor cortex to the nuclei of the corresponding nerves run alongside of one another for a considerable distance.

6. DISORDERS OF THE INVOLUNTARY MUSCLES OF THE EYE AND OF THE SYMPATHETIC.

The muscles in question are the sphincter of the pupil and the ciliary muscle (whose nerve fibres pass through the motor oculi and evidently originate in the groups of small ganglion cells at the anterior extremity of the motor oculi nucleus, *vide* Fig. 6, 1 and 3) and the dilator pupillæ which is supplied by the sympathetic. In addition there are radiating smooth fibres in the eyelids, whose contraction causes moderate dilatation of the palpebral fissure, and the so-called Mueller's muscle, which closes the inferior orbital fissure and whose contraction pushes the eyeball forward (exophthalmus). The two latter muscles are also supplied by the sympathetic, and their nerve supply may be followed experimentally to the sixth and seventh cervical and the first dorsal segment of the spinal cord, the so-called cilio-spinal centre (Budge).

Very little will be said here concerning the muscle of accommodation. Its nucleus, situated close to that median motor oculi nucleus (Fig. 6, 3), which we regard as the centre of convergence, appears to be voluntarily innervated, together with convergence, from the perceptive centre of the visual sphere, and is under the control of conscious vision. There appears to be no direct influence of the cerebral cortex upon the pupillary sphincter. It cannot be stimulated voluntarily, apart from the associated movement in accommodation and convergence, which is, at least in part, a purely mechanical process.

Mention may here be made of Haab's cortical reflex of the pupil. Haab found that, independently of any noticeable change in convergence and accommodation, contraction of both pupils takes place when the attention is directed upon a bright object which has already been present within the field of vision. The brighter the object, the more marked is the contraction of the pupils. Haab assumes very properly that this reflex must be cortical in character because it follows a purely psychical process.

It is evident that in this form of attention there is a general innervation of the cortex of the visual spheres, probably from one of the highest centres, i.e., the frontal lobes. The motor impulse which follows such innervation is distributed to all the ocular muscles and is therefore noticeable only in those which have no antagonists that can be influenced by the will. This is true of the sphincter pupillæ. It is probable that a coincident accommodation impulse can also be demonstrated. It is an interesting fact that the strength of this reflex depends upon the brightness of the object.

The occurrence of the reflex presupposes that the network of the motor corona radiata in the motor oculi nucleus and the ganglion cells of the nucleus of the sphincter are intact. The cells or fibres of the accommodation nucleus may be incapable of function, and Haab reports two cases in which the reflex was retained despite paralysis of accommodation. The reflex will be lost if the internuclear fibres between the nuclei of the ciliary muscle and pupillary sphincter are destroyed, even though both nuclei are intact. In such a case accommodation and the involuntary light reflex of the pupil may be normal, and this has also been observed by Haab.

A cerebral influence on the pupil, possibly in an indirect manner, may also be exerted through the sympathetic. The loss of an entire cerebral hemisphere often produces symptoms of paralysis of the sympathetic upon the same side of the face, *i.e.*, upon the side opposite to the general paralysis. The mydriasis which so often in cerebral hemorrhage is upon the side of the lesion, may be the result of paralysis of the motor oculi, of irritation of the sympathetic, or of both.

For the sake of convenience the anomalies of the movements of the pupils, due to disorders of the motor oculi or of the sympathetic, will be discussed together.

Apart from exceptional cases, such as great differences in refraction, etc., the pupils are normally equal in size, and contract to an equal amount upon the entrance of light into either eye (consensual pupillary reaction), and during convergence and accommodation. The pupils dilate upon irritating the integument in the vicinity of the eye. Consensual pupillary reaction is absent in animals whose optic nerves undergo total decussation.

The light reflex—contraction on the entrance of light, dilatation upon covering the eye—is a function of the motor oculi, resulting from light stimuli which are conveyed from each retina through the optic nerve, chiasm and tractus to the primary optic centres of both hemispheres and thence to the bilateral nuclei of the sphincter pupillæ. The disorders of this reflex may be, *a*, centripetal (sensory visual disorder), *b*, centrifugal (motor), and *c*, central (located between the primary optic ganglia and the muscle nuclei).

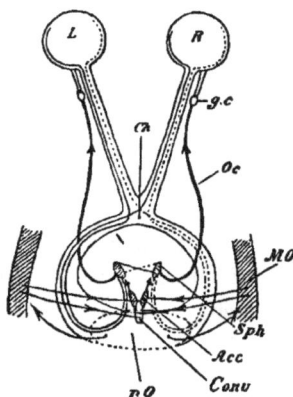

Fig. 15.—Schematic Diagram of the Movements of the Pupils. *R*, right, *L*, left eye; *Ch*, chiasm; *pO*, primary ganglia of the optic nerve; *MO*, macular region of the occipital cortex; *Sph*, *Acc*, *Conv*, nuclei for sphincter pupillæ, accommodation and convergence; *Oc*, oculomotor nerve; *gc*, ciliary ganglion.

a. The centripetal pupillary disorders between the eye and the primary optic ganglia are at the same time visual disorders and hence are bilateral, inasmuch as the optic fibres pass from each eye to both halves of the brain. From the blind parts of the field of vision no light reaction of the pupils can be produced in either eye, from the normal parts the reaction is always equal in both eyes. Hence the pupils are equal in size in unilateral disorders of vision, unless this is prevented by complications.

The reaction of both pupils is lost, after the destruction of one optic nerve, when the light falls upon the blind eye, and is retained and equal when the light enters the sound eye. Jessop's two cases, of double blindness with atrophy of the optic nerves and intact reaction of the pupils to light, have been discussed on page 66.

In destruction of one tractus or the primary optic ganglia of one side there is hemiopic reaction of the pupils, corresponding to the hemianopic disorder of vision. Neither pupil reacts from the blind half of the field of vision, and both pupils react equally from the intact half.

[It may perhaps be well to remark that the existence of the hemiopic pupillary inaction has more value as indicating a lesion at or peripheral to the middle optic ganglia, than the non-existence of the symptom possesses as indicating a lesion higher up in the cerebrum. The former is positive as evidence and, as the editor's experience has shown, can be safely relied upon; the latter is negative and of course has less logical value. We may have a partial impairment of the pupil-controlling fibres and would be led into error should we assume that the absence of the hemiopic pupillary inaction excludes the possibility of the lesion being one whose effect is either peripheral or at the ganglia. It is also conceivable that a tumor dorsad to the middle ganglia and perhaps at some distance may by indirection effect the lesion, either by pressure or by interrupting the fibres.—Ed.]

If the visual disorder has a more central location, the reaction of the pupils to light is not disturbed, although no impressions of light enter consciousness. In all these centripetal pupillary disturbances, the reaction of the pupils to accommodation, convergence, and cutaneous irritants is retained.

b. Centrifugal pupillary disorders are unilateral, unless the causal affection is situated on both sides. In unilateral paralysis of the pupillary sphincter—whether the cause is intraocular (atropine poisoning, trauma), orbital, basal, or nuclear—the pupil of the paralyzed eye does not react despite the perfect perception of light, whether the light enters one eye or the other. In both cases the pupil of the healthy eye reacts properly (complete and partial paralysis of the motor oculi). In centrifugal pupillary disorders the reaction to convergence and accommodation is likewise disturbed in many cases, and this always takes place when the muscle of accommodation is also paralyzed. At the moment of contraction of this muscle the blood of the ciliary body is forced, at least in part, into the iris, and this causes temporary contraction of the pupil. Hence, if the ciliary muscle is still acting, the contraction of the pupil, although considerably diminished, is not entirely abolished in vigorous accommodation and in convergence, which is always associated with an accommodation impulse. The dilatation of the pupil after cutaneous irritation is also present in paralysis of the motor oculi, because it is effected through the sympathetic.

c. Central pupillary disorders are those in which the connection between the primary optic ganglia (anterior corpora quadrigemina, external geniculate body) and the nucleus of the sphincter pupillæ is disturbed or interrupted, but both remain capable of function. Unless complications are present, there is no interference with conscious vision and no paralysis of any voluntary movements; the pupillary reaction in accommodation and convergence may also be intact. There is no reaction of the pupil, however, on the entrance of light (reflex rigidity of the pupil). If the interruption or interference with conduction is unilateral, illumination of the opposite halves of the fields of vision of both eyes causes no reaction of either pupil; on illumination of the corresponding halves of the field of vision the pupillary reaction takes place and in both eyes (hemiopic rigidity of the pupils). Inasmuch as both pupils are equal in size and react uniformly upon the entrance of light, the condition will remain undiscovered unless it is looked for, especially as vision is not necessarily disturbed. Indeed, no case of pure hemiopic reaction of the pupils without disturbance of vision has yet been reported, although it is probable that this symptom is not very rare. But if conduction is interrupted on both sides between the primary optic centres and

8

the nucleus of the sphincter, reflex rigidity of the pupil (Argyll-Robertson symptom) is of frequent occurrence in certain diseases (especially tabes dorsalis) which are to be located in the vicinity of the motor oculi nucleus (*Arch. f. Psych.*, XXIII, 3).   In a case of unilateral congenital ptosis, Siemerling found bilateral changes in the terminal network of the motor oculi nucleus, whose ganglion cells were relatively intact.   This would explain the reflex rigidity of the pupil, which was also present in the case; but it could not possibly have been the cause of the unilateral ptosis.

In reflex rigidity of the pupils we would expect them to be larger than normal, because the light no longer exerts a contracting influence.   In reality, however, this does not often happen.   The pupils are usually narrowed to a considerable extent (myosis).   It appears as if in destruction of the fibres which break up into a terminal network in the sphincter nucleus, a constant irritation is exercised upon the ganglion cells lying within the network.   This view is corroborated by Uhthoff's case (*Berl. kl. Woch.*, 1886, 3 and 4) in which slight pupillary contractions were constantly occurring during reflex rigidity of the pupil.   As this irritation is not necessarily equal on the two sides, the pupils may vary greatly in size.   Cocaine dilates the rigid pupil by stimulating the sympathetic.

Difference in the size of the pupils without any local cause in the eye (inflammation, etc.), and without pronounced diminution in their mobility, is a symptom which points to the region between the primary optic ganglia and the motor oculi nucleus.   The condition may be of the nature of an irritation or a beginning paralysis.

When the pupils differ in size (anisocoria) it is often difficult to decide whether one is too large or the other too small.   In doubtful cases we must regard as normal the one which moves more freely (to entrance of light, accommodation, and convergence), because spasm, as well as paralysis, diminishes the mobility.

From the foregoing remarks it would follow that, in reflex rigidity of the pupils, the condition corresponds, so far as regards its distribution, to homonymous hemianopsia, *i.e.*, the nucleus of both sphincters is or is not innervated from each half of the field of vision. But cases of unilateral reflex rigidity of the pupil have also been described, for example, by Moebius (*Centr. f. Nerv.*, 1888, 23), and moreover the symptom is often much more pronounced in one eye than in the other. This is probably owing to an affection in the immediate vicinity of one sphincter nucleus, which leaves the latter intact but interrupts all its optical connections. In this event the direct light reflex in the eye of the affected side would be wanting, but would be intact on the other side. The pupil on the healthy side would contract on the entrance of light into either eye. At the same time the reflex of accommodation and convergence might be present on both sides. This happened in Moebius' case. As a matter of course, however, such cases will occur rarely and the condition is probably only a temporary one.

Some writers maintain that the nuclei of both pupillary sphincters are connected by fibres and thus explain the fact that both pupils are normally of equal size, even if light enters only one eye. The existence of such fibres is denied by others. Anatomically a mutual connection by fibres between the two nuclei can hardly be denied. But inasmuch as fibres pass to both nuclei from each optic nerve, the existence of connecting fibres is not absolutely necessary to explain the equal size of the pupils.

In the clinical examination of diseases of the region of the nuclei of the ocular muscles greater attention should be paid than has hitherto been done, to the half-sided character of the symptoms (with regard to the field of vision). Motor disturbances (especially of the pupils) may also be homonymously half-sided in regard to the visual field, although no corresponding disturbance of vision is demonstrable. It is evident, therefore, that very accurate local diagnoses may be made in this way.

Heddæus and others apply the term reflex deafness to reflex insensibility of an eye in which, even when vision is intact, the entrance of light does not produce direct or consensual stimulation of the pupil. I do not exactly understand why this term is taken from the organ of hearing. The more appropriate expression would be reflex blindness (reflex amaurosis or amblyopia) or reflex half-blindness. Strictly speaking, the non-production of the light reflex of the pupil, despite its normal mobility, is not a disturbance of movement but of vision, inasmuch as the cause is situated, not in the centrifugal motor paths, but in the centripetal (with regard to the nucleus) paths of optical conduction.

As the pupillary movements are not influenced directly by the will (only indirectly through the impulse of accommodation and convergence), the reaction of the pupils to light takes place through the first reflex arc (*vide* page 87), within which involuntary movements are excited by unconscious sensory impressions. It is true that the impressions of light usually enter the field of consciousness after conduction in the primary optic ganglia, but this is not a necessary occurrence. As we have already seen (page 66), preserved reaction of the pupils to light despite the loss of vision is especially characteristic of cortical visual disorders.

The pupillary movements are discussed in detail by Heddæus, Diss., Halle, 1880, and by Leeser, "Die Pupillarbewegungen," Prize Essay, Wiesbaden, 1881.

A rare disorder of pupillary movement is the so-called hippus, *i.e.*, rhythmical contraction and dilatation without change of illumination; the number and extent of the oscillations may be very variable (*vide* Damsch, *Neur. Centralbl.*, May 1st, 1890). It is observed in recovering paralyses of the motor oculi, and is then associated with nystagmus. It is much rarer as an independent condition and is then found almost always in diseases such as tabes, multiple sclerosis, etc., in which there are frequent lesions in the region of the nuclei of the ocular muscles. In such cases it is the forerunner of reflex rigidity of the pupil. Nystagmus of the pupil is the most appropriate term for this condition. We must assume that, as a result of imperfect irritation or irritability of the muscle nucleus, the innervation takes place explosively, by fits and starts.

Paradoxical reaction of the pupils—dilatation on the entrance of light, contraction on removal of light—is an extremely rare phenomenon (*vide* Oestreicher, *Berl. kl. Woch.*, Feb. 10th, 1890). No sufficient explanation of this symptom can yet be given. Cases like that of Burchardt (*Berl. kl. Woch.*, 1890, 2), in which the pupil was enlarged by contraction of the sphincter, in a case of coloboma of the iris and adhesion of one pillar, do not really belong in this category.

The reaction of the pupil in convergence and accommodation is in part a purely mechanical process. During both acts a part of the blood contained in the ciliary body is pressed into the iris and the pupil is thus temporarily narrowed. In fact, this narrowing of the pupil must be regarded as an associated movement of vigorous innervation due to internuclear connections, in the same way that convergence to a certain distance is accompanied by a corresponding strain of accommodation. In peripheral paralysis or insufficiency of accommodation the increased impulse of accommodation will be manifested chiefly by such associated movements or will intensify them (internal squint in maximum innervation of accommodation).

The presence of the normal reaction of convergence and accommodation with absence of the reaction to light shows that the nucleus of the sphincter pupillæ and the nerve fibres originating from it are intact. On account of the close proximity of the individual groups of ganglion cells of the motor oculi nucleus, it is hardly possible that a lesion can affect only the connecting fibres between these groups.

The dilatation of the pupil in deep inspiration and increased intraocular pressure, and its contraction in expiration and diminished intraocular pressure, are purely mechanical processes.

In spasm of the sphincter pupillæ the pupil is more or less narrowed and reacts less freely to light, convergence and accommodation. The cause may be intraocular (inflammations, myotics), or within the domain of the motor oculi (meningitis), or in the nucleus of the sphincter, although the real cause is situated more centrally.

Dilatation of the pupil after cutaneous irritation is one of the sympathetic reflexes and hence is not lost in paralysis of the motor oculi. It is readily seen only after quite severe irritation in the neighborhood of the eye (head, face and neck). As a rule it is unilateral, or at least much more marked on the side of the irritation than on the other. This reflex is absent in many peripheral anæsthesias of the skin. It is retained in central anæsthesias (corona radiata, cortex, etc.) because the connection between the roots of the trigeminus and their sympathetic ganglia is not interrupted.

We have already stated that the sympathetic supplies the dilatator pupillæ, the fibres in the lids which moderately dilate the palpebral fissure (dilatator palpebrarum), and Mueller's muscle, which closes the inferior orbital fissure and is able to push the globe slightly forward (protrusor bulbi). Hence, irritation of the sympathetic will cause moderate dilatation of the pupil with preserved although slightly diminished reaction to light; spastic dilatation of the palpebral fissure with perfect power of closure; moderate protrusion of the eye, although its mobility is entirely free. At the same time the upper lid does not follow the eye in looking downward to the same extent as normally. Not infrequently the white sclera becomes visible above the cornea (v. Graefe's symptom in Basedow's disease). It is still doubtful whether the intraocular pressure is also increased; diminished power of accommodation is sometimes said to be present (Eulenburg). As a rule this does not hold good. In pure affections of the sympathetic, accommodation, refraction and, usually, the intraocular pressure remain normal.

In paralysis of the sympathetic narrowing of the pupil (myosis) is observed, together with normal reaction to light. The color of the iris often appears somewhat lighter because, on account of the contraction, the coloring elements of the iris are distributed over a larger area. In addition there is slight drooping of the upper lid (ptosis) and slight retraction of the globe into the orbit (enophthalmus). Diminution of intraocular pressure has often been mentioned, and to this is attributed the occasional increase in refraction. On the other hand, diminished accommodation has occasionally been seen on the side of the paralysis. If this is not accidental it is, at all events, very rare. Ocular pressure, refraction, and accommodation are usually unchanged.

In 1869 Horner first gave a distinct clinical history of paralysis of the oculo-pupillary sympathetic fibres. Later his pupil, Nicati, furnished a somewhat schematic account of disease of the cervical sympathetic, beginning with irritative symptoms and passing into paralysis. Such cases are not very rare, and occasionally they may even lead to degenerative trophic changes in the face. Much more frequently there are pronounced paralytic phenomena from the start in some or all the branches. Irritative symptoms may last a long time without terminating in paralysis. The causes are injuries, tumors, inflammations, and suppurations which involve the cervical sympathetic. The cause is generally unknown or the patient attributes the condition to a cold. The affection of the sympathetic is found accidentally during examination for other diseases, as it produces no annoying symptoms.

Abnormalities in the size of the pupils of a central character may be divided accordingly into four classes:

1. Paralytic cerebral mydriasis (paralysis of the sphincter).
2. Spastic cerebral myosis (spasm of the sphincter).
3. Paralytic spinal myosis (paralysis of the dilatator).
4. Spastic spinal mydriasis (spasm of the dilatator) ; the latter form also includes dilatation of the pupil after cutaneous irritation.

Those who deny the existence of a dilatator muscle of the pupil (a simple layer of smooth muscular fibres immediately in front of the posterior iris pigment) must explain the action of the sympathetic by an influence on the calibre of the vessels of the iris, narrowing in spasm and dilatation in paralysis.

In addition to the oculo-pupillary symptoms, vasomotor disturbances in the corresponding half of the face and head are also observed in disease of the cervical sympathetic, viz., narrowing of the vessels in irritation, and dilatation in paralysis of the sympathetic. Kussmaul distinctly saw blanching of the fundus oculi in irritation of the sympathetic. Corresponding changes are more readily seen in the integument of the face, such as unilateral sweating or absence of sweating, unilateral redness or pallor of the skin with several degrees difference in the temperature, and increase or diminution of the lachrymal secretion.

As a rule, no influence is exerted upon the growth and nutrition of the parts, upon inflammatory processes, etc. At a later period, however, " trophic" disturbances are occasionally observed in disease of the sympathetic, but it is evident that another factor has then been superadded. In the same way peripheral anæsthesias usually exert no influence upon the tissue changes, while in other cases visible " trophic" changes also develop.

The so called " neuro-paralytic keratitis" is wrongly included among trophic disorders. It is not a disturbance of the growth and nutrition of the cornea, but an entirely different process due to infection of a corneal ulcer.

After injury or loss of substance of vascular tissues, with or without infection, when healing takes place the vessels play a prominent part, evidently as the result of the direct action of chemical or organized irritants upon them. In non-vascular tissues this direct action upon the adjacent vessels is impossible or is only possible at a late period; it takes place in a reflex manner, through the agency of the trigeminus and sympathetic, and is absent in anæsthesia of the former. Hence the course of recovery, especially of infected surfaces, is modified and prolonged, but there is no real disturbance of nutrition or growth.

If the peripheral sensory tracts are interrupted, the vascular reflexes after sensory irritation, which are not absolutely local, are abolished, and this is particularly noticeable in non-vascular tissues. If the peripheral sympathetic tracts are abolished—and this, in view of their position, must always be associated with disease of the corresponding vascular tract—a substitution may take place along collateral paths. But if the sympathetic ganglion cells are extensively diseased, genuine trophic disturbances make their appearance.

In view of the connections of the sympathetic ganglia with the sensory nerves and the central nervous system, we can understand the fact that, from the start, the affection may assume the character of a sympathetic disorder or may develop from an anæsthesia or even start from the spinal cord (syringomyelia).

The peripheral sensori-sympathetic arc is situated outside of the skull and spinal cavity.

But the sympathetic ganglia also receive centrifugal fibres from the central nervous system through the motor roots, and these exert a tonic or constricting influence on the muscular coats of the vessels. This is true with regard to the brain as well as the spinal cord.

This explains the fact that in disease of an entire cerebral hemisphere unilateral symptoms of irritation or paralysis of the cervical sympathetic are found so frequently. But the local and sensori-sympathetic vascular reflexes are not abolished, indeed they are often exaggerated. This may be the most striking symptom of abolition of the influence of the central nervous system upon the sympathetic.

The higher senses may also exert a tonic influence upon the vessels, especially upon those of the corresponding organ of special sense, and this is really done whenever the organ is engaged in a function which excites the faculty of attention. If, for example, the activity is confined to a single special sense, the central tonic action upon the other senses is abolished. In this way purely functional disorders of the latter may be produced, in the sense of "not seeing and hearing what is going on." If this affects all the senses sleep will result. This condition will again be referred to repeatedly in discussing hysteria, hypnosis, sleep, trophoneuroses, traumatic neuroses, etc.

The sympathetic fibres for the eye probably run along with the motor fibres in the corona radiata and internal capsule, and also along the locality from which Lannegrace has produced unilateral visual disturbances of the opposite eye.

In such visual disorders no abnormal fulness of the vessels can be seen either without or within. For certain reasons it seems probable to me that the effect is produced in the nerves which emerge from the cranial cavity. Interference or facilitation of conduction at the places where the nerves pass through bony foramina would best explain the symptoms in question.

Hence there are three kinds of sympathetic reflexes: 1, the local reflex at the site of irritation; 2, the peripheral sensori-sympathetic reflex arc; and 3, the central reflex arc, which passes through the brain or spinal cord. The reflexes are unconscious in all three kinds, but in the third reflex arc they are voluntary in so far as they consist of associated innervation in motor impulses or in inhibition of such impulses. Under certain circumstances these may be employed for definite purposes, for example, when the whole attention is devoted to visual or auditory impressions.

The very intense innervation of the vessels in violent emotions, such as fright, fear, shame, and anger, are effected through the third sympathetic reflex arc. Such temporary violent innervation after severe fright may pass directly into permanent paralytic symptoms (so-called traumatic neuroses).

In ordinary paralysis of the sympathetic we have to deal essentially with the loss of the central tonic influence of the brain or spinal cord. The ganglion cells of the sympathetic and the centrifugal fibres emerging from them are intact, together with the reflexes in the local and middle sensori-sympathetic arcs. The latter may even be considerably exaggerated.

A very material characteristic of all these symptoms is their unilateral character, but, as a matter of course, they may also occur on both sides.

We have already seen that the oculo-pupillary centre in the spinal cord is situated below the vasomotor centre for the vessels of the head, and that the fibres consequently pass through the nerve roots to the sympathetic ganglia at different levels. If the trunk of the sympathetic is affected the oculo-pupillary and vasomotor symptoms will appear, as a rule, at the same time or in brief succession.

If the lesion is localized in the cord or at the nerve roots, the oculo-pupillary or vasomotor symptoms often appear singly, or only single symptoms make their appearance, such as myosis, or ptosis and myosis. Oculo-pupillary irritative and vasomotor paralytic symptoms may also be combined, or *vice versa*. In diseases of the spinal cord the presence or absence of these symptoms is of great diagnostic importance, and this is also true concerning lesions of the roots of the brachial plexus. In the latter the oculo-pupillary symptoms are a characteristic feature of the "inferior type." In lesions of the brachial plexus proper, however, sympathetic symptoms are no longer observed because the fibres to the sympathetic ganglia are given off before the plexus is formed.

Disturbances in the lachrymal secretion are, in the main, vasomotor sympathetic symptoms, located chiefly in the middle and partly in the central reflex arc (weeping without physical pain, abolition of the secretion of tears in the weeping of many insane, etc.). We are not justified, however in applying the term "tear centre" to parts of the brain from which disorders of the lachrymal secretion may be produced. It is evident that there is merely interruption or disturbance of conducting paths. Nervous epiphora may precede and accompany diseases of the brain and spinal cord.

C. Relations between the Blood-Vessels and the Eye.

1. *Arteries.*—With the exception of a part of the lids the eye is supplied by the cerebral arteries, a fact which accords with the development of the organ from the mid-brain.

Fig. 16 shows the distribution of the large vessels at the base of the brain, according to Merkel ("Handb. d. topogr. Anat.," Bd. 1).

The internal c a r o t i d passes through the cavernous sinus in a double arch and then gives off the ophthalmic artery, which passes through the optic foramen below and to the outside of the optic nerve into the orbit. The central artery of the retina, derived from the ophthalmic, enters the trunk of the nerve with the corre-

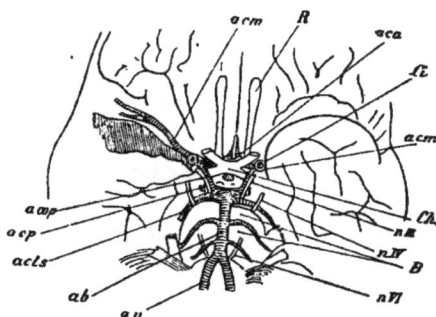

FIG. 16.—Arteries at the Base of the Brain, after Merkel. *R*, Olfactory nerve; *Ch*, chiasm; *nIII*, *nIV*, *nVI*, third, fourth, and sixth cranial nerves; *B*, pons; *Ci*, internal carotid; *av*, vertebral artery; *ab*, basilar artery; *aca*, anterior cerebral artery and anterior communicating; *acm*, median cerebral artery; *acop*, posterior communicating artery; *acp*, posterior cerebral artery; *acls*, superior cerebellar artery.

sponding vein from below and the outside, about 1–1½ cm. behind the eye. A few little twigs pass to the optic tract, chiasm and first part of the optic nerve, from the trunk of the carotid which lies in the outer angle of the chiasm.

The carotid then divides into its two terminal branches. The anterior cerebral artery passes forward at a right angle and communicates at once with the corresponding artery of the opposite side (anterior communicating branch). It passes to the first and second frontal convolutions, to the medial surface of the cerebrum as far as the cuneus (including the præcuneus and excluding the occipital and parietal lobes) and to the inner half of the inferior surface of the frontal lobe. The second branch of the carotid, the middle cerebral artery, passes into the Sylvian fissure and is distributed to the first frontal convolution, insula, convexity of the parietal lobe, first temporal convolution and a part of the second, and the outer half of the lower surface of the frontal lobe.

Posteriorly the carotid sends beneath the optic tract the posterior communicating branch to the posterior cerebral artery. The latter is the terminal branch of the basilar artery which is formed by the union of the two vertebral arteries. The posterior communicating branch runs along the motor oculi nerve and sends branches to the chiasm, infundibulum, hypophysis, and anterior part of the optic thalamus. From it is usually derived the anterior choroideal artery which supplies the choroid plexus.

The posterior cerebral artery supplies the rest of the cerebral hemisphere, viz., the occipital lobe, the inner and lower surface of the temporal lobe, and the largest part of the second temporal convolution.

The areas of supply of the middle and posterior cerebral arteries meet at the parieto-occipital fissure, the cuneus, also known as Nothnagel's perceptive centre, or as the macular locality of other writers (*vide* Fig. 17). It follows that this locality is especially favored as regards its supply of blood. For example, in occlusion of the posterior cerebral artery, this part may still receive nourishment while the remainder of the occipital cortex is destroyed. This explains very simply the persistence of a small central remnant of the field of vision in bilateral homonymous

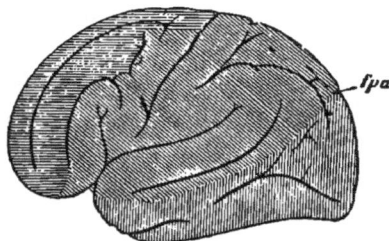

Fig. 17.—Distribution of the Three Large Cerebral Arteries over the Convexity, after Merkel. Distribution of the anterior cerebral artery shaded horizontally; that of the median, obliquely; that of the posterior, vertically. *fpa*, Parieto-occipital fissure.

hemianopsia due to embolism in the occipital lobes (*vide* the cases of Foerster, Schweigger, and others on page 61). I may also remark that the same relation holds good concerning the auditory centre in the second temporal convolution.

According to Heubner all these arteries terminate in the same way. They give off at right angles numerous small branches to the basal ganglia. The anterior cerebral artery supplies the head of the caudate nucleus, the most anterior part of the base, the infundibulum, chiasm, optic nerve and corpus callosum. The middle cerebral artery passes through the lateral perforated space, and in the first centimetre of its course supplies the anterior limb of the internal capsule, the inner part of the lenticular nucleus, etc.; in its second centimetre it supplies the lateral and upper part of the lenticular nucleus, caudate nucleus and external capsule. The posterior cerebral artery passes through the middle perforated space to the cerebral peduncles, corpora quadrigemina, lateral part of the optic thalamus, choroid plexus of the third ventricle, etc.

All arteries of the basal region are end-arteries, i.e., they only anastomose with one another by means of the capillaries. It is not until the arteries have run a course of 2 to 3 cm. that they divide dichotomously in the pia mater, and these branches exhibit numerous anastomoses. They give off numerous lateral branches, short and fine ones to the cortex, longer and larger ones to the medullary substance; these vessels constitute end-arteries. The arteries of the cortex are especially liable to embolism; the small vertical branches of the basal region show a predisposition to diseases of their walls, and to the formation of aneurism and thrombosis.

All other parts of the cerebrum and cerebellum and the nerves from the third to the twelfth are supplied by the basilar and vertebral arteries. Each nerve receives small branches which run in part centrifugally, in part centripetally as far as the nuclei. The ninth, tenth, eleventh and twelfth nerves are supplied by the vertebral arteries, the seventh and eighth nerves by their point of union, and the seventh and third nerves by the basilar artery. From all three a series of vessels passes in the median line as far as the floor of the fourth ventricle; from the vertebral artery they go as far as the lower border of the pons; from the basilar they pass to the pons, the floor of the fourth ventricle, and the beginning of the aqueduct of Sylvius.

Each nerve nucleus thus receives arterial blood from two sides, from the branches just mentioned, and from the centripetal branches in the nerves. The anterior part of the motor oculi nucleus also receives blood from the posterior cerebral artery, the posterior part from the basilar artery. This furnishes the possibility of an independent affection of each division.

The lateral parts of the floor of the fourth ventricle also receive blood from the choroid plexus.

The cerebellar arteries are derived from the basilar and vertebrals, and include the posterior inferior and anterior inferior cerebellar arteries and the superior cerebellar artery. The latter emerges from the basilar at the anterior border of the pons, passes between the motor oculi and trochlearis nerves to the pons, then to the middle cerebellar peduncle, superior vermis, valve of Vieussens, corpora quadrigemina, choroid plexus of the third ventricle, etc.

To give a brief *résumé* of the vascular supply of some important parts of the brain:

The internal capsule receives its supply anteriorly from the middle cerebral artery, posteriorly from the anterior choroideal artery or the posterior communicating artery.

The optic thalamus is supplied anteriorly by the posterior communicating branch, posteriorly by the posterior cerebral artery.

The posterior cerebral commissure and the pineal gland are supplied by the anterior cerebral artery, the latter also by the posterior cerebral and the superior cerebellar artery.

The optic tract is supplied by the trunk of the carotid, the posterior communicating and posterior choroideal arteries; the chiasm and optic nerve are supplied by the carotid, anterior cerebral, communicating and posterior arteries.

The corpora quadrigemina are supplied by the posterior cerebral and superior cerebellar arteries.

The pons, by the basilar and superior cerebellar arteries; the choroid plexus of the fourth ventricle is supplied by the posterior inferior cerebellar artery.

The occipital lobe is supplied by the posterior cerebral artery, its macular portion by the middle cerebral; the region of the anterior cerebral artery also passes, on the inner surface of the hemisphere, to the border of the visual sphere.

2. *Veins.*—The veins in the skull and orbit are destitute of

valves.   The larger trunks are very variable and anastomose freely with one another; the small ones accompany the arteries, the large ones do not.   According to Merkel the small veins within the brain substance no longer anastomose with one another.

The superior cerebral veins from the anterior and upper part of the hemispheres and a part of the median surface empty into the superior longitudinal sinus; the middle veins from the anterior horn, corpus striatum, inferior horn, optic thalamus, peduncles, internal capsule, base of the brain, inferior surface of the occipital lobe, upper and lower part of the cerebellum, pass in the vena magna Galeni to the sinus of the tentorium; the inferior cerebral veins from the temporal and occipital lobes, cerebellum and parts of the base pass to the petrosal, transverse and cavernous sinuses.

The orbit contains two principal veins, the superior and inferior ophthalmic veins, which freely anastomose with one another, particularly in the vicinity of the optic foramen and immediately behind the globe. The superior ophthalmic vein passes through the superior orbital fissure to the cavernous sinus; the inferior ophthalmic empties into the deep facial vein.   Both have numerous connections with the veins of the cheeks,

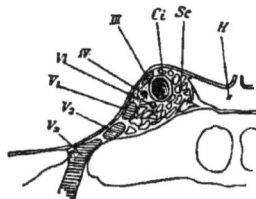

FIG. 18.—Frontal Section of the Cavernous Sinus, after Merkel. *Sc*, Cavernous sinus; *H*, hypophysis; *Ci*, internal carotid; *III*, *IV*, *VI*, oculomotor, trochlearis and abducens nerves; *V* 1, 2, 3, first, second, and third branches of the trigeminus.

lids, nose, and temporal region.   None of them possesses valves.

The central vein of the retina usually empties into the superior ophthalmic, but often passes through the superior orbital fissure directly into the cavernous sinus.

D. Relations between the Eye and the Lymphatics.

The membranes covering the optic nerve pass directly into the cerebral meninges at the optic foramen, to whose upper wall the nerve is quite firmly adherent.   Hence the subdural and subarachnoid spaces of the optic nerve communicate with the corresponding ones within the skull.

The lymphatics in the brain are mainly perivascular and ac-

company the veins. They leave the skull through the carotid canal,
the jugular foramen, and with the vertebral artery to the upper deep
cervical glands. The outflow of lymph in the subarachnoid and
subdural spaces of the optic nerve and the current
of lymph in the nerve itself appear to be directed
toward the cranial cavity.

If an injection is made, under moderate pressure, into the tissue of the nerve in a direction toward the cranial cavity, the fluid does not pass
beyond the chiasm into the tractus, but transversely
across the anterior half of the chiasm to the other
optic nerve, along which it runs in the same way
as if it were directly injected in a centrifugal direc-

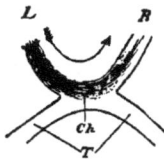

FIG. 19.—Plastic In-
jection into the Left
Optic Nerve, *L*, in the
direction toward the
chiasm, *Ch* ; *R*, right
optic nerve; *T*, optic
tract.

tion. The injected fluid extends farthest forward in the axial bundles of the second optic nerve. (This experiment is not easy and
may at first trial fail; with practice it can be done with certainty.)

This explains the fact that inflammatory processes—whether
infectious or non-infectious—may spread from one eye to the other
without giving rise to notable cerebral symptoms. This is true not
alone of so-called sympathetic ophthalmia but also of an entire series
of inflammatory affections, especially of the choroid.

The lymphatics of the orbit which are connected with Tenon's
capsule empty mainly into the deep glands of the face; the lymphatics of the lids and conjunctiva pass into the glands in front of the
ear and below the inferior maxilla.

The facts and hypotheses hitherto presented furnish the material
from which we may make a local diagnosis of affections of the organ
of sight (in the broadest sense of the term). The subject becomes
more complicated from the fact that diseases within the skull, especially when they develop suddenly, may give rise to remote symptoms which are due to an implication of parts of the brain that are
not directly affected by the disease. The remote symptoms usually
disappear gradually, and it is only those which persist after the lapse
of weeks or months that may be utilized for accurate local diagnosis.
The greatest part of what is known as "the vicarious action of

other parts of the brain" must be attributed to the gradual disappearance of the remote symptoms. For example, the phenomena observed in small circumscribed destructions of the cortex at the periphery of the visual sphere are almost exclusively remote symptoms.

The remote symptoms also include those instances in which, for example, an adjacent or even a distant nerve is pressed against a bony process or a tense artery and is made incapable of conduction. For example, isolated paralysis of the abducens sometimes results from a growth in the opposite hemisphere.

Apart from the focal symptoms, a comparatively small lesion may give rise to irritative or paralytic symptoms of an entire hemisphere or even the entire brain, at another time it may produce only remote symptoms in a certain direction, and in a third case it may produce none at all. Many progressive lesions are surrounded by a zone of irritation which finally gives rise to paralysis. This is observed not infrequently at the base of the brain where the nerves are attacked in succession.

The irritative zone of remote symptoms is usually situated farther away from the original lesion than the paralytic manifestations.

Diseases of the cerebellum, pons, pineal gland, optic thalamus, and other parts near the region of muscle nuclei may or may not be attended with spasms and paralyses of the ocular muscles. In like manner, lesions near the optic tract and chiasm may or may not be associated with disorders of vision. In unilateral lesions of the pons paralysis of the ocular muscles is on the side opposite to the hemiplegia, because the motor parts decussate in the medulla. In diseases of the posterior limb of the internal capsule central disturbances of vision and disturbances of sensation are usually combined.

In addition to the visual symptoms which may be utilized for local diagnosis there are other symptoms of which we may avail ourselves—such as exophthalmus, œdema of the lids, chemosis, choked disc, neuritis, atrophy of the optic nerve, diseases of the choroid and retina, etc. These permit an inference with regard to the character of the disease, rather than its location.

9

E. Individual Diseases of the Brain, Cord, and Nerves.

After this detailed introduction we may be comparatively brief in our remarks on the implication of the organ of vision in the individual diseases of the nervous system. In many of them the eye symptoms are extremely important and characteristic, and determine the diagnosis and treatment. As we shall see, they often throw light on the interpretation of the whole disease.

1. DISEASES OF THE BRAIN.

Anæmia and Hyperæmia of the Brain and its Membranes.

In these two conditions it is to be supposed that the status of the vessels of the retina and the entrance of the optic nerve (which derive their blood supply from the same source as a large part of the brain) would furnish a picture of the vascular condition of the brain, especially of the cortex and the pia mater. This expectation, however, is far from being realized. In the first place, the amount of blood in the brain does not depend solely on the internal carotid (from which the ophthalmic artery takes its origin) but from the entire supply to the circle of Willis. For example, changes in the amount of blood in the carotid may appear only in the retina, where they give rise to hyperæmia or anæmia, while the amount of blood in the brain remains unchanged.

In many cases, however, the amount of blood in the pia mater and in the retina is similar. The slight degrees of change are demonstrable with difficulty or not at all, because the calibre of the vessels varies normally within very wide limits. Hence it is difficult to recognize slight hyperæmia or anæmia of the retina or papilla unless by comparison with the healthy eye. As a rule this is impossible in diffuse affections of the brain.

It may even happen that the retina exhibits the opposite condition to what must be assumed to exist in the brain. The retinal findings, especially when hyperæmic in character, must therefore be relied upon with caution in assuming an analogous condition within the cranium, and with a due regard for all other circumstances.

This is particularly true of acute conditions. In more or less chronic diseases the vascular conditions in the brain and retina gradually appear to become alike, as is seen, for example, in the constant venous hyperæmia of the retina in epilepsy of long standing.

Hyperæmia of the retina is often found, is often absent, in brain diseases of all kinds, but it is constant in none. The apparent hyperæmia of a very hypermetropic eye, especially in young people, or the hyperæmia of the papilla and retina, even amounting to true neuritis, which is a constant symptom at the beginning of myopia, cannot be utilized for diagnostic purposes as regards brain disease. Particularly in threatening myopia is the diagnosis not easy.

Apart from the ophthalmoscopic findings, the symptoms of cerebral congestion are usually said to be restlessness, irritability, diffuse headache and narrow pupils; those of cerebral anæmia are drowsiness by day, sleeplessness at night, circumscribed headache and dilated, sluggish pupils. On the other hand Corning (*N. Y. Med. Rec.*, 13, XI, 87) states that a dilated pupil is found in congestion, a contracted pupil in anæmia, but this is probably not true of the majority of cases.

Cerebral Hemorrhages.

As a rule cerebral hemorrhages develop suddenly, so that, at the start, the remote symptoms predominate. The function of both hemispheres may be abolished quite suddenly. In a little while the symptoms are confined to one hemisphere—hemiplegia or hemianæsthesia with or without disturbance of speech. Very often there is also homonymous hemianopsia in the visual field of the side opposite to the hemorrhage, but this usually disappears in a short time unless the visual zone or the optic radiations in the corona radiata are the site of hemorrhage. Gowers (*Brit. Med. Journ.*, Nov., 1877) first called attention to the frequent occurrence of temporary hemianopsia in apoplectic hemiplegia; this is especially true of the period during which conjugate deviation of the head and eyes is present. The latter symptom is very frequent in cerebral hemorrhage, and takes place toward the side of the lesion. We have already stated that this symptom, in our opinion, is due to irritation of the opposite hemisphere. Hence, in the exceptional cases in which the deviation takes place toward the side opposite to the lesion, the hemorrhage has either been situated far forward in the frontal brain (remote from the visual sphere, which is the most important part of the cortex so far as concerns the motor oculi nerve), or it has been very small (Lépine, *Rec. d'Ophth.*, 1876, p. 280). In these cases the occipital cortex is not included in the paralytic zone, but in the zone of peripheral irritation. In paralysis of one hemisphere without irritation of the other, voluntary associated movements toward the opposite side are abolished to a greater or less extent. This is also a frequent temporary symptom in cerebral hemorrhage.

Sudden homonymous hemianopsia may be the sole symptom, apart from temporary dizziness, of a hemorrhage into the brain. If there is hemiopic reaction of the pupils, the lesion is situated in the optic tract or primary optic ganglia. If the pupils react normally, it is usually situated in Gratiolet's fibres, because such sudden hemianopsias are generally permanent. As a matter of course, a hemorrhage which produces only homonymous hemianopsia must be a small one, and a small hemorrhage into the visual cortex can only give rise to permanent hemianopsia when it destroys the entire perceptive centre. A hemorrhage of such dimensions, however, always causes initial remote symptoms.

The term "signe de l'orbiculaire" is applied by French writers to the inability of hemiplegics to close only the eye of the paralyzed side. This would be a sign of cortical weakness of the orbicularis palpebrarum, *i.e.*, of voluntary closure of the eye. The symptom loses much of its importance, however, from the fact that many healthy individuals are unable to close each eye separately.

Every cerebral hemorrhage forms a focus of destruction, surrounded by a zone of paralyses and, at a greater distance, by a zone of irritation. If the hemorrhage is not followed by inflammation, by which the condition becomes a progressive one, the remote irritative symptoms, which usually include conjugate deviation, first disappear, and then disappear the symptoms of the paralytic zone; under these are included hemianopsia and the inability to perform associated movements of the eyes toward the side opposite to the lesion, unless these are due to direct destruction.   It is not rare that paralytic symptoms disappear with the onset of irritative phenomena in the previously paralyzed part, such as visual hallucinations, spasms and twitchings; for example, associated twitchings toward the side opposite the lesion, when conjugate movements in that direction had been previously paralyzed.

Finally, the symptoms become restricted to those due to the loss of the directly destroyed region, and apparent complete recovery may ensue.   On careful examination, however, we are usually able to demonstrate slight defects in the province of some of the organs of special sense, or of motor activity, or in the general mental condition (memory, change of character, etc.).

From these statements it is evident that an accurately local diagnosis can hardly ever be made except at the very beginning of the attack or after all the remote symptoms have run their course.

If a hemorrhage ruptures outward into the subarachnoid and subdural space, basal symptoms are produced, such as spasms and paralyses, anæsthesias and paræsthesias of the nerves at the base, the chiasm, etc.   The blood may enter the sheath of the optic nerve and produce changes which are visible with the ophthalmoscope (choked disc).   This is often attended with dilatation of the pupils, probably from pressure on the motor oculi nerve.   In one-half of his cases of cerebral hemorrrhage Hutchinson observed rigid mydriasis on the side of the lesion, and believes that this is a very important sign.   It is probably due in most cases to paralysis of the motor oculi from pressure on the nerve, occasionally to irritation of the sympathetic, or to both.   Mydriasis is also observed in meningeal hemorrhages, usually on the same side; more rarely it is bilateral on account of the extension of the hemorrhage; myosis is very rare.   White (*Brain*, 1886, p. 532) and Seeligmueller (*Arch. f. kl. Med.*, XX, p. 101) have called attention to permanent paralysis of the oculo-pupillary sympathetic fibres on the same side as the hemorrhage, *i.e.*, crossed with regard to the general paralytic symptoms.

If the extravasation perforates into the ventricles decided narrowing of the pupils is a striking symptom.   As this is also observed in hemorrhages near the nuclei of the ocular muscles, for example, into the pons, it is probably due to direct irritation of the sphincter nucleus.

Symmetrical hemorrhages may also occur on both sides. Chauffard (*Rev. de Méd.*, 1888, No. 2) reports a case of blindness from bilateral destruction of the optic radiations.

Ophthalmoscopic findings which may be attributed directly to cerebral hemorrhage are very rare. Neuritis or choked disc, with or without retinal hemorrhage, is occasionally observed: it terminates in recovery or in partial or complete atrophy of the optic nerve. Bristowe (*Ophth. Rev.*, March 20th, 1886, p. 88) describes bilateral optic neuritis (probably choked disc) with hemorrhages in a case of sudden hemiplegia with unconsciousness. The diagnosis of cerebral tumor was made, but a hemorrhage, as large as a pigeon's egg, was found in the left optic thalamus. It extended to the posterior border of the internal capsule, to the white substance of the temporal lobe, and to the posterior part of the lenticular nucleus. A few similar cases have been reported. In all such cases the hemorrhage must be large, and the extravasation into the optic sheath, and eventually the choked disc, if unilateral, will be found upon the same side as the hemorrhage.

Choked disc is found somewhat more frequently in pachymeningitis hæmorrhagica, the symptoms of which resemble those of tumor with intercurrent hemorrhages (*vide* Tuczek, *Wien. med. Blaetter*, 1883, 12, and Zacher, *Neur. Centralbl.*, 1885, p. 125).

Complete or partial atrophy of the optic nerve also takes place when the hemorrhage has destroyed the primary optic ganglia, tractus, chiasm or optic nerves, or has exercised upon them a pressure which terminates in degeneration.

On the other hand the eye often exhibits signs which have nothing to do with the existing cerebral hemorrhage but are attributable to the same causes as the latter. For example retinitis, neuritis and albuminuric neuro-retinitis are signs of similar changes in the brain: simple retinal, vitreous, and other hemorrhages are evidences of a wide-spread atheromatous process. Atheroma of the vessels is often seen with the ophthalmoscope (thickening of the walls, narrow column of blood in places, etc.). In rare cases the walls are found to be of a yellowish color in patches, particularly upon the papilla, or miliary aneurisms are found to correspond with similar changes in the brain.

The spontaneous, rapidly disappearing, but easily relapsing hemorrhages of the conjunctiva and retina in old people are of serious significance and betoken possible hemorrhage into the brain.

Embolism and Thrombosis.

An embolus produces the same symptoms, in general, as a hemorrhage and in many cases can hardly be distinguished from the latter. If it is infectious it gives rise to an abscess of the brain, and the symptoms are those of the latter affection. If it is not infectious, a spot of softening will develop. The latter will also develop, although not so suddenly and with fewer remote symptoms, in thrombotic processes which may occur as the result of various diseases of the cerebral vessels.

Spots of softening are by far the most important brain diseases, as regards local diagnoses, because they furnish the purest focal symptoms after the termination of the brief period of reaction. Most of our knowledge of more accurate local diagnosis in man is based upon the clinical symptoms in such cases.

Complications are very rare. In only one case of softening of the occipital lobe is mention made of double choked disc, which disappeared later (Wilbrand, *Arch. f. Ophth.*, XXXI, 3, 319). This is probably not more frequent in softening of other parts of the brain.

A number of cases of bilateral symmetrical softening of the optical centres have been observed (Oulmont, *Gaz. Hebd.*, 1889, No. 38; Bouveret, *Rev. gén. d'Ophth.*, 1887, p. 481; Berger, *Ref. Jahr. f. Aug.*, 1885, p. 289).

Thromboses of the cavernous sinus act differently, according as they are infectious (following caries of the petrous portion of the temporal bone) or non-infectious (marantic thrombosis). In the former event they usually furnish the symptoms of a cerebral abscess together with those of suppuration of the orbital cavity (swelling of the lids, protrusion and immobility of the eye, dilated and rigid pupil, blindness, insensibility of the conjunctiva and cornea, ulceration of the cornea, suppuration of the eye, etc.). So long as the thrombus has not extended to the ophthalmic vein, the ophthalmoscopic appearances may be tolerably normal. But after this has taken place, we find pronounced venous stasis of the retinal veins and more or less abundant retinal hemorrhages, while the entrance of the optic nerve is very red but usually not much swollen.

These ophthalmoscopic appearances are found more frequently in the more slowly progressing marantic thromboses, but only when the thrombus has extended to the ophthalmic vein. Otherwise the current of blood may pass through the posterior facial vein. After injections of wax into the cavernous sinus Ferrari found no stasis of the retinal veins; this took place only when the wax entered the ophthalmic vein itself.

In non-infectious thrombosis of the sinus, we find, apart from general cerebral symptoms, mainly basal symptoms, viz., paralyses of the motor oculi nerves which pass near the sinus (*vide* Fig. 18), rigid and dilated pupil, insensibility of the trigeminus, especially of the cornea and conjunctiva, impairment of sight or even blindness due to interruption of conduction in the optic nerve, etc.

Œdema of the lids and protrusion of the eye then indicate that the process is located in the vicinity of the orbital cavity. All these symptoms may likewise occur in meningitis, and hence the ophthalmoscopic appearances may decide the diagnosis. In meningitis venous congestion and even pronounced neuritis may occur, but the marked stasis of the retinal veins, which is found in thrombosis of the sinus with thrombosis of the ophthalmic vein, is never observed.

Abscess of the Brain.

The symptoms of abscess of the brain consist of those of a tumor combined with those of a purulent infectious inflammation. The destroyed parts of the brain do or do not give rise to symptoms, according to their location, so that a stationary abscess may long remain latent.

In progressive suppurations the destroyed part is surrounded by a zone which gives rise to irritative symptoms, passing later into paralysis. In rapidly advancing suppuration the remote symptoms may even extend to the opposite hemisphere, so that great caution is necessary in utilizing cerebral abscesses for local diagnosis.

Choked disc is found frequently, though not so often as in tumors; neuritis is also frequent, but the most frequent is "obstructive neuritis," *i.e.*, there is slighter swelling but more pronounced inflammatory phenomena than in typical choked disc. This "obstructive neuritis" is the most characteristic ophthalmoscopic finding in abscess of the brain. As a rule the lesion is bilateral though more marked on one side. But if the neuritis is unilateral or much more marked on one side it is generally found on the same side as the abscess. The latter is then located commonly in the anterior parts of the brain, the frontal or temporal lobes. For example, Greenfield (*Brit. Med. Journ.*, 12, II, 1887) reports a case of abscess of the left temporal lobe after otitis, with left optic neuritis and paralysis of the left motor oculi nerve, which recovered after trephining.

The impairment of sight is often very considerable, varying from losses in the field of vision and color disturbances to complete blindness. Vision is rarely normal or nearly so. In one case of cerebral abscess, Nauwerck (*Deutsch. Arch. f. kl. Med.*, XXIX, p. 1) found that the optic nerve was elevated and very much reddened, but there was no disturbance of sight.

In abscesses near the base of the brain, œdema of the lids, protrusion of the globe, pain and photophobia may appear even prior to perforation, as in purulent meningitis.

Conjugate deviation of the eyes and head usually takes place toward the side of the abscess, but there are also exceptions to this rule, particularly when the abscess is remote from the occipital lobe (Roussel, *Prog. Méd.*, 1886, No. 29).

Apart from infectious emboli, suppurations in the ear are the most frequent cause of abscesses of the brain: and about 60 per cent are situated in the temporal or lower part of the parietal lobes, while about one-third are found in the cerebellum. See Hessler in Schwarze's "Handbuch der Ohrenheilkunde," Bd. II, p. 634 (ED.)

In a number of such cases ptosis has been mentioned as a symptom, and must then be regarded as cortical in character. In one case Heinecke (*Muench. med. Woch.*, 1889, p. 571) observed bilateral ptosis and retinal hemorrhages, the latter probably of a septic character.

The rupture of an abscess into the cranial cavity gives rise to the symptoms of a purulent meningitis, viz., basal spasms, paralyses, and sensory disturbances, neuro-paralytic keratitis, etc. Perforation into the ventricles is usually rapidly fatal; this is attended by extreme myosis, probably from direct irritation of the sphincter nuclei.

All remote symptoms, especially the irritative ones (blindness, paralyses, optic neuritis, etc.), may subside after evacuation of the pus, but losses of the field of vision, color disturbances, and more or less pronounced visual disorders may be left. Symptoms which are due to suppuration and destruction of tissues persist after the evacuation of the pus. Irritative symptoms (spasms) prove that the corresponding part of the brain is not destroyed. Hence, restoration of function of the groups of muscles in question may be expected after a successful operation.

In abscesses of the brain, accordingly, the organ of sight may not alone exhibit important local symptoms (hemianopsia, conjugate deviation, paralyses, etc.), but the demonstration of "obstructive neuritis" at the entrance of the optic nerve warrants a very probable diagnosis of the character of the disease.

Tumors of the Brain.

Tumors of the brain may long run a latent course if they develop in a part whose destruction causes no noticeable local symptoms, and if their growth is slow so that the adjacent parts may yield. This may even happen in the immediate vicinity of those parts whose irritation or destruction causes the most striking local symptoms, for example, in the region of the muscle nuclei. In such cases we must assume that the tumor simply displaces the adjacent tissues but does not destroy them or even interfere materially with their functions. It is evident that pressure alone causes no very great disturbance if the parts can escape to one side. Henschen reports a case in which no corresponding homonymous loss of the field of vision was produced by a tumor in the immediate vicinity of the optic radiations, upon which it must have exercised pressure.

A tumor may also grow mainly in one direction while it advances very little or not at all in the opposite direction.

In other cases of rapidly growing tumors which invade neighboring parts, irritate and then destroy the tissues, there are much more extensive symptoms than would correspond solely to the location of the tumor. The remote effects may be manifested over an entire hemisphere, even over both hemispheres. The latter appear usually as symptoms of cerebral compression, viz., headache, vomiting, dulness, slow pulse, somnolence, dilatation of the pupils, etc.

Not infrequently the remote effects are apparently paradoxical. According to Jastrowitz, unilateral and bilateral paralyses of the ocular muscles, especially of the abducens, are frequent. According to Gowers the abducens is especially exposed to pressure on account of its long course in the skull. After artificial increase of pressure in the skull Wernicke found evidences of pressure on the opposite motor oculi nerve. Oppenheim reports a case of tumor of the right frontal lobe with paralysis of the left abducens.

The tumor is often surrounded by a zone of irritated tissue which is destroyed at a later period (irritative symptoms passing into paralysis). The varying amount of blood in the tumor and adjacent parts causes increase and diminution of the pressure symptoms, but as a general thing they increase and finally become constant. Hemorrhages and softenings in the tumor and surrounding parts may develop and give rise to corresponding symptoms. Among the irritative symptoms due to the growth of the tumor are the frequent convulsions and epileptiform attacks, especially if they are unilateral (Jacksonian epilepsy) and terminate in paralysis. The irritative symptoms also include the not infrequent conjugate deviation of the head and eyes.

To this category also belong the paroxysmal attacks of bilateral blindness, which begin with concentric narrowing of the field of vision and must be regarded as rudimentary epileptic seizures. They usually last only a few minutes, but may be repeated a number of times in a single day. Hirschberg observed cases in which the field of vision was constantly contracting and enlarging for half an hour to an hour (*Neur. Centralbl.*, 1891, p. 449).

The growth of a tumor can be best followed in the region of the muscle nuclei and at the base of the brain. Tumors in the latter locality, particularly in the region of the chiasm, the petrous portion of the temporal bone and the sphenoid, often produce unilateral or bilateral exophthalmus from pressure on the cavernous sinus and ophthalmic vein, particularly if the tumor grows into the orbit itself. The ophthalmoscope often reveals single or double optic neuritis, which passes into partial or complete atrophy. Thromboses of the veins of the orbit and the retina may also be produced. In such cases, however, the symptoms may also be due to independent metastases in the orbit.

In tumors of the frontal lobes, unilateral neuritis of the optic nerve or choked disc may be produced by direct implication of the nerve, but this is only observed in a minority of cases.

The local symptoms in tumors have much less diagnostic value than in cases of softening, because they often originate in parts which are quite remote from the tumor and which may be found entirely normal on autopsy. Moreover, such local symptoms may or may not be present in similar tumors of the same region.

If the tractus, chiasm, or optic nerve is directly implicated, a corresponding disturbance of vision develops with partial or complete atrophy of the papilla, visible with the ophthalmoscope. The atrophy may be preceded by irritative symptoms (neuritis), but this is rare. The anatomically visible compression or flattening of the parts does not always correspond to the clinical disturbance of vision.

We sometimes find a perfectly flat optic nerve with almost normal function. On the other hand there may be very great disturbance of function with very slight change of shape. It is evident that the pressure *per se* is less important than the resulting circulatory and nutritive disturbances and the occurrence of inflammatory changes which spread from the tumor into neighboring parts.

Choked disc is much the most important sign in the diagnosis of a tumor. It usually begins quite suddenly, although previously the fundus had been perfectly normal. Within about twenty-four hours—not infrequently after decided remission of previously existing pain in the head—a swelling of the papilla rapidly develops; it may increase to three times the normal diameter, and projects more or less into the interior of the eye. A difference of refraction amounting to 4 or 5 dioptries between the fundus and the top of the elevation is not rare. At the same time the originally hyper-æmic papilla becomes pale, often glassy in appearance, the vessels are sinuous in correspondence with the protrusion, and the veins are often dilated. The border of the elevation often has a bluish or bluish-violet color. In one case of very recent choked disc in a boy, I noticed several concentric rings of reflex around the papilla, evidently due to the formation of folds in the retina which had been pushed aside by the swollen papilla. After a variable length of time the papilla again becomes redder. Small, usually linear hemorrhages develop upon it and in its vicinity, and also whitish or yellowish degenerative foci (similar to those in Bright's disease). The tissue of the papilla usually grows more opaque and the redness then increases, but this is by no means always the case.

At a later period the middle of the papilla fades and hemorrhages and degenerative foci are absorbed; it sinks, and its size is gradually reduced to the normal. The originally blurred border become distinct, and finally the papilla has a uniform chalky-white color without visible details on its surface. Only the sinuosity of the vessels near the previously swollen disc enables us to recognize that the atrophic discoloration (white atrophy) has followed a choked disc. Not infrequently a fringe of pigment is found around the disc, and indicates the circumference of the former swelling. It is only rarely, and usually not in tumors, that the papilla resumes its normal appearance. In such cases swelling predominates, and the opacity of the tissues is almost or entirely absent. I have seen a case of this kind in which the sinuosity of the vessels near the circumference of the former swelling was the sole indication of the previous unilateral choked disc. Central vision and the field of vision were normal, but green blindness remained in this eye.

At least six or eight weeks, often a longer period, elapse before the changes visible with the ophthalmoscope have run their course. The papilla often remains for weeks in a certain middle stage between swelling and atrophy.

As a rule disturbance of vision is observed, but it is not necessarily present in simple choked disc. [Mathewson reports a case of three years' duration—Trans. Fifth Internat. Ophth. Congress, 1876, p. 63.—ED.] Various degrees of impairment of central vision, concentric and sector-shaped narrowing of the field of vision, central and peripheral scotomata passing even into complete blindness, color disturbances, etc., may be present from the beginning or during the progress of a choked disc, and may continue after it has run its course. The ophthalmoscopic appearances and the disturbances of sight are often grossly disproportionate to one another.

As the causal process is progressive the disturbance of vision usually has a progressive character, though often interrupted by more or less complete intermissions. Not all the disorders of vision in tumors of the brain are the result of choked disc, as the visual fibres may also be directly injured in any part of their course. Moreover, disorders of sight also occur in tumors of the brain without any anatomical findings.

In the large majority of cases choked disc is bilateral, although it is not always equal in degree on the two sides. When unilateral, it is usually situated on the side of the tumor, the latter being then located in the anterior part of the brain. It also frequently happens that the choked disc occurs at an earlier period or in a more marked degree upon the side of the tumor.

Typical choked disc is almost the most important symptom in the diagnosis of brain tumor of whatever kind or wherever situated. Sooner or later, often shortly before death, it is added to the other symptoms in a very large percentage of cases (two-thirds or more). It is not one of the early symptoms, but I remember a case in which, from finding a double choked disc, I made a diagnosis of tumor of the brain in a patient who had complained merely of frequent dull headache and occasional vomiting, and had been treated for disease of the stomach. In a few days the autopsy showed the correctness of the diagnosis (glioma of the cerebellum).

Choked disc does not occur with equal frequency in tumors of the different parts of the brain. It is said to be somewhat less frequent in tumors of the frontal lobes, more frequent in those of the cerebellum and adjacent parts. It also occurs in tumors of the spinal cord, though less constantly.

Choked disc is also observed exceptionally in other affections, for example, cerebral hemorrhages, abscesses, and many other diseases. Entirely similar appearances are found occasionally in albuminuria, diabetes, leukæmia, after profuse hemorrhages, etc. However, this hardly impairs the importance of choked disc in the diagnosis of brain tumor.

Anatomical examination shows very constantly an ampulliform swelling of the optic nerve, immediately behind the sclera. It is due to the accumulation of fluid within the sheaths of the nerve. This is often absent in the "atrophic" stage. The ampulla also occurs independently of choked disc, for example, in meningitis (Leber, Broadbent, *vide* "Jahresb. f. Aug.," 1872, p. 359).

At the start, the microscope shows merely œdema of the ocular extremity of the nerve, but finally hemorrhages, swollen nerve fibres, etc., are found. The calibre of the papillary vessels varies. These changes are confined to the ocular end of the nerve; posteriorly no abnormal changes are found, either in the nerve itself or in its sheaths.

After a while proliferative processes develop in the perineurium and interstitial connective tissue, and finally there may even be fatty degeneration and destruction of the nerve fibres. Posteriorly the nerve exhibits no changes, with the exception of ascending atrophy which may develop at a later period. This atrophy, which is attended by the presence of granular cells and amyloid corpuscles, may be followed occasionally into the chiasm and farther back. The ampulla sometimes disappears and the sheath exhibits folds. In other cases, especially when pronounced inflammatory changes in the nerve have been present, proliferation of connective tissue takes place (Fuerstner, *Berl. kl. Woch.*, 25, II, 1889) or endothelial proliferations project into the intervaginal spaces.

In certain cases distinct inflammatory changes (cellular infiltration) occur at the ocular end of the nerve, and are seen with the ophthalmoscope as increased cloudiness of the swollen papilla. We are therefore justified in distinguishing pure choked disc and choked neuritis. Pure choked disc or simple œdema of the ocular end of the nerve very probably causes no direct disturbance of vision, but only indirect disorders from the circulatory changes which are due to the pressure on the vessels and the consequent impaired nutrition. In inflammatory processes of the papilla (choked neuritis) disturbance of vision is more apt to occur, although even then the nerve fibres are affected secondarily, inasmuch as the process is essentially interstitial.

Few writers distinguish between choked disc and choked (obstructive) neuritis. In very many cases a distinction is not even made between choked disc and neuritis. Thus, the English writers often use the term optic neuritis in undoubted cases of choked disc.

Graefe's original opinion that choked disc is due to compression of the cavernous sinus was abandoned as soon as it was demonstrated (Sesemann) that the outflow from the ophthalmic vein takes place in great part through the posterior facial vein. Certain other theories, for example that of Loring, who believed that the condition was due to the circulatory and trophic influence of trigeminal irritation, have also been discarded. At the present time two theories are still maintained, viz., the purely mechanical, transport theory (Manz) and the inflammatory theory (Leber). According to Manz (*Arch. f. Ophth.*, XVI, p. 265), choked disc arises from the entrance of fluid into the subdural and subpial spaces of the optic nerve from the corresponding spaces of the brain, and this compresses the intraocular extremity of the optic nerve. He made experiments by injections of warm fluid into the cavity. According to Leber the tumor causes secretory inflammation and dropsy of the cerebral ventricles; their products act in the intervaginal space as an inflammatory irritant and give rise to the choked disc. Similar views are expressed by Leber's pupil, Deutschmann, who produced typical choked disc, of course with pronounced inflammatory "findings," by the introduction of tubercular material into the cranial cavity.

Anatomical examinations in man prove indubitably that the choked disc of tumor of the brain begins as a pure œdema of the intraocular end of the nerve, without any signs of inflammation.

At a later period interstitial inflammatory changes may be found in the specimen, but in some cases such changes are absent at all stages. In late stages we usually find only secondary atrophic processes. What argument is furnished by the introduction of tubercular matter into the cranium when compared with numerous accurate anatomical examinations in the human subject? In the former event purulent tubercular meningitis develops and, as happens so often, leads to true optic neuritis, which may be associated with a certain degree of swelling. Such experiments prove nothing whatever with regard to typical choked disc in brain tumors. If the theory were true, inflammatory appearances would necessarily be found in the orbital and intracranial parts of the sheaths of the nerve, and this does not happen. Nor does the transport theory explain all the phenomena. According to Ulrich the stasis in the ventricles of the brain extends directly to the lymph spaces of the optic nerve without the agency of its vaginal spaces, and produces œdema of the optic nerve, which leads to choked disc from interference with circulation. Œdema of the trunk of the nerve occurs, for example, in tubercular meningitis, without choked disc.

As a rule, however, œdema of the optic nerve is not found farther back in the orbit in typical choked disc, although it does occur occasionally.

Normally there is a centripetal current of fluid from the eye to the brain within both the nerve and its sheaths, as I can experimentally demonstrate. If this is abolished by increased pressure within the cranium, stasis and œdema must develop. They will become evident, (1) in the optic nerve outside of the cranial cavity where it is not inclosed in a closely fitting sheath, i.e., at the intracular extremity (choked disc), and (2) within the vaginal spaces, where the external sheath is thinnest and most yielding, i.e., immediately behind the eye (ampulla). The latter, however, requires a somewhat increased pressure. Hence, choked disc may occur without an ampulla. The development of choked disc merely requires simple stasis; the "transport" of fluid from the cranial cavity is not necessary.

This simple œdema of the papilla does not give rise to a disturb-
10

ance of vision, in the absence of complications; it may be quite
stationary for a long time and disappear without leaving a trace.
This hardly ever occurs, however, in progressive tumors. With the
increase in pressure intracranial fluid will enter the vaginal sheaths
and will produce the ampulla. In intracranial hemorrhages or
purulent meningitis this entrance of fluid has been demonstrated
anatomically. Even then the fluid does not necessarily produce
inflammation, because nutritive disturbances and destruction of tissue may result from the increased pressure alone (indirectly through
the influence on the vessels!).

As a general thing, however, inflammation does result. Growing
tumors, especially malignant ones, excrete irritating and inflammatory substances, as is shown by the surrounding zone of reactive
inflammation. These substances are also diffused in the cerebro-
spinal fluid, especially the subarachnoid fluid, and enter the sheath
of the optic nerve. They will produce their effect where the œdema
develops most readily, i.e., at the ocular extremity of the nerve. It is
difficult, however, to understand why diffuse inflammations of the
cerebral meninges do not develop, unless the preceding œdema of the
end of the nerve is regarded as a predisposing factor.

The affection of the end of the optic nerve in tumor of the brain
is due, accordingly, to stasis from inhibition of the natural outflow
(œdema of the papilla), to mechanical entrance of fluid into the
sheaths of the nerve under increased pressure (the ampulla is the
result of this factor), and to more or less intense inflammatory phenomena which may, however, be entirely absent.

In this way we can explain the fact that unilateral choked disc
may result from intracranial (at the optic foramen) or intraorbital
compression of the optic nerve.

In cases of tumor of the brain in which sudden disturbance of
vision sets in, after the subsidence of other symptoms (headache),
and fully developed choked disc is found, we must assume that œdema
of the entrance of the optic nerve had already existed and that the
increased intracranial pressure suddenly overcame an obstacle in
the sheaths of the nerves. This obstacle must have been situated in
the region of the canalis opticus and probably acted like a valve.

Simple neuritis, terminating in atrophy, is found less often than choked disc in cerebral tumors. It happens particularly in tumors of the frontal lobes, in which the growth is situated comparatively close to the optic nerve. The affection may be interpreted as a descending neuritis. When unilateral, it may possess local diagnostic value. (Among twenty-five cases of tumors of the cerebrum Oppenheim (*Arch. f. Psych. u. Nervenh.*, XXI and XXXII) observed typical choked disc fourteen times, neuritis five times, and hyperæmia of the papilla once).

More or less complete unilateral or bilateral atrophy of the nerve, with impaired vision, but unattended by inflammatory phenomena, may also occur. In such cases there is probably injury of optic tracts somewhere up to, and including, the primary optic centres.

In very many cases, therefore, the ophthalmoscope furnishes a definite starting-point for the diagnoses. Cases have been observed, however, in which there were pronounced general tumor symptoms with double choked disc, but in which no tumor was found at the autopsy. On the other hand, choked disc may be absent in very large tumors, although we are unable to offer a sufficient reason for this peculiarity.

Bruns (*Berl. kl. Woch.*, 1886, Nos. 21, 22) lays stress upon the absence of choked disc in tumors of the corpus callosum.

When other symptoms are observed, the existence of double choked disc renders the diagnosis of tumor extremely probable. The location of the growth must be decided by the local symptoms, although these may vary greatly in tumors of the same region. Especial importance attaches to progressive irritative symptoms which subsequently pass into paralysis, if the order of succession indicates a definite part of the brain.

The character of the tumor can only be decided in connection with all the other clinical symptoms.

Aneurisms of the cerebral arteries are of frequent occurrence. They are small and give rise to no special symptoms, apart from those of the causal disease of the vessels, which is usually widespread. They may give rise to hemorrhage or thrombosis.

The larger aneurisms are situated most frequently on the carotid and its branches. Special symptoms are commonly produced only after the aneurism has attained about the size of a hazelnut, and these correspond exactly to those of a tumor.

Aneurisms of the carotid within the skull usually cause paralysis of the nerves of the ocular muscles which pass alongside of or through the cavernous sinus (*vide* Fig. 18). They exert a deleterious action on the adjacent chiasm and optic nerve, with a corresponding peripheral disorder of vision and secondary atrophy of the nerve. Unilateral pressure on the nerve may also give rise to unilateral choked disc.

As a large part of the sympathetic fibres of the eye pass through the carotid plexus, the aneurisms produce oculo-pupillary symptoms, occasionally of an irritative but usually of a paralytic character (ptosis, myosis, and slight enophthalmus). The vasomotor fibres to the face take their course along the external carotid.

As the ophthalmic artery is derived from the internal carotid, aneurisms of the latter not infrequently cause changes in the amount of blood in the retinal vessels. This may result in arterial thrombosis, and, if the central artery of the retina is affected, leads to loss of vision with the ophthalmoscopic appearances of embolism of the retina.

In a case of sudden rupture of an aneurism of the anterior communicating artery, Bellamy (*Lancet*, 6, VII, 1889) found blood within the sheaths of the optic nerve and also in the meningeal spaces.

A celebrated case is that reported by Weir Mitchell (*Journ. of Nerv. and Ment. Dis.*, 1889, p. 44). A man aged forty-five years had suffered for five years from headache and disturbance of sight. The ocular muscles were normal. Both nasal halves of the retinæ were entirely blind and there was atrophy of the nasal halves of the papillæ (bitemporal hemianopsia). Central vision was $\frac{1}{10}$ on both sides and fell to $\frac{1}{3}$ before death, which occurred suddenly in coma. An aneurism as large as a lemon was found upon an abnormal communicating branch between the carotids, which passed beneath the chiasm. The latter was divided in the median line from before backward and the sella turcica had been eroded. Two thin atrophic bundles of nerves (the uncrossed bundles of the optic nerves) were found in the place of the chiasm. This finding absolutely demonstrates the partial decussation of the optic nerves in the chiasm.

Tumor of the dura mater may simulate meningitis, as in Unverricht's case (*Centralbl. f. d. med. Wiss.*, 1888, p. 493), in which successive paralysis of all the cerebral nerves and unilateral keratitis neuroparalytica developed. Metastases were found between the dura and the base of the skull; these occluded the foramina. The primary tumor was a round-cell sarcoma, the size of an apple, in the mediastinum.

I have seen a metastasis between the dura and the left frontal bone, starting from a cancer of the stomach, which could not be diagnosed with certainty during life. The symptoms were those of a pachymeningitis hæmorrhagica, and the macroscopic appearances at the autopsy were very similar to this condition. A few days before death the ophthalmoscope showed neuritis of the left optic nerve without notable impairment of vision, but with great photophobia.

Nettleship (*Ophth. Review*, 1887, p. 57) describes post-neuritic atrophy of the optic nerve in a boy of twelve years, suffering from exostoses of the skull.

Marchand (*Virch. Arch.*, Bd. 75, p. 404) reports a case in which multiple cysticerci of the surface of the brain gave rise for years to paroxysmal visual hallucinations; later blindness developed suddenly in both eyes, and in two and one-half months the fatal termination occurred. Engler (*Prag. med. Woch.*, 1888, No. 2) observed double choked disc without subjective disturbance of vision in a girl of twenty-three years who had numerous cysticerci of the subcutaneous cellular tissue and muscles, and who suffered from periodical headaches, occasional paræsthesiæ, nausea and vomiting. Whether these symptoms were due to a cysticercus in the brain, must remain doubtful, inasmuch as no post-mortem examination was made.

Tumors of the eye sometimes grow along the optic nerve into the skull and brain. This is particularly true of retinal gliomas, while choroidal sarcomata develop metastases usually in other parts, such as the liver, lymphatic glands, etc. Tumors of the optic nerve and other malignant tumors of the orbit may also proliferate into the cranial cavity, but such growths are rare.

Meningitis.

The eye is extensively involved in the different forms of meningitis (simple, epidemic, tubercular, acute and chronic; internal and external hydrocephalus; leptomeningitis and pachymeningitis). General eye symptoms are unimportant. During diffuse cerebral irritation the eye takes part in the general hyperæsthesia by great photophobia and sensitiveness to light. When there is increased intracranial pressure, the pupils are dilated and react poorly, but there are numerous exceptions to this rule.

Much more importance attaches to the local symptoms, which vary according as the meningitis affects the convexity or the base of the brain.

In meningitis of the convexity, disturbance of the function of the visual cortex may give rise to double blindness in which the reaction of the pupils to light may remain intact and the ophthalmoscope show no changes. In such a diffuse disease it is only for a short time that the lesion will act upon one side alone (homonymous hemianopsia with intact pupillary reaction to light), although this is certainly not infrequent. If the process comes to a standstill, recovery may occur.

Conjugate deviation of the eyes may also appear. Jaccoud ("Jahresb. f. Aug.," 1879, p. 243) observed conjugate deviation to the right in purulent meningitis of the left hemisphere and congestion of the right hemisphere. As a temporary symptom conjugate deviation is not infrequent.

Disturbances of vision are usually peripheral and due to changes at the base of the brain. These and other basal symptoms are among the most important diagnostic signs of basilar meningitis.

The exudation at the base causes irritation and inflammation of the nerves imbedded in it. It is usually most abundant in the region of the chiasm. Hence we find a great variety of irritative and paralytic symptoms in the province of the basal cerebral nerves, the abducens, motor oculi, trochlearis and facial. Leichtenstern (*Deutsch.*

med. Woch., 1885, 31), who reported twenty-nine cases of epidemic meningitis, found that the abducens was attacked most frequently, the motor oculi very rarely (in the hydrocephalic stage there was slight bilateral ptosis in one case, weakness of an internus in one case, and inequality of the pupils in one case). In tubercular meningitis the pupil is often dilated and very sluggish; in the epidemic form, he observed this symptom in only two cases. In one case there was isolated abducens paralysis, although all the nerves were imbedded in exudation and both motores oculorum nerves were very much reddened. In one patient, who was still conscious, there was paralysis of all the ocular muscles, including the levatores palpebrarum, but the reaction of the pupils to light was normal. Nystagmus was observed several times.

Trigeminal hyperæsthesiæ, paræsthesiæ, and anæsthesiæ are quite frequent. Anæsthesiæ of the cornea may give rise to so-called neuroparalytic keratitis, for example, Robinson (*Lancet*, 1880, II, p. 612), Spierer (*Monatsbl. f. Aug.*, 1891, p. 222).

Spasms and paralyses have a peripheral character. Unless due to irritation of the cortex of the convexity, they are not conjugate and associated, but affect individual nerves or parts of nerves and soon lead to permanent degeneration of nerve and muscle.

The nerves of special sense, the acoustic, optic tract, chiasm and optic nerve are often imbedded in the more or less cellular exudation. At first this gives rise to irritative symptoms, later to destruction of the nerve with corresponding disorders of sight and hearing. After a while the ophthalmoscope shows more or less atrophic discoloration of the papilla.

The optic nerve is often implicated directly. It and its sheaths take part in the inflammatory infiltration, and the ophthalmoscope shows the picture of optic neuritis, one of the most important aids in the diagnosis of meningitis. The optic nerve is more or less reddened, its borders obliterated, its tissue cloudy. The latter sign is the most important in the differentiation from simple hyperæmia. Pronounced hyperæmia of the vessels is not always visible. It is very often present, however, and is made manifest by the fact that a large number of otherwise invisible little vessels become visible upon the papilla. In many cases the location of the papilla can only be recognized by the entrance of the vessels. The papilla is not at all or very slightly prominent; hemorrhages and exudations are comparatively rare.

A microscopic examination in this stage shows that the intervaginal space is not dilated, but the connective-tissue septa and the pial sheath are infiltrated with cells. This infiltration increases toward the periphery. At the lamina cribrosa the impression is created as if the centrifugally wandering cells had been filtered into that locality, and indeed this may be the actual anatomo-pathological occurrence.

At a later period the papilla grows paler. It finally becomes as white as chalk, is sharply defined, and shows almost no details upon its surface. The vessels are more or less narrowed (post-neuritic atrophy). The findings are similar to those seen after a choked disc has run its course with marked inflammatory symptoms. In this stage the microscope shows connective-tissue increase in the nerve and more or less complete loss of nerve fibres.

Optic neuritis in meningitis is generally bilateral, although it does not always appear at the same time on the two sides or with equal intensity. The termination in atrophy is usually observed only in those cases which do not prove fatal.

The disorders of sight may vary greatly but are usually considerable. They include impairment of vision, of the recognition of colors, concentric and sector-shaped narrowing of the field, central and peripheral scotomata, sometimes for colors alone, and even complete blindness. Complete restoration of sight is rare. Even if blindness does not ensue, there is usually more or less impairment of vision; disorders of the color sense, defects of the field, and also defects of the other organs of special sense, particularly of hearing, remain.

It is difficult to furnish definite statements concerning the frequency of optic neuritis in the different forms of meningitis, especially as many writers include choked disc, choked neuritis, and neuritis under the latter term. But it is undoubtedly found in the majority of cases, although often not until a late period when the diagnosis has been assumed from other symptoms. When it appears at a comparatively early period it may be of the greatest importance in excluding typhoid fever, pneumonia, etc. In the latter affections optic neuritis is extremely rare.

In purulent meningitis the inflammation extends not infrequently through the superior orbital fissure to one or both orbits. We then find œdema of the conjunctiva (chemosis), which may run its course without any further visible sign of inflammation and is occasionally one of the first symptoms. In other cases infiltration and suppuration of the orbit set in with protrusion of the eye, œdema of the lids, etc. The œdema of the lids exhibits one feature which is characteristic of orbital suppuration, *i.e.*, it ceases abruptly at the bony rim of the orbit. Notable disturbance of vision is not necessarily present even in orbital abscess, but it usually occurs sooner or later from implication of the optic nerve and may lead to permanent blindness. Interference with the movements of the eye in one or another direction is not uncommon.

As a rule the inflammation extends into the orbit along the veins. Exceptions are sometimes observed, as in the following case reported by Hofmann (*Neurol. Centralbl.*, 1886, p. 357). After furunculosis of the neck meningitic symptoms developed, attended by violent headaches and slowing of the pulse. The left eye was pushed forward, blind and immovable: there were ptosis, dilated pupils and choked disc. No pus was found on opening into the orbit, but the optic nerve was swollen to the size of the little finger, and pus was evacuated from the ampulla. After drainage recovery set in, but ptosis, blindness and atrophy of the optic nerve were left.

A sero-plastic-purulent choroiditis is another frequent symptom during and after meningitis. It begins with ciliary injection, discoloration of the iris, distortion of the pupil and cloudiness of the fundus, the details of which can no longer be recognized. A yellowish hypopyon and a similar pupillary exudation not infrequently develop after a little while. There may be parenchymatous opacity of the cornea (by immigration from its borders), but true suppuration is not often seen. Panophthalmitis with termination in suppuration and phthisis bulbi is also a comparatively rare symptom. Blindness usually occurs at a very early period, and the eye is soft from the start.

At the end of a few weeks the ciliary injection, hypopyon and pupillary exudation commonly disappear, the cornea clears up, the iris is found superficially adherent to the lens, is usually lighter in color than that of the other eye, and later is distinctly atrophic; the pupil remains irregular and immovable. The blindness is generally permanent, and it is only in mild, not fully developed cases that some visual power remains.

The eye subsequently grows softer, becomes considerably smaller, and often appears quadrate under the pressure of the recti muscles. The lens may remain transparent for a long time and enable us to recognize in the vitreous a dense, whitish-yellow mass which often extends to the lens. In this exudation we often see new-formed vessels, also slight hemorrhages, so that occasionally it is hardly possible to avoid mistaking it for glioma of the retina prior to its "glaucomatous stage." This is so much more apt to occur because both meningeal exudative choroiditis and glioma are prone to occur in children, and because the previous symptoms of meningitis have not been correctly interpreted. As the eye is always blind, the error in doing enucleation is not very disastrous, in view of the great malignancy of retinal glioma. Diagnosis is based upon a greater or less degree of probability, and the opposite mistake would be much more serious in its consequences.

After the lapse of years calcifying cataract gradually develops, often also band-shaped keratitis with development of callosities and calcifications.

As a rule the disease is unilateral. It occurs usually in simple meningitis of young children, but it may also appear at a late period, for example, during epidemics of meningitis. The disease may appear at the onset of the meningitis or some weeks (six to eight) after the latter has subsided. Recoveries with fair or good sight do occur but are rare. Knapp (*Zeitschr. f. Ohrenheilk.*, XIV, p. 241) reports a case of bilateral deafness and exudative choroiditis after cerebrospinal meningitis in a boy of six years. The left eye recovered, the right eye became phthisical.

On anatomical examination of such cases it has been found that the optic nerve and its sheaths were free from inflammation (Modl,

Wien. med. Woch., 1880, No. 29; Oeller, *Arch. f. Aug.*, VIII, p. 357). Berlin assumes, accordingly, that orbital inflammations extend from the cranial cavity solely through the medium of phlebitis of the ophthalmic vein (through the superior orbital fissure). The simplest hypothesis is that of a real metastasis, but this will not hold good for the cases in which meningitis and choroiditis are almost simultaneous in their onset. Here we must assume the coincident deposit of the same morbific substance in the pia mater and choroid. This theory is the more plausible in view of the fact that both structures are developmentally alike.

Bull (*Jahr. f. Aug.*, 1873, p. 304) observed two rudimentary cases in epidemic meningitis. In one case there was swelling of the papilla with numerous yellowish, prominent patches in the choroid; in the other case there was a coherent mass of choroidal exudation in the fundus. In the latter case atrophy of the optic nerve subsequently developed. There are also all degrees of transition to milder forms of choroiditis and cyclitis such as are observed frequently after typhoid affections, especially after relapsing fever.

As the result of purulent meningitis following aural suppuration, Nettleship observed purulent choroiditis on the left side, and only neuritis on the right side. It seems as if, in such cases, neuritis alone develops when the chemical and not the living agents of inflammation enter the optic nerve and its sheaths. Purulent choroiditis develops in a discontinuous manner, although, as Hofmann's case shows, the direct extension of the process cannot be excluded in all cases.

Purulent inflammations of the eye may be the starting-point for metastatic purulent meningitis. Occasionally it is merely a purulent conjunctivitis (Politzer, *Jahr. f. Kinderheilk.*, 1870, p. 335), but usually it is a traumatic septic panophthalmitis with or without a foreign body in the eye. Some writers advise, though wrongly, against the enucleation of an eye affected by panophthalmitis. If, after the removal of such an eye (as a matter of course, under antisepsis), meningitis sets in, the operation was performed too late, because there is no doubt that the eye is the source of infection. Purulent meningitis also occurs in panophthalmitis without enucleation, for example, after cataract operation (*Jahrb. f. Aug.*, 1888, p. 217). Rolland (*ibid.*, 1885, p. 365) performed seventy enucleations in panophthalmitis without meningitis, nor have I had a fatal result in numerous cases. [About forty fatal cases of meningitis after enucleation of the eye have been published, yet even with acute suppurative panophthalmitis as a frequent cause of enucleation (about 14 per cent), the ratio of death is about 1 in 4,000. See "Text-book of Diseases of the Eye," by Noyes, 1894, p. 545.—ED.]

In epidemic cerebro-spinal meningitis conjunctival catarrh often appears at the beginning. An important symptom is an early œdema of the conjunctiva, from extension of the inflammation in the orbit along the veins and interference with the return flow, especially in the ophthalmic vein.

According to Foerster (*l.c.*, p. 105) deep subepithelial infiltrations of the cornea, which may be entirely reabsorbed, also occur in epidemic cerebro-spinal meningitis. Among twenty-eight cases of the disease Jedrzejewicz (*Jahrb. f. Aug.*, 1880, p. 249) found two cases of the above-described form of choroiditis.

In tubercular meningitis the symptoms of simple meningitis are supplemented by those of tuberculosis of the eyeball or brain. While choked disc is very rare in pure meningitis (Bramwell, *Jahr. f. Aug.*, 1879, p. 244; Panas, *Rec. d'Ophth.*, 1886, p. 651), it is more frequent in the tubercular form, especially if the brain contains a solitary tubercle which *per se* gives rise to all the symptoms of a tumor.

In the typhoid terminal stage of meningitis, as in all diseases which present a typhoid state, a desiccation keratitis is not infrequently observed. On account of the imperfect closure of the lids and arrest of winking, the exposed parts of the cornea and conjunctiva desiccate and infectious inflammation develops beneath the crust. Cessation of the pupillary reaction to cutaneous irritants is said to be a terminal symptom, but it merely proves peripheral anæsthesia of the irritated parts of the integument.

Senator (*Charité Ann.*, XI, p. 248) saw nystagmus and iritis in meningitis. Gairdner and West (*Jahrb. f. Aug.*, 1878, p. 241) report rapid alternation of myosis and mydriasis in tubercular meningitis. In several cases of basilar meningitis, Kahler (*Prag. med. Woch.*, 1887, No. 5) observed maximum dilatation of the pupil when the patient sat up; this disappeared on lying down.

Bock (*Wien. med. Woch.*, 23, XI, 1889) reported five cases of cataract after meningitis in persons in the thirties, which were cured by operation. The diagnosis of meningitis is by no means undisputed, and the connection between the two conditions appears obscure. Perhaps a part is played by the spasms which occur in the ciliary muscle and which might produce nutritive disturbances in the lens from interference with the circulation. It is a striking fact, however, that no other writer speaks of the development of cataract after meningitis—of course apart from the secondary form due to exudative choroiditis—although violent spasms are hardly ever absent in the more acute forms.

In chronic hydrocephalus the symptoms of compression predominate. Choked disc and choked neuritis and simple optic neuritis are observed not infrequently. They are almost always bilateral and associated with marked disturbance of vision. They terminate in atrophy of the optic nerve and, as a rule, in blindness. Widely staring eyes (so-called "hunger for light") are often mentioned, but are common enough in blindness from any kind of total atrophy of the optic nerve. Nystagmus is not uncommon.

In one case of chronic leptomeningitis Veronese (*Wien. klin. Woch.*, 1889, No. 24) observed sudden blindness with normal ophthalmoscopic appearances, without any other disturbances of consciousness or motion. This was evidently a central visual disturbance in the cerebral cortex or the corona radiata. Subsequently two cerebral apoplexies occurred, and the fatal termination ensued one and a half years later. At the autopsy the optic nerve, chiasma, and tractus appeared normal, and purulent, bloody fluid was found in the lateral ventricles.

The oft-mentioned affections of the optic nerve in deformities of the skull (Manz, Ber. d. Heidelb. ophth. Ges., 1889, p. 18; Vossius and Stood, *Jahr. f Aug.*, 1884, p. 354; Michel, *Arch. f. Heilk.*, 1873, p. 39; Schneller, *Jahr. f. Aug.*, 1881, p. 253) consist of more or less complete atrophy, usually after preceding neuritis, with

corresponding disturbance of vision. They are due to chronic meningitic changes which lead, on the one hand, to premature ossifica-
tion of the cranial bones; on the other hand, to constriction of one or
both optic nerves at the optic foramen. Michel found, in such
cases, endothelial proliferation into the constricted intervaginal spaces
of the optic sheath.

Circumscribed meningitis, of extra-uterine origin, must be regarded as the cause of the large majority of cases of so-called hereditary affections of the optic nerve. They run their course as a neuritis with subsequent atrophy (usually partial, rarely complete) of the
optic nerve. At the start we often find marked changes in vision,
as nyctalopia, photophobia, color phantasms, etc. More or less impairment of sight, central or eccentric scotomata and other defects of
the field, color disturbances, etc., are usually left (Leber, *Arch. f.
Ophth.*, XVII, p. 249). Daae's cases (*Jahr. f. Aug.*, 1870, p. 379) of
frequent occurrence of hemianopsia in one family, or of progressive
binasal hemianopsia, point to the region of the tractus and chiasm.
In a typical case of hereditary disease of the optic nerve I found
normal vision despite atrophic discoloration of the papilla, although
green blindness was present.

The presence of neuritis, however slight, and its slow course, distinguish this disease from Graefe's retrobulbar neuritis with sudden
blindness, with at first negative ophthalmoscopic findings and later
atrophy of the papilla.

Hemorrhagic pachymeningitis gives rise to the symptoms both
of meningitis and of a growing tumor with periodical hemorrhages.
As the disease is located mainly on the convexity, basal symptoms
are usually absent. If the base of the brain be involved blood may
be found within the sheaths of the optic nerve and meningitic changes
will also appear at the base. Local cortical symptoms may be produced, such as hemianopsia, conjugate deviation toward the side of
the disease or toward the opposite side, ptosis, nystagmus, etc.
These symptoms may disappear during the intervals between the
attacks. Changes in the pupils also occur (Fuerstner, *Arch. f.
Psych. u. Nerv.*, VIII, 1, p. 1).

When the pachymeningitis is located in the anterior part of the skull we may find unilateral or bilateral hyperæmia and stasis of the optic nerve, neuritis, choked disc with or without impairment of sight, and choked neuritis, according as the symptoms of tumor or meningitis predominate. If the ophthalmoscopic appearances are unilateral they are usually found on the same side as the disease, or at least upon the side on which the lesion is most pronounced.

We have already spoken (p. 149) of the possible hemorrhagic pachymeningitis with superficial metastatic tumors between the cranial bones and the dura mater.

Insanity.

We have already explained that so-called mental diseases must be regarded as extensive diffuse or localized affections of the frontal cortex, to which centripetal associative fibres pass from all directions and from which centrifugal fibres pass to all other parts of the brain.

The purer the mental disease, the more strictly must the process be confined to the frontal lobes and the more completely will local symptoms be absent. On account of the frequent implication of other parts of the cortex, sensory and motor symptoms are also found in very many cases.

The fact of the affection of other parts of the cerebral cortex, the presence of hallucinations of feeling (motor cortex), hearing (acoustic cortex), sight (visual sphere), etc., is employed as an important principle of classification of mental diseases. It is well known that hallucinations of hearing are more frequent than those of sight, corresponding to the shorter distance between the temporal and frontal lobes. Such hallucinations may occur long before or during the disease. In the first event the insanity begins at the moment when the hallucinations are no longer recognized as such but are regarded as objective. Illusions or false interpretations of actual objective sensory stimuli are also found in many cases. We will soon enter more in detail into different forms of unilateral hallucinations and illusions of sight.

Numerous ophthalmoscopic examinations have been made in the insane (Jahresb. f. Aug., 1874, p. 428, and 1883, p. 338), but no constant findings have been discovered, apart from general paresis and complications, such as tabes.   Not even conditions of depression and exaltation furnish corresponding ophthalmoscopic appearances. The frequent injection of the conjunctiva, often amounting to conjunctivitis, in maniacal and excited patients is a result of this condition and of the insomnia rather than a direct equivalent of congestion of the cerebral cortex.   Laudenbach and Bennett's statistics (*N. Y. Journ. of Nerv. and Ment. Dis.*, 1886, No. 13) of the ophthalmoscopic appearances in 707 insane showed that no constant findings correspond to any definite form of insanity.   Whenever a demonstrable affection of the eye is present—apart perhaps from hyperæmia or anæmia of the fundus—it is always due to some material cause such as tabes, sclerotic foci, syphilis, albuminuria, etc., in the course of which insanity develops from involvement of the frontal lobes.

Occasional findings of all kinds have been reported.   According to Manz (*Berl. kl. Woch.*, 1884, No. 50), the insane often exhibit congenital anomalies of the eye, such as different colors of the iris of one or both eyes, accumulations of pigment in the choroid and retina, partial poverty of pigment, albinism, unusual shape of the pupils, abnormalities in the origin and course of the retinal vessels, etc. These conditions are probably signs of degeneration.

According to Raehlmann (Volkmann's Vortraege, No. 128), maniacal patients often have enlarged pupils; narrow pupils are characteristic of diminution of the cortical functions.   This may be true as a general thing, but exceptions to the rule are very frequent.

Injuries to the eye are not uncommon in insanity.   The patient may gouge out one or both eyes, pull out the lashes, place sand and other articles into the conjunctival sac, etc.

Psychoses also develop not infrequently after diseases of the eye, and especially after operations upon the organ.   This occurs almost always in predisposed individuals (heredity, alcoholism, etc.), in whom the operation acts as an exciting cause.   Repetitions may happen in successive operations on the eye.

Among thirty-one cases of post-operative insanity Frankl-Hochwart (*Jahrb. f. Psych.*, IX, 1 and 2) distinguished the following forms: *a*. Psychoses with hallucinations (without alcoholic causation), 15 cases. In 6 it began in the first twenty-four hours; in 9 cases after an interval of from two days to three weeks. The condition consisted of frightful hallucinations and terror; confusion was present almost constantly but was not always very pronounced. In 4 cases recovery occurred in a few days, in the others the course was a protracted one. *b.* Simple confusion in old people, senile conditions of excitement with no knowledge of the surroundings, 6 cases. Three of these rapidly improved, the other three terminated in dementia. *c.* Psychoses in alcoholics, viz., delirium, 7 cases. They began one to two days after the operation, usually at an earlier period than in non-alcoholics; they ran a rapid course and soon terminated in recovery. *d.* Psychoses in marantic individuals, mental confusion from irritation, 3 cases. All proved fatal.

The greater frequency of mental disturbances after operations on the eye than after other operations is due in part to the absolute rest, darkness and seclusion with closed eyes which are often required. These favor the occurrence of hallucinations and may thus arouse a latent mental affection in a predisposed individual. Some part is also played, at times, by the use of atropine which may give rise to hallucinations, especially of sight. Of course the psychoses may also develop independently of the use of atropine.

Valude (*Ann. d'Ocul.*, 115, p. 242) observed in a woman of sixty-five years a maniacal attack follow an iridectomy in which atropine had not been employed; it was relieved by the removal of the bandage. Parinaud (*ibid.*) reported a case of aphasia after a cataract operation; right hemiplegia developed soon afterward. Here the operation on the eye and the attendant excitement were the exciting cause of a cerebral hemorrhage.

On the other hand Moulton (*Ann. d'Ocul.*, Nov., 1891) claims to have cured a melancholic patient, aged forty years, by the removal of an eyeball which had been phthisical since the age of six years.

An interesting observation was made by Soltmann (*Neur. Centralbl.*, 1890, p. 749). Children who are requested to write with the left hand usually employ the ordinary script, but in a slow and awkward manner. A small number, however, write backhanded, *i.e.*, from right to left; these individuals suffered from a psychopathic

11

hereditary taint. Deaf-mutes practise "mirror writing" when the deafness has been acquired in utero or in infancy, and ordinary writing when the deafness is acquired in later life. This is also true with regard to the blind. Of 16 imbecile children 13 employed "mirror-writing." Soltmann explains this fact on the ground that the control of the left hemisphere over writing is lacking, but this explanation does not seem to us to be sufficient. Perhaps it happens that one hemisphere does not, as is usually the case, acquire a more marked controlling influence over the function of writing.

Hallucinations may be of various kinds (Mendel, *Berl. kl. Woch.*, 1890, 26, 27). There may be associate sensory impressions (photisms in the visual organ, phonisms in the auditory organ); for example, audible sounds produce a sensation of color. High-pitched sounds usually evoke light colors, low-pitched sounds dark colors, but the greatest individual differences are observed. Occasionally definite olfactory or gustatory impressions also produce the sensation of color; for example, the smell and taste of vinegar produce red vision (*olfaction et gustation colorée*, olfactory and gustatory photisms). In like manner the perception of color may be roused by the sight of certain numbers, letters, etc. This is said to occur particularly in hysterical psychoses, melancholia, and paranoia. Analogous associated sensations also occur in all the other special senses.

Protracted after-sensations may be experienced after great mental excitement, the administration of cannabis indica, and at the beginning of insanity (Mendel).

Subjective visual sensations may be peripheral, as, for example, the phantasms of fire and sparks in inflammatory diseases and congestions of the optic nerve, retina and choroid, the yellow vision of santonin poisoning, etc. They are distinguished from central hallucinations and illusions in that the latter are false judgments about real sensorial phenomena. Hallucinations are either simple and elementary, such as fire, colors, lightning, etc., or compound, such as the vision of shapes, animals, processions, either in a shadowy or perfectly distinct manner. In hallucinations of the muscular sense in the ocular muscles, the subjective visual impressions undergo movement, enlargement or diminution in size.

Such a condition may lead to a refusal to take food; for example, in a lunatic who saw everything enlarged to an enormous extent and was frightened by the tremendous size of the articles of food (*Neurol. Centralbl.*, 1888, p. 445). Illusions include those conditions in which the patient recognizes acquaintances in every one (delirium palingnosticum), or sees everything changed (delirium metabolicum), even his own shape, as in hypochondriasis, etc. Entoptic phenomena or the central scotoma of alcoholics may also give rise to illusions. Hallucinations and illusions are sometimes seen only when the eyes are open, sometimes when they are closed.

Several senses are often affected at the same time. The hallucinations may be unilateral or different on the two sides; for example, a man is seen on the right side, a woman on the left side. Bilateral hallucinations may become unilateral, and *vice versa*.

Hallucinations are rare in focal diseases; it is said that they are then more apt to be elementary. Bennet's epileptic patient, who had a lesion in the angular gyrus, had red vision as an aura. In a paretic dement with unusually prominent visual hallucinations, Mendel (*Neur. Centralbl.*, 1882, No. 58) found a remarkable amount of disease in the occipital lobes. The presence of visual hallucinations proves that the visual cortex is still performing its function. Hence if a defect in the visual field is cortical in origin, this is shown by the hallucinations, *i.e.*, the latter do not appear within the province of the defect in the field. Pick has observed a case of this kind. If both visual spheres are destroyed hallucinations of sight are no longer possible.

Hallucinations have been observed in a child of fifteen months (stramonium poisoning).

The main causes of hallucinations are conditions of weakness, inanition, the period between waking and sleeping, many poisons (alcohol, cannabis indica, opium, stramonium, belladonna, santonin, etc.), hysteria and hystero-epilepsy, hypochondriasis, all the psychoses with the exception of idiocy.

The origin of visual hallucinations is direct or indirect irritation of one or both visual spheres, as a result of which the visual cortex, as well as the associated parts of the cortex (frontal lobes), experiences the same changes as if the visual impression corresponding to the hallucinations had been conveyed from the retina.

According to Feré (*Rev. de Méd.*, 1890, p. 758) the pupils contract or dilate in visual hallucinations according as the apparent objects approach or move farther away. This shows that the visual cortex is also centripetally (motorially) active. Manifestly corresponding (cortical) convergence and accommodation take place. It is well known that conjugate deviation of the head and eyes toward the situation of the apparent object takes place very often. It occurs when the hallucinations are not situated at the point of fixation but in the periphery of the field of vision.

Paretic Dementia.

In paretic dementia eye symptoms are important. They possess great diagnostic and therefore prognostic significance because they often occur at an early period or supplement other doubtful symptoms.

They consist, apart from certain trophic or vasomotor disorders, of disturbances of sight with or without ophthalmoscopic findings, or, more commonly, of disorders of the ocular muscles.

The peripheral disorders of vision, in which ophthalmoscopic changes appear sooner or later, are not characteristic of the disease. There is usually a simple gray atrophy of the optic nerve, which is observed more frequently in tabes. It affects one or both eyes and leads to gradual impairment, concentric narrowing of the field, interference with color-conduction, and in the end usually terminates in complete blindness. One difference from tabetic atrophy is the fact that there is often a slight preliminary cloudiness of the papilla and the adjacent parts of the retina, which does not lead necessarily to atrophy of the nerve. Uhthoff (Ber. d. Heidelb. ophth. Ges., 1883) found this cloudiness in 28 per cent of the cases of paresis, Siemerling in only 8 per cent. This change in the fundus points to an inflammatory condition, however mild it may be, and other signs of the same character are not infrequent, viz., hyperæmia of the retina, neuritis of the optic nerve, retinitis, etc.

Klein (*Wien. med. Presse*, 1877, No. 3) observed 2 cases of atrophy of the optic nerve among 42 cases of paresis (nearly 5 per cent); Uhthoff (*l.c.*) found 8⅔ per cent among 150 cases; Siemerling (*Charité Ann.*, XI, p. 339) observed it 9 times among 151 cases (6 per cent).

Unlike the ordinary form of atrophy with concentric narrowing of the field, Hirschberg (*Berl. kl. Woch.*, 1883, No. 39) reports a case which began as a progressive central scotoma, while the boundaries of the field and color perception at the periphery were normal. Boediker (*Neurol. Centralbl.*, 1891, p. 187) has published a similar case. This finding warrants us in inferring, with a certain degree of probability, neuritic processes in the optic nerve. In such a lesion this disorder of vision is very frequent, as in the "axial" neuritis of so-called toxic amblyopia.

Atrophy of the optic nerve is usually a late symptom of progressive paralysis, but it may also occur at an early period or even precede the mental disturbance (Wigleworth and Bickerton *Brit. Med. Journ.*, 21, IX, 90). Optic atrophy may be the sole eye symptom, but it is usually associated with others, especially with disorders of the ocular muscles.

Among the less frequent ophthalmoscopic findings Siemerling (*l.c.*) reports slight optic neuritis in 2 per cent of the cases. Uhthoff and Moeli (*l.c.*) found, among 150 cases, 3 cases of hyperæmia of the papilla without pronounced opacity, 1 case of retinal hemorrhage and 2 cases of spindle-shaped dilatation of a small artery near the papilla. The latter condition reminds us of Klein's retinitis paralytica, which he observed 18 times among 42 cases. In addition to opacity of the retina the arteries, more rarely the veins, exhibit spindle-shaped dilatations, often two or three in the course of the same vessel. These appearances have also been observed, however, in other cerebral diseases. Klein also found 4 cases of retinitis, 1 of hyperæmia of the fundus and 1 of choked disc.

Among the central disorders of vision hallucinations are very common during the course of the disease and often present at the start. In one case, in which they were especially striking, Mendel found at the autopsy an unusual amount of disease in the cortex and pia mater of the occipital lobe. The hallucinations of sight may also be unilateral.

Homonymous hemianopsia, paroxysmal or permanent, is also observed; sometimes it occurs in succession on the two sides and then it leads to complete cortical blindness. In one case of this kind Stenger (*Arch. f. Psych. u. Nerv.*, XIII, 1, p. 218) found pronounced atrophy of the occipital lobes. Hemianopsia is often associated with hemianæsthesia or cortical paralyses.

Psychical blindness also occurs with comparative frequency; it may be paroxysmal, lasting for days and then disappearing. It is always temporary; but diminution of sight and finally blindness usually develop at an early period.

Riegel has investigated the disorders of reaction in progressive paralysis, and the results may be found in various Wurzburg theses (Rabbas, 1884; Kirn, 1887; Kraemer, 1888. The higher grades of such disorders among the insane are found only in paralytics, but are not constant. The characteristics are: a large number of "slips of the tongue," which every one makes occasionally, repetition of certain words and syllables, substitution of words which are not related in sense, sound or print, foreign-sounding words, a senseless gibberish of words, and usually the inability to recognize what has been falsely read. Another characteristic appears to be the frequency with which particular words are employed in all possible combinations.

Riegel regards these disorders as allied to aphasia and Berlin expresses a similar opinion. The latter assumes the " reading centre" to lie in the cortex in the vicinity of the third frontal convolution, the superior parietal gyrus, angular gyrus and the upper temporal convolution. These are regions of the cortex beneath which run the associative fibres between the occipital cortex and the motor speech centre. The disorders of reading in progressive paresis are evidently due to lesion of the motor speech centre and its associations with the higher frontal cortex, and also to disturbance of the associative connection of the visual sphere with the cortical regions just mentioned.

The most important and characteristic eye symptoms of progressive paralysis are the muscular disorders. They may affect all the muscles of the eyes or only single ones. Anatomically these affections are internuclear (reflex rigidity of the pupil) or nuclear in character, or they involve the nerve roots or even the peripheral nerves themselves. Not even a lesion of the nerve terminations or the muscles themselves can be positively excluded.

The paralyses of the muscles are often temporary, but apt to relapse. They may occur very early during apparently perfect health, or they appear during the course of the disease. Apart from the pupils, the abducens and accommodation are affected most frequently. According to Moeli accommodation is paralyzed in about 1½ per cent of all cases.

Simple or combined paralyses of the ocular muscles may be the very first symptom. Boediker (*Arch. f. Psych. u. Nerv.*, XXIII, 2, p. 313) saw a case which began with bilateral abducens paresis. One year later reflex rigidity of the pupil appeared, and four and a half years later atrophy of the optic nerve developed. It was only at this period that the knee-jerk on the right side was diminished and there was slight difficulty in speech. Death occurred eight years after the onset of the disease. The muscular disorder was due, in this case, to an interstitial neuritic process in the muscle nuclei and nerves. Siemerling reports a case which began with unilateral mydriasis and parálysis of accommodation.

A characteristic feature is the so-called reflex rigidity of the pupil (*vide* p. 113), which is present in about half the cases of paresis and is rarely observed in other diseases, with the exception of tabes. At first the pupil reacts very little and later not at all to the entrance of light (Argyll-Robertson symptom); still later it does not react to convergence and accommodation. The pupils may be of normal size, contracted or dilated, but narrowing (myosis) is observed most frequently. The pupils are often unequal, especially at the beginning. They usually dilate after the introduction of cocaine (contraction of the vessels). When this condition is not fully developed, the pupil may contract unequally, so that it changes its shape on the introduction of light, or constantly exhibits an irregular shape. According to Salgó (*Wien. med. Woch.*, 1887, 45, 46), irregular shape and reaction of the pupil are more frequent than inequality and rigidity of the pupil.

Despite the reflex rigidity of the pupil, it may continue to dilate for a long time after cutaneous irritation. Later, this reaction grows slower and is finally lost. It may also be abolished from the start.

In 64 per cent of the cases Siemerling found absence of the reaction of the pupil to light, usually on both sides; Uhthoff and Moeli found it absent in about half the cases. Moeli found reflex rigidity of the pupil in 47 per cent (among 500 cases), doubtful reaction in 4 per cent and sluggish reaction in 10 per cent.

Uhthoff (*Berl. klin. Woch.*, 1886, No. 3) found reflex rigidity of the pupils 492 times among 4,000 insane; of these 421 (85.5 per cent) were paralytic dements. Thomsen (*Charité-Annal.*, XI) found 172 cases of general paresis (83 per cent) among 205 patients who exhibited reflex pupillary rigidity. Moeli (*Centralbl. f. Aug.*, 1885, Sept.) found this symptom in only 1.6 per cent of non-paretic insane. Moeli (*Arch. f. Psych. u. Nerv.*, XIII, p. 621) also made a special examination of the reaction of the pupil to cutaneous irritants. When the reaction to light was good in general paresis, the reaction to cutaneous irritants was almost always present. When the former was impaired the reaction to cutaneous irritants was often absent, and was absent as a rule when the former was abolished. The knee-jerk was usually intact when the pupillary reaction to light was good, but was only present in about half the cases when the latter was abolished.

Reflex rigidity of the pupil may long be the sole prodromal symptom of general paresis. But the evil result will not invariably follow.

Motor disorders of a cortical nature (convulsions, speech disturbances, etc.) are very common; they rarely affect the eye (ptosis). Conjugate deviation of the head and eyes is quite frequent during the "attacks." Zacher (*Arch. f. Psych. u. Nerv.*, XIV, p. 463) observed this symptom in 12 cases. The head and eyes were usually turned (7 cases) toward the side of the predominant irritative symptoms; in 2 cases without such symptoms they were turned toward the side opposite to the paralysis. In 2 cases the head and eyes were turned in opposite directions, evidently dependent on the situation of the localities of greatest cortical irritation.

Bechterew (*Jahr. f. Aug.*, 1881, p. 316) saw conjugate deviation of the head and eyes toward the right; right hemiplegia and deviation toward the left subsequently developed. Foerstner (*Arch. f. Psych. u. Nerv.*, VIII, 1, p. 182) observed conjugate deviation toward the left which lasted two weeks. He also reports unilateral disturbance of vision with reacting pupil and normal ophthalmoscopic findings in the eye of that side of the body which was subject to the more severe seizures.

According to Blocq (*Arch. de Neurol.*, 1889, No. 54) ocular migraine (scintillating scotoma) is a frequent herald of general paresis. Pick (*Prag. med. Woch.*, 1899, No. 1) makes a similar statement. Graff (*Jahr. f. Rec.*, 186, p. 294) reports a complication with facial hemiatrophy and neuro-paralytic keratitis. In another case hemorrhages into the conjunctiva suddenly developed; the secretion of tears was checked, the cornea became dry but remained clear. The patient died soon afterward from a hemorrhage into the locus cæruleus and the small descending root of the trigeminus.

In general paresis, according to Sgrosso (*Psychiatria*, Bd. V), a conjunctivitis *sui generis* occurs, dependent upon nervous stases in the palpebral conjunctiva; it is said to appear particularly in the late stages. It is true that conjunctivitis often occurs at this time, but its causes are much simpler.

According to the teachings of pathological anatomy, what is known as general paresis is a symptom rather than a definite disease. It is a progressive degeneration of the cortex which is most marked in the frontal lobes, but may vary in character. Apparently either the vessels, nerve fibres or interstitial tissue can be primarily affected. Moreover, it is often complicated with other focal and systematic diseases of the brain and spinal cord, even with multiple neuritis, or it occurs as a complication during their course. The majority of the eye symptoms are probably due to these frequent combinations, all of which owe their development to a common cause. For example, Lawford (*Lancet*, 1883, II, p. 1090), among 7 cases of atrophy of the optic nerve in general paresis, found spinal sclerosis in 5 cases, tabes in 1 case and lateral sclerosis in 1 case. While the insanity implies disease of the frontal cortex, the visual cortex displays no signs of irritation such as hallucinations, etc. The pupillary symptoms which are most characteristic indicate focal affections in

the neighborhood of the muscle nuclei.   These also vary in character
and are probably co-ordinate with the main disease.

In Paralysis Agitans

(Parkinson's disease), which is probably due to insufficient cortical influence upon the motor organs, Galezowski (*Rec. d'Ophth.*, Feb.,
1891) usually found no eye symptoms except slight tremor of the eyelids (very rarely unilateral).   He sometimes found amblyopia with
negative ophthalmoscopic appearances and occasionally defects in the
field of vision.   Galezowski (*Neurol. Centralbl.*, 1891, p. 224) also
reports the sudden loss of the inner, upper and lower quadrants of
one eye, without ophthalmoscopic findings but with slow reaction
of the pupils.   This was evidently a unilateral peripheral affection
of the optic nerve, similar to that in Graefe's retrobulbar neuritis.

Gray atrophy of the optic nerve, bilateral ptosis, fixity of gaze,
etc., are also reported occasionally in paralysis agitans.   According to
Galezowski the imperfect movement of the head and eyes interferes
with close work.   According to Debove (*Jahr. f. Aug.*, 1878, p. 252),
reading was impossible because, after reading one line, the eyes could
not be carried to the next but were compelled, by a sort of imperative
movement, to continue along the line already read.

Nystagmus is very rare in paralysis agitans.   Wille (*Corresp. f.
Schweiz. Aerzte*, 1888, No. 8) observed it in one case.

Diffuse Encephalitis,

according to Graefe's original idea, was said to be the cause of the
corneal softening of poorly nourished, marantic infants, because
numerous granulo-fatty corpuscles were found in all parts of the brain
in such patients.   It was soon learned, however, that such findings
are normal at this age.   The affection in question is a simple desiccation keratitis associated with insufficiency or abolition of the lid
movements.   If the causal disease, usually cholera infantum, subside the corneal affection may recover, after exfoliation of the
necrotic parts.

Among three cases of diffuse sclerosis of the brain Erler (Diss.
Tübingen, 1881) found a case of bilateral abducens paralysis,

although the paralysis was not uniform on the two sides.   The symptoms of diffuse cerebral sclerosis are essentially those of increased intracranial pressure and irritation of the cortex.   Sluggish reaction of the pupils is present as a rule; nystagmus is very often noticed.

Schmaus (*Neurol. Centralbl.*, 1888, p. 626) also observed a left abducens paresis in a girl of three years.   The pupils were dilated and sluggish; there was considerable impairment of vision and horizontal nystagmus; the left facial nerve was slightly paretic.

Quincke (*Deutsch. Arch f. kl. Med.*, XXVII, p.193) saw choked neuritis in a case of hemorrhagic encephalitis.

According to Leube, Ziemssen and Merkel (*vide* Foerster, *l.c.*, p. 125),

Cheyne-Stokes Breathing,

which is more or less frequent in numerous diseases of the brain, is attended with characteristic pupillary symptoms.   During the respiratory interval the pupils are narrow and do not react to light. Moderate dilatation sets in with the renewal of respiration.   When hemiplegia was also present the contraction of the pupils was attended by horizontal associated movements of the eyes, which were most marked toward the paralyzed side.   According to Murri (*Jahr. f. Aug.*, 1888, p. 446), the respiratory intervals may be shortened voluntarily by opening the lids, but according to my experience this does not succeed in all cases.   I was unable to discover visible changes in the fundus during Cheyne-Stokes breathing.   The contraction of the pupils during the respiratory intervals depends probably upon paresis of the sympathetic.

Injuries to the Brain

may produce eye symptoms in various ways: *a.* By direct destruction or irritation of those parts which produce local symptoms, for example, hemianopsia or cortical blindness in injury to the occipital region, or conjugate deviation to the right in injury to the left side of the skull, etc.   Ball's case (*Jahr. f. Aug.*, 1883, p. 329) is very instructive in regard to the occurrence of nystagmus.   After injury to the parietal bone, left hemiparesis ensued with epileptoid attacks; during the latter the eyes were first turned to the right, then nystagmus occurred, then deflection to the left (left-sided irritative symptoms, then diminution with enfeebled cortical innervation, passing later into right-sided irritative symptoms). Beger (*Jahr. f. Aug.*, 1880, p. 432) observed temporary nystagmus after injury in the region of the eyebrows.

b. By indirect injury, as in the frequent fractures at the base from contrecoup. All the nerves at the base may be torn or their function greatly impaired, giving rise to manifold paralyses (the abducens is attacked with comparative frequency) and sensory disturbances. Neuro-paralytic keratitis may develop in anæsthesia of the first branch of the trigeminus.

Special importance attaches to the injuries of the optic nerve which are sustained in this manner. Berlin has shown that fissures at the base, due to contrecoup, pass through the optic foramen with remarkable frequency, and not rarely through both foramina. In such cases the injury, which is often apparently trivial, is followed immediately by unilateral or bilateral blindness with abolition of the reaction of the pupil to light and, for a time, no ophthalmoscopic findings. This may be the sole symptom. For example, I had under observation a railway official who, during a trip, struck the rim of the orbit lightly against some obstruction. The patient was not unconscious for a moment, but from that time on he was blind upon the side of the injury.

Berlin has shown that in fissures through the optic foramen the nerve is also torn. But a hemorrhage into the optic nerve or its sheath, as the result of the concussion, is also sufficient to produce blindness.

In accordance with the peripheral character of the injury atrophy of the optic papilla always develops. According to Berlin, ophthalmoscopic examination in daylight shows beginning atrophy at the end of two weeks (the color is a pure yellow). At the end of three weeks it is demonstrable with artificial illumination. More or less pigmentation of the atrophic nerve subsequently develops in not a few cases. Knapp (*Arch. f. Ophth.*, XIV, 1, p. 252) was the first to describe two cases of this kind. The blindness and atrophy of the optic nerve are usually complete. But Capron (*Arch. f. Aug.*, XVIII, p. 407) observed, after a fracture at the base, that only the inner three-fourths of the nerve was shining white and atrophic; quantitative perception of light was still retained. Such cases also exhibit a decided tendency to pass into complete atrophy. In Rieger's case (*Jahr. f. Aug.*, 1887, p. 299) the capillary hyperæmia of the nerve, moderate impairment of vision with concentric narrowing of the field and color disturbance, must be attributed to a fissure at the optic foramen with comparatively slight implication of the optic nerve.

It is well known that these basal fissures and fractures may be far distant from the point of injury. Thus, Hoog (*Jahr. f. Aug.*, 1887, p. 203) saw paresis of the right facial and abducens nerves after a blow upon the intersection of the coronal and sagittal sutures. Vossius (*Mon. f. Aug.*, 1883, p. 284) observed, after a fall upon both tubera ischii, blindness of the right side followed by atrophy of the optic nerve, which could hardly be explained in any other way than by a fissure at the base of the skull. After a fracture of the frontal bone, Tuffier (*Jahr. f. Aug.*, 1884, p. 353) observed nasal hemianopsia, evidently from implication of the anterior angle of the chiasm.

In indirect fractures of the base the bony walls of the orbit may be directly involved. Hemorrhages into the orbit, exophthalmus, hemorrhages into the eyelids, pressure on and rupture of the optic nerve, muscular paralyses, etc., may be produced in this way.

c. Eye symptoms may result from a hemorrhage which may give rise to general symptoms (compression of the brain) and also, according to its situation, to local symptoms. In injuries to the skull, blood is often found in the sheaths of the optic nerve, extending forward to the sclera. Talko (*Mon. f. Aug.*, 1873, p. 341) explained a hemorrhage into the vitreous as the result of compression of the retinal vessels consequent upon an extravasation of blood into the intervaginal space, but this explanation is far-fetched. If the blood is extravasated between the dura and the bones it cannot enter the intervaginal space of the optic nerve.

Hutchinson (*Ophth. Rev.*, 887, p. 97) states that in compression from meningeal hemorrhage there is usually mydriasis on the corresponding side, but it may also be bilateral; myosis is very rare. He explains the mydriasis by direct pressure on the motor oculi.

The symptoms of compression of the brain (unconsciousness, coma, slow pulse and respiration, etc.) also include dilatation of the pupils, which do not react to light.

d. If the injury is followed by meningitis, the corresponding symptoms are produced. Probably all cases of neuritis or neuroretinitis after injuries are due to secondary meningitis (*vide* p. 150). Cerebral abscesses which are due to trauma, and often do not give rise to symptoms until the lapse of years, show no peculiar feature (*vide* p. 131).

e. A series of symptoms remain which are attributed to cerebral concussion and are supposed to be purely functional disorders without anatomical findings. Macroscopic findings are not infrequently absent in cases which were evidently due to material injuries—for example, in basilar paralyses of the nerves. In individual cases, however, careful microscopical examination occasionally shows lesions (capillary hemorrhages, changes in the vessels, especially in the small and smallest ones, corkscrew-shaped axis cylinders, *i.e.*, torn nerve fibres, etc.) which suffice to explain the irritative or paralytic symptoms.

According to Hutchinson (*l.c.*) the signs of cerebral concussion include a certain sluggishness in the reaction of the pupil, without pronounced myosis or mydriasis. Unilateral or bilateral mydriasis is rare, and myosis occurs only when inflammatory processes develop.

Nystagmus, a sign of insufficient cortical activity or interference with the conduction of the innervation impulse which starts from the cortex, is seen not infrequently.

In addition there may be symptoms which correspond, in the main, to those of hysteria: concentric narrowing of the field with and without disturbance of central vision and the color sense, often associated with sensory disturbances on one side of the body. Such a condition may be called traumatic hysteria. It is also known as traumatic neurosis. It is assumed that the symptoms are central in character and that they have no anatomical substratum. According to our present anatomical knowledge, however, unilateral concentric narrowing of the field of vision can only be peripheral in character. The subject will be further considered in the discussion of hysteria.

Heatstroke or sunstroke may be regarded as a peculiar form of cerebral trauma. According to Spalding (*Jahr. f. Aug.*, 1887, p. 519) only one case has been reported in which blindness occurred after sunstroke, but there have been six cases of optic neuritis with restoration of vision (these are probably Holz's cases mentioned below). Kesteren (*ibid.*, 1882, p. 442) reports a case in which yellow vision existed for three months after exposure to the intense heat of the sun. The ophthalmoscope showed a slight neuritis. Hotz (*Jahr. f. Aug.*, 1879, p. 255) describes six cases of unilateral and bilateral neuritis, with notable disturbance of vision, which he attributes to heatstroke. The cause is said to be an effusion at the base of the brain with secondary disease of the optic nerve sheaths.

Finally, we will consider a few cerebral diseases in which we have to deal with more or less sharply defined processes.

Porencephaly.

In case of cavities in the brain either congenital or acquired, and which are usually the result of large spots of softening, symptoms arise from absence of the affected parts. Among them may be visual disorders and paralyses. If the cavity is not located centrally as regards the primary ganglia and muscle nuclei, there will, however, be secondary degeneration of the interrupted conducting paths. In the optic nerve this is visible as simple gray atrophy. If it is congenital or arises in early childhood, such atrophies may be due to arrested development of fibres or to the fact that they are not medullated. Nystagmus may be observed (Otto, *Arch. f. Psych. u. Nerv.*, XVI, 1, p. 215), and is attributable to imperfect development of the cortex and its peripheral connections.

Fuerstner and Stuehlinger (*Arch. f. Psych. u. Nerv.*, XVII, p. 1) observed the following eye symptoms in porencephaly: 1, unilateral reflex rigidity of the pupil and gray atrophy on the same side; 2, bilateral gray atrophy and ptosis of the left side; 3, bilateral gray atrophy and rigidity of the pupil; 4, bilateral gray atrophy of the optic nerve.

The diagnosis of porencephaly during life is hardly possible and the findings in the eye furnish no data. The local diagnosis of the site of the affection is sometimes possible, and the eye symptoms may then prove very useful.

Bulbar Paralysis and Allied Diseases.

In the typical forms of bulbar or glosso-pharyngeo-labial paralysis eye symptoms are wanting. The disease presents an ascending and a descending form, and may be complicated with eye symptoms in two ways. In the descending form we find sympathetic nerve lesions with oculo-pupillary and vasomotor symptoms, passing from irritation into paralysis. In the ascending form there are sometimes sensory disturbances (trigeminus) and paralytic symptoms referable to the external and internal ocular muscles and the ocular facial nerve, and occasionally disturbances of vision. In these paralyses the nuclei, the roots of the nerves, or even the peripheral nerves within the cranium may be primarily diseased. Apart from the continuous extension of the principal focus, isolated sclerotic and atrophic foci may develop, not alone in the brain but also in the peripheral nerves.

In typical bulbar paralysis such symptoms are extremely rare. Ritchie (*Glasgow Med. Journ.*, XII, 1888) saw bilateral gray atrophy of the optic nerve, with considerable impairment of vision and dilated, immobile pupils. Heller (*Petersburg med. Woch.*, 1882, No. 9) found right abducens paralysis which recovered. Eisenlohr (*Zeitschr. f. kl. Med.*, I, 3, p. 435) observed difference of the pupils in one case, paresis of the orbicularis in another case. Fischl (*Prag. med. Woch.*, 1879, No. 4) found in acute bulbar paralysis, paralysis of the left abducens, levator palpebræ superioris, and the ocular facial; the paralyses recovered. Kahler (*Wien. med. Presse*) reports in a woman of thirty-seven years paralysis of the frontal and orbicularis palpebrarum, while sight and ocular movements were normal, and he mentions similar cases reported by Birdsall and Remak. Minot (*Jahr. f. Aug.*, 1879, p. 249) reports left-sided ptosis. Difference in the size of the pupils, nystagmus and conjugate deviation are found somewhat more frequently.

Wernicke was the first to differentiate from ordinary bulbar paralysis or inferior polio-encephalitis, a special form known as superior polio-encephalitis, in which eye symptoms constitute an essential feature. This disease is either acute (hemorrhagic inflammation of the floor of the fourth ventricle, usually in alcoholism) or chronic, as in the majority of· cases. It may run an ascending or descending course, may spread in a continuous manner or skip certain places.

On account of the predominant affection of the eye muscles the disease is also known as progressive paralysis of the ocular muscles or progressive ophthalmoplegia. In *Arch. f. Psych. u. Nerv.*, XXII, Supplement, Siemerling has written a detailed monograph on the subject.

The chronic form often shows nuclear disease with degeneration of nerve and muscles (fatty and connective-tissue degeneration); there may also be degeneration of the muscle and nerve, leaving the nucleus intact; or sclerotic foci interrupting conduction in the nerve, while the remainder of the nerve, the muscle and nucleus are intact. The findings may also be entirely negative (Eisenlohr, *Neurol. Centralbl.*, 1887, Nos. 15–17; Bristowe, *Brain*, VII, p. 313).

The clinical symptoms consist of progressive paralyses of the ocular muscles. They are almost always bilateral. They may be complete or incomplete, and affect the internal or external ocular muscles, or both. The paralyses are usually irregular; ptosis is comparatively rare. They may disappear entirely or in part, and then relapse, but on the whole the process advances. All the symptoms which may result from disease of the nuclear region, whether nuclear, perinuclear, or internuclear (*vide* p. 71), are observed more or less frequently—for example, reflex rigidity of the pupil, paralysis of convergence, divergence paralysis (more correctly, spasm of convergence), and not infrequently nystagmus. In one case Siemerling observed protrusion of the right eyeball; Lichtheim noted bilateral protrusion, more pronounced on the left side.

It is only in exceptional cases that progressive ophthalmoplegia

12

is an independent disease. In some cases the process is a descending one, and bulbar symptoms develop to which the patient succumbs, or the disease is a part of multiple sclerosis, tabes, and other diseases of the spinal cord. It is also a forerunner of insanity, particularly of general paresis.

Acute superior polio-encephalitis is much rarer (*vide* Salomonsohn, *Deutsch. med. Woch.*, 1891, p. 849). It begins with pronounced somnolence, although the patient is entirely conscious. Then advancing paralysis of the ocular muscles develops, progressing to more or less complete ophthalmoplegia. Death occurs in somnolence within one or two weeks. Salomonsohn's case is the only one in which recovery occurred.

In Wernicke's three cases the ophthalmoscope showed neuritis (redness of the papilla, with or without swelling) and in two cases retinal hemorrhages. Fever was never observed, and even subnormal temperatures occur. Of Wernicke's three cases two occurred in hard drinkers, the third was due to sulphuric-acid poisoning. Thomsen's two cases (*Arch. f. Psych. u. Nerv.*, XIX, p. 185) and Kojewnikoff's case (*Prog. Méd.*, 1887, Nos. 36 and 37) were also hard drinkers. Salomonsohn's patient, who recovered, did not suffer from alcoholism.

The anatomical lesion is an hemorrhagic encephalitis of the gray matter at the floor of the fourth ventricle and the aqueduct of Sylvius; numerous punctate and a few larger hemorrhages; the small vessels and capillaries are distended, but their walls are normal; granular corpuscles are found in the vicinity of the hemorrhages.

Somnolence serves to distinguish the disease from simple nuclear paralysis. If there be no paralysis the symptoms are those of so-called nona and the allied somnolence of negroes, which is attended with great general emaciation and increasing muscular weakness and apathy, without rise of temperature or disorder of sensation and motion, and which terminates in death. The benign maladie de Gerlier (vertige paralysant), an evidently infectious disease which is manifested by vertigo, staggering gait, great muscular weakness and double ptosis, also points to a localization of the lesion in the region of the muscle nuclei.

Multiple Sclerosis.

In multiple sclerosis the eye is very often affected and, indeed, it may exhibit symptoms which are characteristic of the disease.

The principal clinical symptoms consist of tremor when fixing an object, nystagmus, peculiar slowness of speech and spastic-paretic phenomena of the limbs. There may also be headache, vertigo, psychical disorders, convulsions, apoplectiform attacks, visual disturbances, etc. These symptoms grow worse, walking and standing become impossible, the tremor becomes general, intelligence is impaired, and speech becomes unintelligible. Finally, there is paresis of the abductors of the glottis, incontinence of urine and feces, etc., and death occurs either from increasing bulbar symptoms, in an apoplectiform attack, or from an intercurrent disease. Sensibility is undisturbed; the electrical reactions to both currents are normal.

All these symptoms may appear occasionally without finding sclerotic patches at the autopsy. For example, Carter Gray (*Jahr. f. Aug.*, 1889, p. 526) found only œdema of the brain and extensive leptomeningitis of the convexity. As a rule, however, gray foci of degeneration are distinctly visible, sometimes in almost all parts of the central nervous system. Oppenheim (*Berl. kl. Woch.*, 1887, No. 41) also found numerous microscopic foci in the pons, medulla oblongata, chiasm and optic nerve, but these had not given rise to symptoms.

The most important ocular symptoms are the nystagmus and twitchings. We have already seen that this associated disorder is to be regarded as insufficient cortical innervation of the nuclei of the ocular muscles. In multiple sclerosis the nystagmus or tremor is due to an interruption of conduction between the oculomotor cortex and the nuclei of the ocular muscles. Voluntary innervation may be conveyed by fits and starts, even when the interference with conduction is located in the peripheral motor nerves. Hence arises the possibility of unilateral nystagmus, which has actually been observed (Gordon Norris, *Jahr. f. Aug.*, 1888, p. 434). As the favorite site of the sclerotic patches is the region of the pons, the nuclei of the ocular muscles and the medulla, the nystagmus is probably due in the main to perinuclear foci in the fibres of the corona radiata (*vide* p. 84).

The frequency of nystagmus in multiple sclerosis is variously estimated. Uhthoff found 12 per cent of typical nystagmus (pendulum movements around the position of rest) and 46 per cent of nystagmus-like twitchings (twitching movements of the eyes toward the desired position, either during movements in all or only in certain directions). True nystagmus is extremely rare in other diseases of the brain. Nystagmus-like twitchings also occur in a great many other nervous diseases, but in none with such frequency as in multiple sclerosis.

Distinct restriction of the mobility of the eye in one or more directions occurs not infrequently, either with or without nystagmus. In these disorders we must also assume that the site of the interruption of conduction is on the central aspect of the muscle nuclei (probably perinuclear), although it is also possible that the cause might attack the nuclei or nerve roots. In the two latter cases, however, we would be much more apt to expect complete abolition of motion in certain directions, especially upward or downward.

In addition to these disorders of associated movements, which Uhthoff observed in 3 per cent of his 100 cases, other paralyses of the ocular muscles are not infrequent. They are in great part nuclear in character but may also be peripheral. For example, Leube (*Arch. f. kl. Med.*, VIII, p. 1) saw both motor oculi nerves converted into thick gray bands. In a number of cases sclerotic foci have been demonstrated in the peripheral nerves. Uhthoff noted 6 cases of abducens paralysis (unilateral in 4 cases), 3 cases of partial motor oculi paralysis, 3 cases of paralysis of convergence, and 2 of ophthalmoplegia externa, *i.e.*, muscular paralyses in 17 per cent. These were distinct paralyses, but a less degree of impairment of mobility is much more common. The paralyses almost always recover, but relapses may occur. In only three cases (abducens paralysis) did they appear at the beginning of the disease or as prodromata.

Changes in the pupils are rare in multiple sclerosis. Uhthoff (*l.c.*) found 1 case of reflex rigidity of the pupil, 4 cases of myosis with slight reaction to light and convergence, in 1 case the reaction to light very much diminished but without myosis, 3 cases of unequal size of the pupils, 2 cases of slight convergence reaction despite comparatively good reaction to light, *i.e.*, anomalies of the pupils were present in 16 per cent of the cases. In the majority there were only slight deviations from the normal condition. Parinaud (*Prog. Méd.*, Aug. 9th, 1884) also mentions myosis with considerable increase of the reaction to light and marked contraction during convergence and fixation, *i.e.*, increased reflex excitability of the pupil. Thus, the pupils show no characteristic changes in multiple sclerosis, as they do in progressive paralysis.

Much more importance attaches to the disturbance of vision and the ophthalmoscopic appearances. The disturbance of vision is peripheral,—a central scotoma, occasionally for colors only, irregular narrowing of field of vision with or without disturbance of central vision, or both combined, or in comparatively rare cases concentric narrowing. Both eyes or one alone may be affected. It often develops very rapidly, may undergo improvement, and even disappear entirely. The patient usually has a sensation of a more or less dense mist, without subjective light sensations. Impaired sight may appear at the beginning of the disease and may even be the sole manifestation for a long time. It increases occasionally with the exacerbation of other symptoms or with bodily strain, and on the whole runs a very changeable course. The central scotoma is rarely absolute and complete blindness is rarely observed. Hemianopic symptoms are infrequent and occur only in complications with other diseases of the brain.

The visual symptoms point to a partial retrobulbar affection of the optic nerve. Characteristic changes are found sometimes immediately behind the eye: foci of granular degeneration of the medullary sheaths, while the axis cylinders remain intact. There is sometimes proliferation of the interfibrillary connective tissue which becomes finely granular and fibrillated in addition, proliferation of nuclei in the larger septa of connective tissue, especially near the vessels and in the finest interfibrillary bands (interstitial neuritis). Dilatation and proliferation of the small vessels may also be found in the intracranial portion of the optic nerve or in the chiasm.

Pathologically the disease of the optic nerve in multiple sclerosis stands midway between pronounced neuritis and simple atrophy. The process begins in the finer connective-tissue septa, then passes to the coarser ones; the atrophy of the nerve substance is secondary. The atrophy and destruction of the medullary sheaths are rapid and complete, while the axis cylinders often remain permanently intact. According to Uhthoff the vessels are not primarily diseased. Whether this is true of the optic nerve affection in all cases which belong clinically to multiple sclerosis is very doubtful, to judge from the experience in other diseases of the brain and cord. The clinical characteristics are the occurrence of focal symptoms and the maintenance of conduction, although this is interfered with to a greater or less extent.

As the majority of nerve fibres within the foci of disease remain intact, ascending or descending degeneration of the optic nerve does not develop at all or is very slight. Nor does the retina show any atrophic changes. In one case Uhthoff observed a wedge-shaped atrophy of the retina with pigmentation as the result of obliteration of an arterial twig.

This corresponds with the ophthalmoscopic findings. Pronounced atrophy of the optic nerve is rare (3 per cent), incomplete and partial atrophy is much more common (19 per cent). The ophthalmoscopic appearances are often (18 per cent) very similar to those of toxic amblyopia (nasal half of the nerve grayish-red, cloudy and often obliterated, temporal half pale "like dull porcelain"); in 6 per cent Uhthoff found neuritis, and this, as a matter of course, may terminate in white atrophy (five cases in literature). Only 4 per cent showed normal ophthalmoscopic findings.

Very often there is a striking disproportion between the appearance of the optic nerve and the disturbance of vision, the latter being much more marked than the ophthalmoscopic appearances would lead us to expect. It may even be considerable when the ophthalmoscopic appearances are normal. This corresponds entirely to the anatomical fact that the axis cylinders in the patches of sclerosis are intact in great part, but probably conduct poorly on account of insufficient isolation.

In the course of the visual tracts the degenerative foci are found most frequently in the optic nerve, rarely in the chiasm and tractus. They also occur in the primary optic ganglia (Schuele, *Arch. f. kl. Med.*, VII, p. 259).

The patches which are situated in the visual tracts correspond entirely to those in other parts of the brain. Marie and Rindfleisch assume a primary disease of the vessels, Charcot (and Uhthoff) a primary disease of the neuroglia, Adamkiewicz a primary affection of the nerve fibres. In fact, all three anatomical processes really occur. When they develop in multiple patches and the axis cylinders remain intact for a long time so that conduction is not completely interrupted, the clinical picture of multiple sclerosis is produced. We will again refer to analogous conditions in other diseases of the brain and cord.

2. DISEASES OF THE SPINAL CORD.

Diseases of the spinal cord *per se* give rise to eye symptoms only when the region of the cilio-spinal centre is affected directly or indirectly. Other eye symptoms occur only when the disease of the cord extends to the cranial cavity or in case of complications. Both are very frequent.

Tabes Dorsalis.

The eye symptoms are very important in tabes dorsalis. They may decide the diagnosis and not infrequently constitute the first symptom of the impending disease. This is true particularly of gray atrophy of the optic nerve and of the muscular disorders, but vasomotor and secretory anomalies are not uncommon.

Tabes is often associated with general paresis, insanity, bulbar paralysis, multiple sclerosis, or with other systemic diseases of the cord.

Atrophy of the optic nerve in ataxia appears ophthalmoscopically as a gray discoloration of the papilla, which is first visible on the outer half, then extends slowly over the inner half of the nerve until the papilla has a uniform gray color and is slightly excavated. The details on the surface of the papilla, especially the tendinous network of the lamina cribrosa, remain clearly visible as in glaucomatous atrophy, and in contrast with post-neuritic white atrophy in which the meshes of the lamina cribrosa are invisible or very indistinct. When there is pronounced physiological excavation, it is quite possible to mistake it for glaucomatous excavation, particularly in the last stages. This is especially true when the gray atrophy occurs in short-sighted eyes, and the myopic meniscus (crescent), together with the atrophy of the choroid which has been drawn over on the inner side of the optic nerve, simulates the appearance of the glaucomatous "halo."

Inflammatory phenomena are entirely absent in gray atrophy. As a rule, the calibre and degree of fulness of the retinal vessels are also unchanged, because the original process is located behind the entrance of the central vessels. The retinal hemorrhage in gray atrophy reported by Auscher (*Jahr. f. Aug.*, 1887, p. 528) must be regarded as an accidental finding.

The absence of inflammatory phenomena is also anatomically characteristic of gray atrophy. The affection is a primary atrophy of the nervous elements of the trunk of the nerve, perhaps following granular degeneration.

At a later period atrophy also occurs in the finer connective-tissue bands within the larger meshes. The coarser septa retract and become more homogeneous and sclerosed. The walls of the small vessels within them also become sclerotic. On the whole the structure of the optic nerve with its regular alternation of nerve substance and septa is retained. Signs of interstitial processes and nuclear proliferation are never present, and the subsequent thinning of the nerve is never so great as when there have been preceding interstitial inflammatory changes (retrobulbar neuritis). The entire process thus appears to be a simple descending atrophy. The retina constantly exhibits atrophy of the nerve-fibre and ganglion-cell layers; the other layers appear essentially normal. In only one case did Uhthoff, from whom we copy this description, observe rarefaction of the outer part of the external granular layer.

The atrophic nerve fibres, which are often varicose at the beginning, usually remain intact with the medullary sheath; in very rare cases the latter also degenerates.

The disturbance of vision consists of diminution of central vision, concentric narrowing of the field and disorder of the color sense. Central vision usually possesses a definite relation to the narrowing of the visual field, and is so much more impaired the more the boundaries of the field approach the point of fixation. The impairment is usually very slow, and years elapse (even ten to twenty) before complete blindness. Temporary improvement may occur, especially at the beginning, but on the whole the process is progressive. An arrest, either temporary or permanent, may occur, however, at any period, but this is not common. Cases in which sight is lost within a year are included among the rapid ones. Hirschberg's case ("Klin. Beobacht.," 1874), in which vision was almost entirely lost in eight weeks, is extremely rare.

It will often be found that in feeble illumination or in artificial light, vision becomes disproportionately poorer, although no anomaly of the light sense can be discovered. The disorder of the field consists essentially of concentric narrowing, often with entering angles which slowly enlarge and approach the point of fixation. The latter is not usually preserved until the last; an eccentric part of the retina is at the end still sensitive to light. On the whole the contraction is concentric to the blind spot, not to the point of fixation (Foerster).

Despite the impairment of central vision the visual field may remain normal, and the latter may be very much narrowed although central vision is good.

A central scotoma hardly ever occurs and always rouses the suspicion of a complication or a false diagnosis. Gowers (*Jahr. f. Aug.*, 1881, p. 315) reports a case of combined sclerosis of the posterior and lateral columns with gray atrophy of the optic nerve, and bilateral, oval, central scotoma, especially for red and green. This finding is the typical one in so-called pseudo-tabes (multiple neuritis). To this category probably belonged the two cases of tabes in children of eleven and fourteen years, in whom Bouchut (*Gaz. des Hôp.*, 1874, p. 297) observed neuritis and hyperæmia of the optic nerve.

The color disturbance is characteristic of interference with conduction (*vide* p. 29). Real exceptions are very rare, apparent ones are more frequent, especially in comparatively rapid cases. Prognostically the color disorder is important, because those cases in which the color boundaries are very much narrower than those for white belong to the rapidly progressive cases. But if the color boundaries are only narrowed to correspond with the boundaries for white, the process is either very slowly progressive or stationary. Usually the color boundaries are somewhat more narrowed than those for white, corresponding to the slowly progressive course; or an entering angle for the color boundaries is the forerunner of a similar contraction of the field of vision in general.

There are all kinds of exceptions to the ordinary course which cannot be entered into here.

There is very often a disproportion, in both directions, between the visible atrophy of the nerve and the disturbance of vision. As a general thing the atrophic discoloration of the nerve is visible with the ophthalmoscope before a disturbance of vision is demonstrable. When the visible atrophy appears to be much greater than we might expect from the amount of disturbance of vision, the case is usually a slowly progressive or stationary one. There are even cases in which the ophthalmoscope shows the appearances of total gray atrophy, although there is hardly any disturbance of vision. When the latter is disproportionately greater than the visible atrophy of the nerve, the case is generally a rapidly progressive one. There are numerous exceptions, however, to this rule.

Atrophy of the optic nerve may begin in any stage of tabes, and even precede the first symptom for years. Kahler observed it seven years, Charoct ten years, Gowers fifteen and twenty years prior to any other tabetic symptoms. The older the individual, so much slower is its course. It has been claimed that when blindness from atrophy of the optic nerve occurs at an early period, especially when it runs a rapid course, a slow course of the tabetic symptoms proper may be expected (Martin, *Arch. f. Aug.*, 1890, p. 122; Benedict, *Wien. med. Woch.*, Aug. 14th, 1887). The latter writer says: "It is a rule, without exceptions, so far as my experience goes, that motor tabetic symptoms subside, however severe they may have been, as soon as atrophy of the optic nerve begins. The latter symptom, however, presents a very unfavorable prognosis." I have seen exceptions to both these statements. But a very rapid course of the visual disturbance always arouses in me the suspicion of a wrong diagnosis, because this is much more common in neuritic and sclerotic processes than in primary atrophy of the nerve substance. As a general thing, the prognosis of those diseases which may be mistaken for tabes is more favorable, and only those cases in which an autopsy is obtained may be regarded as conclusive.

Both optic nerves are usually attacked, but often at different times, so that years sometimes elapse between the affection of the two nerves.

The frequency of optic nerve atrophy in tabes is variously estimated. For evident reasons the statistics of ophthalmologists furnish a higher percentage than those of neurologists. Gowers mentions 13.5 per cent (35 per cent in cases attended with psychical disturbances, 8 per cent in those unattended with such symptoms), Marina 10 per cent, Walton 21 per cent, Leber 26 per cent, Berger 33.7 per cent, Dillmann 2 per cent, Uhthoff (Nerve Clinic) only 20 per cent.

According to Galezowski about two-thirds of all optic nerve atrophies are tabetic; Peltesohn found a still higher percentage, viz., 78 out of 98 cases. According to other writers the proportion is lower, and it is evident there are great differences in the clinical material of different observers. At all events every genuine gray atrophy of the optic nerve raises the suspicion of tabes, although a long time (ten to thirty years) may elapse before the appearance of further symptoms.

The treatment is not very promising; the best plan is the hypodermic injection of strychnia into the temples (0.001 to 0.004 at a dose). In some cases the improvement is merely subjective, for example, in clear weather, etc. To this category belongs the apparent improvement from the internal use of santonin; the yellow vision simulates clearer vision. Caution must be exercised

in the application of electricity to the optic nerve. I have seen
this followed in several instances by an acute exacerbation.
Careful cold-water treatment often has a favorable effect on the
optic nerve trouble as well as upon the tabetic symptoms proper.
According to Abadie and Desnos suspension has effected an improvement of vision, while Eulenburg, Mendel, Galezowski, and
Clarke detected no influence.

There is often a certain independence in the course of the optical symptoms as compared with the spinal symptoms, and this may
also be noticeable in the therapeutic results. For example, Kahler
(*Prag. med. Woch.*, 1878, No. 36) found that all the ataxic symptoms
retired under the administration of nitrate of silver, while the visual
disorder remained unaffected.

We must enter a caution against vigorous antisyphilitic treatment despite the well-known relations of tabes to syphilis. [At the
pressent time great stress is laid upon the value of hydropathic
and hygienic treatment of tabes as conducted in the best modern
establishments—not excluding iodide of potassium.—ED.]

Foerster emphasizes the carelessness and even cheerfulness with
which patients endure their affliction. Among my patients with beginning atrophy of the optic nerve at the onset of tabes, two committed suicide in one summer. One case was interesting from the
fact that under the influence of strychnine injections vision remained
tolerably fair for a long time. The patient lost patience, however,
and contrary to my advice electricity was applied by another physician. The consequent rapid impairment of sight led the patient to
commit suicide.

It is evident from the statements already made that the process in
the optic nerve begins as a disseminated one, and that the atrophy
extends in an ascending and descending direction. The middle
fibres of the nerve remain longest intact, and we may therefore
infer an action starting from the periphery. This may take place
most readily at the place where the nerve passes through the unyielding optic foramen. The gray atrophy is also noticeable in the
chiasm, tractus, and as far as the primary optic ganglia, but there
is no direct connection with the degeneration in the spinal cord.

Other visual disorders such as homonymous hemianopsia, symmetrical defects of the field (according to Berger, these occur frequently), central scotomata, etc., are exceptional, but may occur in pure dorsal tabes, usually as complications of the later stages. The homonymous defects are probably due to foci of degeneration in the tractus or the primary ganglia.

Muscular disorders of all kinds occur with equal frequency in tabes. Berger (*Arch. f. Aug.*, 1888) found them in 38 per cent of the cases, Dillmann in 41 per cent, Uhthoff in only 20 per cent.

The muscular paralyses preceding and during tabes are always peripheral or nuclear; at the most they are perinuclear or internuclear. The nuclei and nerve roots have been repeatedly found intact on post-mortem examination. On the other hand, true conjugate or associated paralysis is never observed, except in complications. In the case of muscular paralysis, as in tabes itself, the lesion is chiefly a primary atrophy of the nervous elements of the nerve fibres and ganglion cells. Pronounced inflammatory or hemorrhagic inflammatory phenomena, for example, inflammation of the ependyma of the fourth ventricle and hemorrhages into the subjacent gray matter (Buzzard), do not belong to tabes proper, but to multiple neuritis, the symptoms of which are well known to be very similar to those of tabes. However, the possibility of the combination of these two processes is not excluded.

The paralyses of the external ocular muscles are often bilateral but not symmetrical, often unilateral, and frequently affect only single muscles. Mydriasis, ptosis and other partial, as well as complete, motor-oculi paralyses, and abducens paralysis are the most frequent. As a rule they develop suddenly and usually disappear after a longer or shorter period either with or without treatment. Relapses are frequent. Their duration varies from a few hours to a year or more. According to Hutchinson spontaneous recovery of a ptosis awakens suspicion of tabes.

According to Fournier (*Bull. de Méd.*, 1887), syphilitic paralysis of the motor oculi involves a larger number of muscles; headache, vertigo, epileptiform attacks, aphasia, mental disorders, etc., are present at the same time. In the tabetic form, in which the nerve roots are chiefly involved, a single muscle is affected in many cases. In syphilis accommodation suffers at an early period, in tabes it remains intact for a long time. Tabetic paralyses are often temporary, sometimes lasting only a few hours, and often relapse. Syphilitic paralyses are more permanent and develop gradually. True as these statements are in general, there are also frequent exceptions.

Rarer symptoms are complicated ophthalmoplegias, such as unilateral and bilateral ophthalmoplegia externa and interna (Danillo, *Central. f. Aug.*, 1891, p. 128; Dillmann, *l.c.*; Hoffmann, *Arch. f. Psych. u. Nerv.*, XIX, 2, p. 438), or bilateral ptosis (Déjérine, *Prog. Méd.*, 1884, No. 43). In the latter case the muscle and nerve were completely degenerated.

The muscular paralysis sometimes assumes the character of progressive ophthalmoplegia (Galezowski). Pel (*Berl. kl. Woch.*, 1890, No. 1) observed seven relapses of motor-oculi paralysis precede the development of ataxia in an epileptic patient with a neuropathic heredity.

Paralysis of accommodation is frequent, occasionally even at the beginning of tabes. Asthenopia from paresis of accommodation is also often observed (Landolt, *Neurol. Centr.*, 1891, p. 307).

According to Galezowski unilateral paralysis of accommodation without mydriasis is often the first sign of beginning tabes, and is associated in a characteristic manner with anæsthesia of patches of the skin in the temporal region.

Nystagmus is rare; it is probably perinuclear in origin. Vierordt (*Berl. kl. Woch.*, 1886, 21) observed bilateral horizontal nystagmus. Berger saw true nystagmus in only one case, but often observed slight tremor of the eyes. Dillmann had one case of nystagmus and several of nystagmus-like twitchings.

These rarer forms are perhaps always due to complications (multiple sclerosis, general paresis, etc.) which are very often observed.

According to Moeli (*Charité-Ann.*, 1881) the "head symptoms," *i.e.*, gray atrophy and paralyses of the ocular muscles, are much more frequent when psychical disturbances are present (gray atrophy with such symptoms, 35 per cent, without them 8 per cent; muscular paralyses with psychical symptoms 47 per cent, without them 15 per cent). According to Berger (*Deutsch. med. Woch.*, 1885, Nos. 1 and 2) muscular paralyses are more frequent in tabetics who have syphilitic antecedents. The spinal symptoms predominate in other cases.

Pronounced paralysis of the orbicularis palpebrarum is rare, but paresis is frequent, and is manifested by tremor on closing the lids.

Berger found 6 cases of facial paralysis in 109 cases, but in only one was the ocular facial also affected.

The following figures give an approximate idea of the frequency of the individual muscular affections. Among 100 cases of tabes Dillmann found the motor oculi affected 26 times (9 cases of paresis of all branches, 3 cases of paresis of the external branches, 6 of ophthalmoplegia interna, 1 of accommodation alone, 5 of individual branches); the abducens 12 times; the trochlearis 3 times. There was also one case of nystagmus. The paralyses are more often complete and permanent than in multiple sclerosis.

According to Kahler the most frequent early symptom is abducens paralysis, next ptosis; the trochlearis is affected with least frequency. Paralyses of the associated movements (especially of convergence) occur rarely and usually at a later period. In our opinion paralysis of convergence is a nuclear or at least a perinuclear symptom. De Watteville (*Neurol. Centralbl.*, May 15th, 1887) also states that paralysis of convergence may appear at the very onset. Grainger Stewart (*Brain*, 1880, II, p. 182) also observed double vision during convergence. Some of the bilateral abducens paralyses which, according to Kahler, are so frequent, may perhaps be interpreted as spasm of convergence.

Early paralyses usually recover, the later ones often persist. In my experience, the older the individual, the longer the duration of the paralyses which eventually disappear.

Every paralysis of an ocular muscle which occurs suddenly in a healthy person (without injury, apoplexy or other brain symptoms, diabetes mellitus or insipidus, syphilis, albuminuria, etc.)

arouses the suspicion of a beginning tabes, especially if it recovers in a comparatively short time or subsequently relapses.

The pupillary symptoms are also very frequent and possess great diagnostic importance. We have already noted that mydriasis from oculomotor paralysis, with or without implications of accommodation, is not a very frequent symptom. Much more frequent are the sympathetic symptoms which indicate irritation or paralysis of the cilio-spinal centre, especially myosis, which may be extreme. The reaction to light may be retained or absent, but it is always very much diminished. The myosis may be associated with ptosis sympathetica. But the latter is often absent and, on the other hand, may be present without myosis.

Dillmann found failure of reaction of the pupil to light (Argyll-Robertson symptom) in 76 per cent of the cases; in 74 per cent the knee-jerk was also absent. Myosis was observed in 23.7 per cent and inequality of the pupils in 34.2 per cent of the cases. In about one-fourth of the cases the rigidity of the pupil to light occurred very early, perhaps as the very first symptom. Among 41 cases with muscular paralysis, rigidity of the pupils to light was noted 35 times. The reaction of the pupils to cutaneous irritation seems to disappear later than the reaction to light. If the myosis is marked, dilatation of the pupils after cutaneous irritation is absent.

In 31.6 per cent of Dillmann's cases the pupils reacted neither to light nor to convergence (total reflex rigidity of the pupil, reflex deafness).

In only 4 cases among 109 did Berger find a normal condition of both pupils. In 8 cases the reaction to light took place by means of oscillating movements, which must be regarded as the very beginning of the paralysis (Gowers' symptom); in 2 cases the reaction to light was absent on one side alone.

The light reaction of the pupil may be absent where the latter is normal in size, in myosis and mydriasis; in spinal myosis the action of mydriatics is diminished, in spinal mydriasis that of myotics. Kahler, Mendel and Oppenheim found the oculomotor nucleus intact in absence of the light reaction. Hence the lesion must be sought in the centripetal fibres to the motor oculi nucleus or in their origins, unless the condition is due to a toxæmia. A peripheral cause is extremely improbable. In Moebius' case (*Centr. f. Nerv.*, 1888, No. 23), in which there was unilateral light rigidity of the pupil, the latter did not exhibit consensual reaction (*vide* p. 115) although the other eye did. In addition to the interference with conduction in the centripetal optic fibres to the nucleus of the sphincter pupillæ, we might also assume, in this case, an interruption in the connections between the two sphincter nuclei, although this explanation is not absolutely necessary (*vide* p. 115).

The absence of the pupillary reaction to light, which is followed by loss of the reaction to accommodation and convergence and to cutaneous irritants, is one of the most characteristic signs of impending or beginning tabes and forms an integral part of the further clinical history of the disease. It is true that this symptom also occurs very often in other diseases of the nervous system, for example, in general paresis. In this affection, however, the mental disturbance or the characteristic disorder of speech will aid the correct diagnosis. Moreover, a combination of these two diseases is observed not infrequently and has also been demonstrated anatomically.

Inequality and irregular shape of the pupils, with or without reflex, are found very often, occasionally even the so-called paradox reaction of the pupils (p. 116).

All the pupillary symptoms of tabes cannot be explained by a lesion of a single part. In addition to spinal spastic mydriasis and paralytic myosis, which point to the cilio-spinal centre, there are also nuclear and perinuclear processes (reflex rigidity of the pupil, etc.) as a result of which the centripetal fibres of the first reflex arc (*vide* p. 116) are interfered with or divided. The nucleus itself may be involved at a later period. This explains the comparative independence of the spinal and cerebral pupillary disturbances which may be absent or present in various combinations, as the result of irritation of one part and paralysis of another. This does not imply that occasionally the process may not first attack the

13

sphincter nucleus itself or its emerging roots. The latter seems to be the rule in the other disorders of the ocular muscles.

In contrast with the rapid onset and disappearance of the cerebral paralyses, the spinal symptoms develop gradually and remain permanent. In Klinkert's case (*Jahr. f. Aug.*, 1885, p. 312) a reflex rigidity of the pupil disappeared after antisyphilitic treatment. Rumpf (*Berl. kl. Woch.*, 1838, No. 4) relieved this symptom, associated with myosis, by the aid of the faradic brush.

Epiphora is a very frequent sympathetic symptom. Berger claims to have seen it in half his cases. It is generally a direct disorder of secretion, but it may also be the result of imperfect action of the orbicularis palpebrarum (Berger). In one case I observed obstinate epiphora without objective findings as the initial symptom of tabes. True "crises lacrymales" sometimes develop.

Joffroy (*Un. Méd.*, 1888, No. 156) observed slight exophthalmus. According to Berger diminished tension of the globe is present in a full third of the cases, but this appears to me to be an exaggerated statement.

Sensory disturbances, such as insensibility of the cornea with or without anæsthesia of other parts of the skin supplied by the trigeminus, are rare (Westphal, *Arch. f. Psych. u. Nerv.*, VIII and IX). These are not pure cases of tabes but combined diseases of the columns of the spinal cord. Neuralgia of the trigeminus, shooting pains and abnormal sensations in the eyes, years before the first prodromal symptoms, have also been reported, but it is doubtful whether they are directly connected with the tabetic process.

Insensibility of patches of skin near the eye and in the temporal region was noted by Galezowski (*Rec. d'Ophth.*, 1888, p. 85) at the beginning of tabes. In one of my cases atrophy of the optic nerve was preceded by subjective coldness of the eye, first upon one side, then upon the other; further symptoms of tabes have not yet developed in this case.

It is characteristic of the ataxia in tabes that the tottering of the patient increases upon closure of the eyes. The imperfect innervation of the motor organs on account of defective conduction of peripheral sensory stimuli may be replaced entirely or in part by voluntary impulses under the control of the sense of sight. But if the cause of the ataxic symptom is central (medulla oblongata or cerebellum), or is situated on the motor side of the first reflex arc, it will not be affected by voluntary motor impulses. The ataxia will not increase on closing the eyes. This is also true of hereditary ataxia (Friedreich's disease) in which other symptoms (nystagmus) also indicate imperfect conduction of the motor impulses.

As is well known, all the symptoms of tabes may occur not only in primary, non-inflammatory degeneration of the posterior columns of the cord, but also in other chronic systemic and even focal diseases of the same locality. Cerebellar diseases, hemorrhages into the pons (Mendel, *Berl. kl. Woch.*, Oct. 10th, 1887), etc., may produce similar symptoms. But it is particularly chronic multiple neuritis of the posterior spinal roots which produces exactly similar symptoms. In such cases the eye symptoms often decide the diagnosis. Moreover, anatomical findings are sometimes entirely negative in an apparent tabes dorsalis.

Hereditary ataxia begins generally at the period of puberty, often appears in several members of the same family, and is regarded by Friedreich and the majority of other writers as a developmental disorder of the spinal cord. Its principal symptoms are: *a*, disorders of movement, like those of ataxia, in the hands and feet; *b*, ataxic disorders of speech; *c*, deformity of the feet, viz., equino-varus and permanent extension of the great toes; *d*, absence, rarely increase, of the patellar reflex; *e*, absence of disorders of sensation, of the visceral reflexes, of the movements of the pupil, of cerebral disturbances and spastic symptoms. A characteristic ocular symptom is nystagmus or rather nystagmus-like twitchings which are almost always present. According to Charcot their significance is decisive in the differential diagnosis from the tabes of young people. These twitchings (I would prefer to call them motor nystagmus) are irregular movements of the eyes when an object is fixed or followed with the eyes. Choreic movements of the head may develop at the same time. Mendel (*Berl. kl. Woch.*, Nov. 24th, 1890, cases 1 and 3) could produce the symptom by revolving the patient three or four times on his axis, although it was absent during rest; this does not always succeed (Case 2).

Other eye symptoms are rare. Joffroy (*Gaz. Hebd.*, 1888, No. 10) observed slight ptosis and temporary diplopia, and a similar case is reported by Ormerod (*Brit. Med. Journ.*, 1885, I, p. 435). Bernabei (*Riform med.*, May, 1888) observed optic neuritis; Wharton Sinkler (*Med. News*, July 4th, 1885), beginning atrophy of the optic nerve and color disturbance; Mendel (*Berl. kl. Woch.*, Oct. 10th, 1887), bilateral paresis of the abducens in a patient aged four and a half years.

Titubation is not increased in this disease by closing the eyes.

From these statements it is evident that we have to deal with a motor ataxia, *i.e.*, the motor involuntary and voluntary innervation is enfeebled, while the centripetal sensory condition is disturbed very slightly or not at all. The nystagmus (probably also the speech affection) of multiple sclerosis develops in a similar manner. Hence voluntary motor innervation cannot aid the involuntary innervation, as it does in ordinary tabes. In the latter the voluntary innervation under the control of sight partly supplies the imperfect involuntary and voluntary innervation after sensory impressions and muscular sense; if the former is removed by closing the eyes, the ataxia will be increased. In Friedreich's disease the centripetal tracts are, in the main, unimpaired, but the centrifugal tracts are not sufficiently innervated, probably on account of insufficient development of central fibres. The muscle nuclei (anterior horns of the cord) and peripheral nerves exhibit no qualitative anatomical changes. Paralyses of the ocular muscles are rare in the different forms of progressive muscular paralyses apart from the oculo-pupillary symptoms in affections of the upper dorsal and lower cervical cord. Occasionally the abducens or motor oculi nerve is attacked. For example, Zacher (*Neur. Centralbl.*, 1886, No. 23) observed inequality of the pupils and right convergent strabismus (probably cured abducens paralysis) in amyotrophic lateral sclerosis. The bulbar and cerebral nerves were also attacked finally in a case of anterior chronic poliomyelitis (Nonne, *ibid.*, 1891, p. 439).

According to Erb (*Deutsch. Zeitschr. f. Nerv.*, Bd. I), the progressive muscular dystrophies in their different forms (juvenile pseudo-hypertrophic, infantile and hereditary) are to be regarded as a clinical unity, and differ only in regard to the earlier or later beginning of the disease and the more marked implications of the upper or lower part of the body. Early implication of the face is characteristic of Duchenne's infantile form (Landouzy-Déjérine type). The mimetic muscles (ocular facial) are almost always affected, very rarely the ocular muscles proper. Oppenheim (*Charité-Annal.*, XIII, p. 38) has observed nystagmus in a case of juvenile muscular dystrophy.

Moebius' "infantile disappearance of the eye muscle nuclei" (*Muench. med. Woch.*, 1891, Nos. 3 and 4) may be regarded as a peculiar form of atrophy which has become stationary. According to Moebius, frequent implication of the facial and constant escape of the internal ocular muscles are characteristic features.

In acute ascending (Landry's) paralysis—which is only in part an acute myelitis and may be due to multiple neuritis, and even furnish negative anatomical findings—Schwarz (*Zeitschr. f. kl. Med.*, XIV, p. 293) observed secondary contraction in abducens paralysis (he calls it convergent strabismus), and Hoffmann (*Arch. f. Psych. u. Nerv.*, XV, 1) noted slight ptosis. Hun and Pellegrino also reported diplopia in one case.

Achard and Gunion (*Jahr. f. Aug.*, 1889, p. 546) observed blindness within six days, sixteen days later ascending paralysis (acute diffuse myelitis), and death in five months. Three large sclerotic foci were found; first, in the optic nerve and tract; second, in the cervical cord; and third, in the dorsal cord, in addition to numerous secondary degenerations.

In acute myelitis Knapp (*Jahr. f. Aug.*, 1885, p. 273) observed bilateral ophthalmoplegia and choked neuritis; Dreschfeld (*Lancet*, Jan. 7th, 1882), double neuritis in two cases; Noyes (*Arch. f. Aug.*, X, 3, p. 331), also double neuritis with varying vision; Chauvel (*Prog. Méd.*, 1880, No. 32), optic neuritis with considerable hyperæmia, first on the right side, then on the left.

In transverse myelitis Rumpf (*Deutsch. med. Woch.*, 1881, No.

32) found diminished vision and a condition "midway between neuritis and choked disc;" the faradic brush was attended with surprising therapeutic effects. Steffan (Ber. d. Heidelberg ophth. Ges., 1879, p. 90) observed temporary amaurosis, first on the left side, then on the right, with slight descending neuritis followed by permanent hemianopsia, as forerunners of an acute transverse myelitis.

Eye symptoms have often been reported in caries of the spine and its sequelæ. Among 38 cases in children, Bull (*Jahr. f. Aug.*, 1875, p. 343) saw 40 cases of neuritis, 32 of very noticeable hyperæmia of the fundus, and 2 of marked anæmia of the papilla (the children likewise were probably anæmic). In 36 cases the pupils were large and very sluggish, indicating spinal irritation (spastic spinal mydriasis). Really positive ophthalmoscopic findings do not appear to be frequent. Abadie (*Ann. d'Ocul.*, 1876, p. 85) observed double optic atrophy in a case of Pott's disease, probably an accidental complication.

The eye symptoms which are often seen in syphilitic affections of the cord are probably complications due to a similar affection within the cranium.

In tumors of the spinal cord choked disc is found in the same form as in brain tumors but in a smaller percentage of cases. Hirt (*Berl. kl. Woch.*, 1887, No. 3) found rigidity of the pupils, ataxia, left-sided ptosis and abducens paralysis in an individual who had fifteen to twenty cysticerci in the upper part of the cord, beneath the pia mater.

In the spinal affections which are attended with paralysis, the same thing happens as in diseases of the nuclear region of the ocular muscles. Primary disease of the nucleus, nerve roots and basal cerebral nerves produces the same symptoms as the paralysis attending degeneration of the nerve, nucleus and muscle (poliomyelitis superior). It is only after considering all the other circumstances, especially the anatomical relations of adjacent parts, that a definite diagnosis can be made of disease of the nucleus or nerve roots, or of multiple neuritis, and even then the diagnosis may be unexpectedly corrected at the autopsy.

It may also be difficult to decide, in spinal-cord cases, whether we have to deal with multiple neuritis or with primary disease of the motor nerve nuclei (anterior horns). Even the possibility of a primary disease of the muscles must be considered. If the motor fibres are destroyed above the nuclei (anterior horns) but the cells of the latter are intact, the spontaneous voluntary movements will be abolished and the involuntary reflexes retained or exaggerated. The disorder of movement is unchanged by opening or closing the eyes, as the motor impulse from the cortex is no longer conveyed to the cells of origin of the motor nerves.

If the centripetal spinal paths are destroyed or no longer capable of conduction while the centrifugal tracts are intact, the voluntary movements under the control of the higher senses (sight) may replace in part the automatic movements. If this control is lost, the disturbance of movement becomes much more pronounced. In such cases, also, the differential diagnosis between primary disease of the posterior columns and multiple neuritis of the nerve roots (tabes and pseudo-tabes) may be difficult. The ophthalmoscopic findings (gray atrophy or axial neuritis of the optic nerve) may prove decisive.

Injuries of the Spinal Cord.

Wharton Jones states that affections of the optic nerve are frequent after injuries to the spinal cord. Among seventeen rapidly fatal cases Allbutt (*Lancet*, 1870, I, p. 76) saw no changes in the fundus; among thirteen chronic cases eight showed varying degrees of hyperæmia of the papilla; this appeared so much earlier, the higher the site of injury to the cord. Vision was impaired very little and despite the long duration there was much more tendency to recovery than to a transition into atrophy. Hence it was due probably to a vasomotor paralysis. Almost every extensive injury to the cord at once produces vasomotor paralysis and rise of temperature upon the side of the motor paralysis; this may terminate in trophic disturbances. Allbutt never found true neuritis, but Firth (*Practitioner*, May, 1886) claims to have seen double optic neuritis in an injury to the lower cervical and upper dorsal spine, with temporary paresis of the right arm. Fowler (*Journ. f. Ophth.*, Jan., 1891) observed the gradual development of double optic-nerve atrophy after an injury to the spine, followed first by paralysis, later by weakness of both legs. This was possibly a mere coincidence.

In concussion of the spine (railway spine) eye symptoms are quite frequent. In part, they are unconnected with the cord (retinal hemorrhages, atrophy of the optic nerve, etc.); in part they belong to the so-called " traumatic neurosis;" in part they are due to irritable weakness of the sympathetic system, especially of its vasomotor functions. Schmaus (*Virch. Arch.*, 122, 3, p. 487) has found notable anatomical changes in spinal concussion, such as necrosis of the nervous elements with or without destruction of the neuroglia, and later, columnar degenerations, softenings, formation of cavities, etc. In addition larger and smaller blood-vessels may be torn, so that the term concussion is insufficient in many cases.

Injuries of the lower cervical and upper dorsal cord produce corresponding oculo-pupillary and vasomotor symptoms of irritation or paralysis. The latter are more frequent and may develop from preceding irritative symptoms. These phenomena may be important in making an accurate local diagnosis, especially in stab wounds.

Trophoneuroses.

The trophoneuroses form a transition between the central and peripheral diseases of the nervous system. They are often due to peripheral diseases, but may also be central in origin. The latter is undoubtedly true, for example, of syringomyelia (formation of cavities in the cord) in which, apart from trophic disorders, spastic muscular paralyses, general vasomotor and secretory disturbances and the characteristic partial disturbances of sensation, there are also sympathetic (vasomotor and oculo-pupillary) ocular disorders, such as increased or diminished secretion of tears, one-sided sweating (rarely abolition of sweating) of the head. The presence or absence of oculo-pupillary symptoms may possess a local diagnostic value. Syringomyelia sometimes exhibits bulbar symptoms, such as abducens paralysis and nystagmus, and in the later stages optic neuritis and impairment of vision. Concentric narrowing of the field of vision has also been noticed. In seven cases Déjérine and Tuiland (*Semaine Méd.*, 1890, No. 30) found considerable narrowing of the visual field, especially for green (up to 10°); the ophthalmoscopic appearances were normal. They believed that hysteria could be positively excluded in these cases.

Morvan's disease—symmetrical panarities (paronychia) terminating in necrosis—appears to be merely a variety of syringomyelia.

In the other trophoneuroses the symptoms on the part of the sympathetic or the sensory nerves may predominate. There may or may not be definite anomalies or diseases of certain organs, such as the thymus, thyroid and pineal glands, although we are unable to offer any explanation of the supposed connection.

In acromegaly or hypertrophy of the terminal parts of the body (Pierre Marie's disease), the eyelids are occasionally affected, in addition to the fingers, toes, nose, ears, lips, cheeks, tongue, chin, penis, clitoris, uvula, xiphoid process, nipples, etc. The affection occurs in various combinations but is almost always symmetrical. It consists essentially of a thickening, and to a less extent of elongation, of the affected bones and soft parts. This distinguishes it from partial giant growth, which is usually not so symmetrical and is generally noticeable at an earlier age. Acromegaly does not begin until the age of thirty or forty years, after general growth has ceased. Atrophy of the thyroid and hypertrophy of the hypophysis cerebri are often mentioned; persistence of the thymus gland is frequent but not constant. There are often pains and other nervous disturbances in the growing parts, in which peripheral neuritis and disease of the vessels have been found.

In this disease Surmont (*Centr. f. Nerv. u. Psych.*, 1891, p. 28) observed double choked disc terminating in blindness. Minkowsky (*Jahr. f. Aug.*, 1887, p. 305) found slight exophthalmus on both sides and considerable disturbance of vision with defects in the field; the ophthalmoscopic appearances were normal. Minkowsky believes that these symptoms were due to pressure of the enlarged hypophysis on the chiasm. Morax (*Arch. de Neurol.*, XVII, p. 436) saw optic neuritis in acromegaly. He also mentions a number of hysterical symptoms (concentric narrowing of the field of vision, monocular polyopia, micropsia, macropsia, etc.).

Maisonneuve (*Prog. Méd.*, May 16th, 1891, p. 413) observed double exophthalmus, but the elongated and thickened lids still covered the eyes. The eyes were situated nearly in front of the orbit and could be easily luxated in front of the lids. In addition the pupils, which were of normal size, did not react to light, while the reaction to accommodation and convergence was intact. The fundus was hyperæmic; central vision was slightly impaired, the visual field and color sense were normal.

It is evident that the enlarged hypophysis cerebri may occasionally produce all the eye symptoms of a non-infectious brain tumor and local symptoms corresponding to its situation, *i.e.*, visual disorders which point to an affection of the optic tract and chiasm. This occured in Bignami's and Bury's cases (*Neurol. Central.*, 1892, p. 20).

A condition similar to acromegaly, but more diffuse, less confined to projecting parts and unilateral, is found in the very rare affection known as hemi-facial hypertrophy. The eye, orbit and lids may also take part in this hypertrophy. Hankel (*Berl. kl. Woch.*, 1884, No. 34) observed a total staphyloma corneæ. But as the lids could not be closed this would naturally result from an infectious keratitis and have no relation to the general hypertrophy. Ziehl (*Virch. Arch.*, 1891, p. 92) found in right-sided facial hypertrophy marked myopia and convergent strabismus on the same side. I have observed a similar case: extreme myopia and extensive choroidal changes in the enlarged eye on the side of the facial hypertrophy. Shieck (*Berl. kl. Woch.*, 1833, No. 45) observed hypertrophy of the lower lid alone, while the eye was normal.

Progressive facial hemiatrophy is a much more frequent disease than the above. It begins, as a rule, with the symptoms of unilateral irritation of the cervical sympathetic, which gradually passes into paralysis. Then progressive atrophy develops in all the tissues of one side of the face with the exception, perhaps, of the bones. In addition to vasomotor and oculo-pupillary paralytic symptoms (ptosis, myosis, enophthalmus), we find that the integument of the lids grows constantly thinner, finally cicatricial in appearance, and is sometimes pigmented; the eyelashes and eyebrows become gray and fall out. On account of the cicatricial retraction of the integument of the lids, the eye is not sufficiently covered and is exposed to all sorts of injuries. In other cases the orbital fat atrophies and the eye sinks deep into the orbit (Henschen, *Jahr. f. Aug.*, 1883, p. 319; Kuester, *Berl. kl. Woch.*, 1882, No. 10; Virchow, *ib.*, 1880, Nos. 20 and 36).

Sometimes, however, oculo-pupillary irritative symptoms are found in advanced hemiatrophy. For example, Seeligmueller (*Deutsch. Arch. f. kl. Med.*, XX, p. 101) found enlargement of the left pupil and the palpebral fissure in left facial hemiatrophy; both symptoms were intensified by pressure on the superior cervical ganglion.

Total or partial paralyses of the trigeminus with corresponding anæsthesia occur with comparative frequency in facial hemiatrophy. They may even lead to neuroparalytic keratitis (Graff, *Jahr. f. Aug.*, 1886, p. 263). Graff's case, however, was complicated with progressive paresis on a syphilitic basis, and began with conjunctival hemorrhages. Peripheral anæsthesia of branches of the trigeminus (usually traumatic or due to inflammatory processes in the vicinity) may also be the starting-point of the disease (Ruhemann, *Centr. f. kl. Med.*, Jan. 5th, 1889).

The beginning of the hemiatrophy may be attributed not infrequently to traumatism, either with injury to the sympathetic (Moebius, *Berl. kl. Woch.*, 1884, No. 15; Epéron-Nicati, *Arch. d'Ophth.*, 1883, pp. 193, 423) or a fall upon the head (Delaware, *Jahr. f. Aug.*, 1880, p. 276). The atrophy sometimes occurs in the course of a single nerve, for example, the left supra-orbital nerve (Karewski, *Berl. kl. Woch.*, 1883, p. 549) or is confined to one temple, with coincident hypertrophy of the upper lid (Estor, *Rev. de Méd.*, 1888, p. 200).

The following symptoms have been occasionally observed: choroiditis and myopia in a girl aged twelve years (Kalt, *Jahr. f. Aug.*, 1889, p. 517); ptosis, divergent strabismus, cataract, impaired mobility of the eye, contracted pupil as the result of an inflammation of the maxilla which was also the cause of the hemiatrophy (Ruhemann, *Central. f. kl. Med.*, 1889, No. 1); small light patches in the fundus of both eyes in a case of left hemiatrophy (Spitzer, *Wien. med. Blaett.*, 1885, No. 1). [These patches were probably a congenital anomaly. I have found the same condition accidentally in a patient who consulted me concerning atropine mydriasis.] Epéron (*l.c.*) saw, upon the side of the hemiatrophy, diminished vision and concentric narrowing of the field, with an atrophic spot in the macular region. Kahler (*Prag. med. Woch.*, 1881, No. 6, 7) describes a hemorrhage into the vitreous; Hirschberg (*vide* Virchow, *l.c.*) observed in the retina the remains of an inflammatory affection and sudden cylindrical swelling of the two veins which pass outward, with rosary-like expansions.

Flashar (*Berl. kl. Woch.*, 1880, No. 31) describes a double neurotic facial atrophy; upon the side which was most affected the reaction of the pupil was lost, and there was divergent strabismus, with amblyopia and atrophy of the optic nerve.

According to Nothnagel (*Neur. Centralbl.*, 1891, p. 320) there are two forms of facial hemiatrophy, one due to lesions of the sympathetic, the other to lesions of the trigeminus; the latter have been demonstrated anatomically. In my opinion both nerves are always affected, although the lesion of one sometimes predominates, sometimes that of the other.

3. DISEASES OF THE NERVES.

Diseases of the peripheral nerves rarely affect the eye, unless they belong to that organ or are adjacent to it.

Multiple Neuritis,

i.e., more or less acute interstitial inflammation of the nerves, giving rise to secondary degeneration and finally atrophy of the nervous elements, occurs occasionally after all forms of infection. It develops after poisoning, especially of a chronic character (arsenic, alcohol, lead, etc.), after acute (typhoid fever, influenza) and chronic infectious diseases (syphilis, general carcinosis, etc.), as the result of constitutional anomalies (diabetes) or organic diseases (cirrhotic kidneys). It may even constitute the principal symptom of an infectious disease, as in beri-beri or kakke. The symptoms vary according as the anterior or posterior roots, the nerve plexuses or the peripheral nerves are chiefly affected.

When the posterior roots of the spinal nerves are most involved, especially if the inflammation is ascending in type, the symptoms may be very similar to those of tabes (pseudo-tabes). As a general thing the course is more rapid than that of true tabes. If it begins acutely with fever and with paræsthesiæ and pains which pass rapidly into anæsthesiæ and motor disorders, a mistake is hardly possible. But if the inflammatory process in the nerve roots runs a very chronic course, a differential diagnosis may be very difficult, unless aided by the etiology or certain eye symptoms.

Any disturbance of vision which may be present depends mainly upon an axial neuritis of the optic nerve, and consists of central scotoma with color disturbance which is so characteristic of toxic amblyopia. This does not occur in tabes, or, at least, very rarely. Other forms of visual disturbance are rare. The ophthalmoscope also shows the findings which are characteristic of toxic amblyopia, viz., grayish-red opacity and obliterated borders of the inner (nasal) half of the optic nerve, with atrophic discoloration of the outer half where the fibres to the macula lutea are located. These findings may also be present without any notable disturbance of vision.

The absence of true spinal symptoms, particularly myosis and reflex rigidity of the pupil, is also characteristic of multiple neuritis. Other paralyses of the ocular muscles may be present if the nerves at the base of the brain are attacked. Nystagmus is rare because it does not follow inflammation of the nerves and must result from complications.

If the anterior roots are mainly attacked, we find stationary and progressive (amyotrophic) muscular paralyses, which are distinguished with difficulty from those due to an affection of the anterior horns of the spinal cord. This category includes acute ascending paralysis (Landry's paralysis), which may occur as a primary affection of the cord or as a primary multiple neuritis. Ocular symptoms are only present accidentally and cannot be utilized in differential diagnosis.

Oculo-pupillary symptoms occur in the inferior type of combined amyotrophic paralysis of the arm (paralysis of the hand, forearm and triceps, with pronounced affection of sensation) if the nerve roots or the spinal cord itself have been implicated. The oculo-pupillary fibres from the cord to the sympathetic pass through the seventh and eighth cervical and first dorsal nerve roots. If these become impermeable, ptosis, myosis and enophthalmus, either separately or in combination, will be present on the same side. If they are the site of spasm or neuralgia, spastic mydriasis is often noticed, with or without dilatation of the palpebral fissure and exophthalmus. If the affection is located peripherally in the brachial plexus, these symptoms will be absent because the plexus proper no longer contains oculo-pupillary fibres.

In the superior type (shoulder-arm paralysis, without notable disturbance of sensation), vasomotor disturbances are often seen in the face, if the nerve roots have been involved. They are absent in pure paralysis of the plexus.

Diseases of the Trigeminus.

a. Inflammations of the first branch of the trigeminus, extending to its terminal ramifications, may give rise to herpes. In addition to neuralgic pains there is more or less disturbance of sensation and even complete anæsthesia. For the sake of convenience the different forms of herpes of the eye will be discussed in the chapter on diseases of the skin. The infectious inflammations of the cornea which often attend herpes may also present the signs of neuro-paralytic keratitis.

In one case of herpes zoster ophthalmicus Daguenet (*Rec. d'Ophth.*, 1877, p. 117) found optic neuritis with swelling, exudation and very sinuous veins, later, atrophy of the optic nerve with $V = \frac{1}{6}$. It is possible that the ophthalmoscope would furnish positive results more often than is generally assumed. But hyperæmia of the fundus must not be regarded as of special importance, because it is due to the diminution of intraocular pressure which is usually present. Paresis and paralysis of accommodation and mydriasis have been observed repeatedly. Uhthoff (*Jahr. f. Aug.*, 1886, p. 459) noticed cessation of the lachrymal secretion in neuritis of the trigeminus.

b. The ocular branches may take part in trigeminal neuralgia, and then redness of the conjunctiva, increased lachrymal secretion, slight swelling of the lids, photophobia, etc., are frequent symptoms. On the other hand the pains (ciliary pains) may radiate into other branches of the trigeminus.

In all neuralgias, but especially in those of the trigeminus, dilatation of the pupil is often found corresponding to the mydriasis after cutaneous irritation. In a case of severe neuralgia of the first and second branches of the nerve, Gerhardt (*Arch. f. kl. Med.*, XVI, 1 and 2) found slight hyperæmia of the retina and an arterial pulse in the eye on that side.

c. A very interesting affection is the so-called neuro-paralytic keratitis which occurs in anæsthesia of the trigeminus. This is a form of progressive infectious corneal ulcer which is peculiar in several respects. Many regard it as a proof of the existence of special trophic fibres in the first branch of the trigeminus.

In anæsthesia of the cornea it is very often found that ulcerative keratitis develops spontaneously, after a longer or shorter period, in the exposed part of the cornea. The ulcer becomes infected, enlarges slowly but constantly, and may lead to complete destruction of the cornea. It may also recover at any stage, particularly when the anæsthesia disappears. The course is peculiarly asthenic; the inflammatory reaction of surrounding parts is absent or abnormally slight, and this is evidently the cause of the peculiar progressive course. Similar affections are observed in herpes corneæ, starting from the ulceration, and occasionally in Basedow's disease. In both cases the cornea is usually insensible. A similar condition is found in the marantic or desiccation keratitis after the profuse diarrhœa of childhood, cholera, etc.

Special trophic fibres in the trigeminus have been very generally assumed in explanation of this peculiar course of neuro-paralytic keratitis. It was found that the inflammation did not develop when the innermost fibres of the trigeminus were spared on section of the nerve (Meissner). According to Gaule the condition does not develop if any fibres escape; he assumes that this is owing to the fact that under such circumstances the sensibility of the cornea is subsequently restored.

Neuro-paralytic keratitis occurs only in peripheral anæsthesias of the trigeminus and also after destruction of the Gasserian ganglion. It does not develop when sensory conduction is interrupted centrally (Magendie), even after the cornea has been entirely insensible for years.

Snellen advanced the first weighty argument against the existence of special trophic nerves. He showed that the disease does not develop when it is possible to protect the cornea against all external irritants and injuries.

Ferrier (*Lancet*, 1888), who observed neuro-paralytic keratitis in anæsthesia dolorosa, regards the former as an irritative condition analogous to the latter. But we are hardly warranted in regarding a purely functional disorder (pain) as co-ordinate with a very material anatomical change.

Gaule (*Centralbl. f. Physiol.*, 1891, p. 15) has recently described visible changes in the corneal corpuscles and circumscribed necroses and foci of proliferation of the epithelial cells, which follow immediately upon division of the trigeminus. He attaches great importance to the consequent disturbance of the nutritive currents in the cornea, giving rise to an insufficient supply of fluid to the cornea and to relative dryness. This is favored still more by the insufficient winking movements. He denies the existence of special trophic fibres, because no centrifugal fibres pass to the cornea. He regards the corneal corpuscles as the terminal organs (preferably organs of origin) of the sensory fibres. After division of a sensory nerve its cells of origin at the periphery are destroyed in the same way that the cells of origin in the muscle nucleus degenerate after division of a motor fibre.

Two points must be kept in mind: 1. The interruption must be situated in the peripheral nerve. 2. The disease does not develop if an external injury to the cornea with loss of substance can be avoided. We may state, accordingly, that neuro-paralytic keratitis is an infectious inflammation resulting from a traumatic loss of substance in the cornea, and which runs a peculiar course on account of the interruption to conduction in the centripetal nerve tracts.

The nutrition of the uninjured cornea is evidently not impaired on account of its insensibility, but under pathological conditions the corneal tissue reacts differently. The absence of vessels is the ground of the difference between the cornea and other tissues. If infection occurs after a loss of substance in a vascular tissue, its products act directly on the vessels and produce the well-known changes which lead to stasis, emigration of white blood globules, etc.

In the cornea, when rendered insensible from peripheral causes, this takes place only to a slight extent, and only becomes more marked when the process approaches its margin (demarcating inflammation). In the normal cornea the stimuli which are conveyed centripetally evidently act upon sympathetic ganglion cells, whence a direct influence is exerted upon the vessels of the adjacent tissues, particularly the conjunctiva. In normal nutrition this influence is less prominent because it takes place mainly from the intraocular tissues (uvea); the vascular reflexes in question are perhaps presided over by the ciliary ganglion. In pathological processes, on the other hand, in which the vascular tract of the conjunctiva is chiefly affected, the case is different. The vascular changes in the conjunctiva which occur in infectious inflammation of the cornea are greatly diminished or abolished when the sensory conduction between the cornea and sympathetic is interrupted.

Moreover, the corneal corpuscles from which the regeneration of the destroyed tissue must take place become necrotic, as Gaule has shown, after division of the trigeminus. Hence, the regeneration is only possible from the cells of the conjunctiva, and as a matter of course takes place very slowly, particularly in complete anæsthesia of the entire cornea. It will take place much more readily if conduction alone is interrupted while the nerve fibres remain intact. In the latter event the corneal corpuscles probably also remain intact.

Whether the centripetal stimuli reach consciousness is immaterial. It is sufficient that they be conveyed to the ganglion cells of the sympathetic. An equally slight influence is exerted when centrifugal sympathetic conduction is interrupted centrally from the sympathetic ganglion, as happens in the oculo-pupillary and vasomotor sympathetic paralyses which are located in the cord, the roots of the spinal nerves, or the sympathetic ganglion. In such paralyses the nutrition may remain unimpaired for years, and the reaction of the tissues to pathological irritants likewise appears to be unchanged. It is evident that the conduction between these "central" sympathetic centres and the terminal ramifications of the sympathetic nerves is interrupted by ganglion cells of a lower order.

The case is different when the sympathetic ganglion cells are
14

diseased but not destroyed. A direct, continuous and anatomical change in the cellular elements of the vessels—with which the endothelium of the corneal corpuscles is co-ordinate—is then produced. In such cases the pathological findings are: dilatation of the vessels, changes in the endothelium together with the tissue changes observed after inflammatory irritants (accumulation of round cells, proliferation of the tissue cells, and formation of cicatricial tissue). If vascular tissues become anæsthetic, the altered reaction to injuries and infected lesions is less striking, but often distinctly noticeable.

These statements harmonize very well with our knowledge concerning the so-called trophoneuroses:—the frequent initial symptoms of irritation of the sympathetic nerves, often terminating in paralysis; the onset with anæsthesiæ and paræsthesiæ; or the predominance of sensory or of sympathetic symptoms in the same disease (unilateral facial hemiatrophy, starting rarely from the trigeminus, more frequently from the sympathetic) ; or the fact that in the same disease (syringomyelia) trophic disorders may be either present or absent (according as the process has involved the intervertebral ganglia or has remained confined to the spinal cord), etc. Such trophic disturbances and abnormal reaction to external (probably also internal) noxious influences are found whenever the sensory (centripetal) conduction to the sympathetic ganglion, the latter itself, or its centrifugal fibres to the tissue cells of the vessels (macular fibres and endothelium) are interfered with. Symmetrical affections will develop when a spinal lesion extends to the nerve roots and to the spinal ganglia of both sides at the same level; for example, symmetrical gangrene in syringomyelia. It is well known that in primary disease of the vessels the developmental and nutritive conditions are abnormal, and the reaction to inflammatory and other irritants is different from that in other localities in which the vessels are healthy. In all these cases there is either an imperfect transmission of sensory stimuli to the centrifugal sympathetic fibres or a disturbance in the latter, including their origin (ganglia) and terminations (walls of the vessels).

If the term trophic nerve fibres is to be used at all, it should be applied to the peripheral ramifications of the vasomotor nerves, and these are absent in the cornea. This is the very reason of its peculiar reaction after peripheral interruption of its centripetal sensory conducting tracts.

It is evident that desiccation of the surface plays no great part in ordinary neuro-paralytic keratitis because the latter is unilateral in the great majority of cases.

Winking occurs uniformly on both sides, even if one eye is insensible. But in the so-called marantic keratitis, in the corneal ulceration of Basedow's disease, the insufficient moistening and superficial desiccation are important.

Collins' case (*Brit. Med. Journ.*, June 23d, 1888) of unilateral cataract upon the side of a complete trigeminal anæsthesia, while the conjunctiva and cornea remained normal, is unique. Perhaps it was a mere coincidence in an hysterical anæsthesia.

Diseases of the Facial Nerve.

In peripheral facial paralyses and spasms, not confined to single branches, the external ocular muscles supplied by the nerve are constantly affected. In central affections the so-called ocular facial (orbicularis palpebrarum and frontalis) is often not involved, in others it alone is affected (page 23).

Guinon (*Rev. de Méd.*, 1887, No. 6) found concentric narrowing of the field of vision in convulsive tic. The former is not due to the latter, but both are probably symptoms of the same nervous affection.

4. FUNCTIONAL NEUROPSYCHOSIS.

This term includes a series of diseases of the nervous system in which material anatomical findings are absent, or, if present (as in chorea), act only as an indirect cause by remote action upon the nerve fibres and cells. The findings in the eye have an unmistakable similarity in all these diseases. We will devote our attention mainly to concentric narrowing of the field of vision with or without color disturbance and with or without disturbance of central vision. This is evidently a peripheral disorder of conduction in the optic nerve.

Hysteria.

As a matter of course eye symptoms are not wanting in the varied clinical history of hysteria. They include disturbances of sight, paralyses and spasms, disorders of sensation and secretion, peculiar associated sensations (*audition colorée*) and the like.

a. Disturbances of sight; bilateral or, much more often, unilateral amaurosis. The reaction of the pupil to light may be retained or lost; its size often changes without known cause. The ophthalmoscopic appearances are normal. In unilateral blindness with intact reaction of the pupil to light we may be led to suspect simulation, especially if simulation tests show that the apparently blind eye possesses normal vision when the patient thinks he is seeing with the good eye, and also that the good eye is blind when he believes that he is using the blind eye (for example, in tests with prisms or the stereoscope). In such cases the presence of other undoubted hysterical symptoms may be very important. Such cases undoubtedly occur, but it may be very difficult to make a differential diagnosis between unilateral hysterical blindness and simulation, even if the individual has not the slightest object in simulation.

Unilateral amblyopia is much more common than complete blindness, but double amblyopia is much rarer. The amblyopia occurs as retinal anæsthesia (page 32), a more or less marked concentric narrowing of the field with or without impairment of central vision and with or without disturbance of the color sense corresponding to the narrowing of the field of vision. This may even proceed to color-blindness and total color-blindness, but then central vision is usually more or less impaired.

In hysterical disorders the boundaries for the different colors may be entirely abnormal; red is recognized farther toward the periphery than yellow, etc., or the curves of the different colors cross one another. A large part of these anomalies is explained by the sudden development of the affection. If, coming suddenly, red, orange yellow and yellowish-green are perceived in the same intensity of shade, it makes no difference which name I apply to any of these colors; if the same disorder of color sense develops slowly and gradually, certain data for differentiation are furnished (viz., optic atrophy). But this explanation does not suffice for all cases. Some resemble the "disorder of the perception of colors" (page 33), and others mock all efforts at explanation.

According to Charcot a characteristic sign of hysteria is the unilateral concentric narrowing of the field of vision with more or less impairment of central vision and of color preception (v. Graefe's anæsthesia retinæ, *vide* Leber, "Handb. d. Aug.," V, p. 980; Foerter's kopiopia hysterica, *ibid.*, VII, p. 89). It is usually only a feature of a general hemianæsthesia of the same side; occasionally this is confined to the face and neck. As a rule it is associated with anæsthesia of the conjunctiva and cornea, but the latter never terminates in neuro-paralytic keratitis. According to Féré, anæsthesia of the conjunctiva is always present, but this statement is probably exaggerated.

If the amblyopia is bilateral it is more pronounced, according to Charcot, on the side of the hemianæsthesia. According to Gilles de la Tourette, hysterical disorders of vision are always bilateral, but are almost always more marked on one side.

The same remarks hold good in regard to examining for simulation in hysterical amblyopia as have been made above concerning hysterical amaurosis. According to Moravesik (*Neurol. Centralbl.*, 1890, p. 230) peripheral stimuli, such as dropping ether upon the arm, etc., enlarge the visual field by a few degrees; joy enlarges, sorrow contracts it, etc. Vision can sometimes be improved by wearing any kind of glasses (if they do not diminish vision for distance), evidently by arousing a certain amount of innervation on the part of the patient; plane glasses, weak prisms, etc., are often sufficient for this purpose.

According to Charcot hemianopsia does not occur in hysteria; hence scintillating scotoma (ophthalmic migraine) is only an accidental symptom; it occurs in the hysterical as often as in the non-hysterical. Landolt has observed hemianopsia, however, and in one of his cases, unilateral amblyopia had been diagnosed by others. His cases were not pure, however, because the ophthalmoscope showed abnormal findings. Glorieux (*Jahr. f. Aug.*, 1887, p. 295) reports a case, in a boy aged sixteen years, of right hemianæsthesia and temporary right hemianopsia. Rosenstein (*Centralbl. f. Aug.*, 1870, p. 351) observed a bitemporal hemianopsia. But such cases are so rare that we are inclined to suspect a complication.

Hallucinations also occur and are very similar to those of alcoholism. They are, correctly speaking, illusions and are due to a false interpretation of objects perceived entoptically.

A rarer hysterical symptom is unilateral or bilateral diplopia or monocular polyopia without corresponding findings. No sufficient explanation has yet been offered, although there is much in support of the assumption of partial or irregular contraction of the ciliary muscle (Charcot and Parinaud). Astigmatism may be produced in the same way. (Borel, *Arch. f. Aug.*, 1886, p. 253). [A case of monocular polyopia in each eye, sometimes passing into tetropia, because the two eyes did not always co-ordinate in their movements, came under my notice. It was due to astigmatism, combined with latent strabismus convergens, in an extremely hysterical woman. By suitable glasses, both cylindric and prismatic, and general treatment the symptoms were removed.—ED.] Photopsia, hemeralopia, unilateral hemianopsia (?), micropsia (paresis of accommodation), macropsia (spasm of accommodation), scotoma, etc., have also been reported, but they are not at all characteristic of hysteria. Finkelstein (*Jahr. f. Aug.*, 1886, p. 292) claims to have seen two cases of paracentral scotoma, similar to that of Hirschberg's alcoholic amblyopia.

Occasional mention is made of colored hearing (*audition colorée*), *i.e.*, certain letters, tones, or words, when heard, produce a coincident sensation of color. A still rarer condition may be called "sounding vision," in which the sight of certain colors produces a certain perception of sound.

Hysterical disorders of sight almost always begin suddenly, may last for any length of time, and usually disappear suddenly.

b. Among the muscular disorders of the eye, spasms are frequent while paralyses are remarkably rare. Duchenne (*Gaz. des Hôp.*, 1875, p. 682) observed only one case of hysterical paralysis of an eye muscle; in addition to other paralyses there was temporary paresis of the right, and later of the left abducens. Double abducens pare-sis is also mentioned in Roeder's case (*Monatsbl. f. Aug.*, Nov., 1891). Moebius denies the occurrence of paralysis of the eye muscles in hysteria, and regards them as traumatic.

Besides paresis of the abducens, uncomplicated ptosis occurs in hysteria. Schaefer finds it (*Arch. f. klin. Med.*, Bd. V) a frequent symptom of hysteria in children.

This restriction of hysterical paralysis to unilateral or bilateral abducens paresis and ptosis furnishes food for thought. It appears much more probable to me—although I have not seen such cases— that the abducens paresis should be called a spasm of convergence. If we may also regard, as is extremely probable, the ptosis as the re-sult of paralysis of the sympathetic (perhaps of spasm of the orbicu-laris) then we would be justified in maintaining that paralysis of single external muscles of the eye does not occur in pure hysteria. Slight paresis of all the external muscles may in reality be a condition of slight spasm; in both cases the mobility in all directions is mod-erately impaired.

Pareses and paralyses seem to occur in the smooth internal muscles of the eye, especially the ciliary muscle, although the diagnosis is not always easy. For example, in marked concentric narrowing of the field with diminution of central vision and spasm of the sympa-thetic, it is not justifiable to conclude forthwith that immobility of the iris to light is due to paralysis of the sphincter pupillæ.

Tonic and clonic spasms of the orbicularis palpebrarum (blepha-rospasm) are often observed; they may or may not be attended with pain. Some of the cases described as ptosis are probably due to spasm of this muscle. Spasm of accommodation often combined with ma-cropsia is also observed.

An occasional symptom is spastic contraction of both internal recti (spasm of convergence). We have already observed that many so-called abducens paralyses may be attributed to the same condition.

Constant twitching of the upper lids, repeated about twice a second, is said to be a very characteristic symptom. It is probably due, like nystagmus, to weakening of the motor cortical innervation to the muscles.

Apart from hystero-epileptic seizures, conjugate deviation is rare (Griffith, *Jahr. f. Aug.*, 1888, p. 416, 2 cases; Forst, *ibid.*, 1884, p. 680).

Variations in the size of the pupils, without known cause, is said to be quite frequent. In Donáth's case (*Neurol. Centr.*, 1892, p. 156) there was periodical hysterical paralysis of the pupils and accommodation, which was cured by hypnotism. It is doubtful whether the case belongs to this category, because atropine was found in the possession of the patient. Rapid changes in the fulness of the vessels of the iris may be the cause of this pupillary symptom.

The membranes of the eye may also be "hysterogenic" and contact with them may provoke attacks. According to Gilles de la Tourette (*Ann. d'Oc.*, Oct., 1891, p. 266) this is true of the conjunctiva, cornea, inferior lachrymal duct and the mucous membrane of the lachrymal sac.

c. Among the disorders of sensation, anæsthesias of the integument of the temples and eyelids, of the conjunctiva and cornea, are quite frequent, and may be associated with retinal anæsthesia. Contact with the insensible cornea may provoke an abundant secretion of tears (Gilles de la Tourette). It has been maintained that the pupil on the anæsthetic side is larger than on the other, but this is not true of all cases. According to Féré the iris on the anæsthetic side has a darker color, and this does not change on transfert (that is, when by the influence of suggestion abnormal sensations are transferred to the opposite side of the body), but there are also frequent exceptions to this rule.

Paræsthesiæ of all kinds are often observed in the eye. Hyperæsthesia of one side of the face, including the lids, conjunctiva and cornea, and with or without photophobia, is also a frequent symptom. Vision in the hyperæsthetic eye is sometimes increased.

Pains around and within the eyes are very common. Apart from all the forms of migraine, which possess no characteristic features, the most frequent are the so-called ciliary pains, *i.e.*, pains in the eye radiating toward the forehead, temple, teeth, etc., and occurring spontaneously or after attempts at accommodation. The region of the ciliary body is sometimes tender to the touch, as in cyclitis. Although these pains render almost every function of the eye impossible and are extremely intractable to treatment, nothing is found objectively. These "ciliary pains without findings" constitute a part of Foerster's kopiopia hysterica. They are identical with Donders' painful accommodation, Nagel's hyperæsthesia of the ciliary muscle, Schenkl's hysterical eye pain, and Horner's neuralgia bulbi.

d. Among the principal disorders of secretion may be mentioned: epiphora without discoverable cause (frequent), disorders of the sweat secretion, blue sweat (chromhidrosis), particularly of the eyelids, etc. The so-called *œdème bleu des hystériques* is œdema combined with dilatation of the vessels.

These symptoms may occur in every possible combination. They usually begin suddenly and disappear in the same way; they change from the right side to the left, and may last a very short or very long time. When unilateral, they may usually be made to disappear and be transferred to the other side (transfert) by means of various manipulations, such as contact with a magnet, with certain metals, etc.

Eye symptoms, particularly the unilateral concentric narrowing of the field of vision, may be produced by touching certain parts of the skin with vibrating tuning-forks, or, when present, these symptoms may be cured by looking through colored glasses, etc.

While the absence of objective findings is characteristic of all hysterical affections, it is to be noted that occasionally we find trophic disturbances, such as hemorrhages, vesicles on the skin, falling out of the nails and hair, etc. This furnishes a certain relationship to diseases of the sympathetic, in which these symptoms occur more frequently. Atrophic processes in the muscles, which could hardly be attributed to disuse, have also been reported.

Leber has also found objective changes in the optic nerve in the amblyopia with concentric narrowing of the field which, according to Charcot, is characteristic of hysteria. On account of the importance of these findings with regard to the theory of hysteria, I here reproduce Leber's report:

"A woman, æt. 45 years, from Griesinger's clinic. Amblyopia hysterica without ophthalmoscopic changes. R., hardly fingers at 1'; L., M. ⅓, Jaeger No. 5, with difficulty. Concentric narrowing of both visual fields. Slight double abducens paresis (spasm of convergence?). Anæsthesia of the left side of face, at times complete left hemianæsthesia, and weakness of the left side of the body. Death from septicæmia after removal of a small tumor in the left axilla which was falsely regarded as a neuroma. Post-mortem findings negative as regards the primary disease. The optic nerve, chiasm and tractus appeared entirely normal to the naked eye (!). After hardening, transverse section of both optic nerves immediately in front of the chiasm showed a narrow gray band, which did not stain with gold, and consisted of nerve bundles which in part were entirely atrophic, in part contained a few medullated fibres among the non-medullated, atrophic fibres. The other nerve bundles stained uniformly with gold, but on teasing, a few atrophic fibres were found, and also some amyloid corpuscles. Nearer to the eyeball a series of sections showed a somewhat lighter, more yellowish color, starting at the periphery, and this was gradually lost. Hardly any changes were found here with the microscope. In the tractus a moderate number of amyloid corpuscles were imbedded in a very thin, super-ficial layer, composed of fibres running in a somewhat circular direction; in this region the pial sheath was infiltrated quite abundantly with lymph corpuscles, which also filled the perivascular spaces. On account of the high grade of these changes it was difficult to ascertain their extent and mode of development. The impression produced was that of a nutritive disturbance of the superficial bundles starting from a slight perineuritis. This was entirely inadequate to explain the marked amblyopia." (The changes found by Leber evidently exceed the slight a'rophy of the peripheral bundle of the optic nerve which, according to Fuchs, is constantly present).

I regard these positive anatomical findings as extremley impor-tant. While the motor, sensory and secretory symptoms of hysteria permit only a qualified local diagnosis, a positive diagnosis may be made of the location of the visual disorder which, according to Char-cot, is characteristic of hysteria. This disorder is identical with that observed in interference with conduction in the optic nerve, and its frequent unilateral occurrence proves that it is peripheral in origin. In the entire central organ beyond the entrance of the optic nerve into the chiasm there is no locality in which a lesion can produce unilateral disturbance of vision.

As hysterical symptoms have been generally regarded as central, Charcot and his pupils have been led to assume a decussation of the uncrossed optic nerve fibres behind the chiasm and the primary optic ganglia, in order to explain unilateral visual disorders by central causes. Such an assumption is anatomically indefensible. Lanne-grace also assumes an action upon the vessels of the eye, not upon the optic fibres, in the unilateral visual disorders which he produced experimentally by certain injuries of the brain.

Even if the cause of the visual disorder *per se* is central, it acts peripherally upon the organ of sight at the spot close to the chiasm where Leber found the periphery of the optic nerve atrophic and many nerve fibres destitute of medulla. This finding appears to me to be entirely sufficient to explain the visual disorder. The absence of visible ophthalmoscopic changes is explained by the fact that the fibres are not destroyed but that the medulla is wanting in places; in addition the central artery of the retina enters the optic nerve much farther forward. As a result of the absence of the medullary sheath the conduction to the central organs is interfered with or abolished. The condition is similar to that of the degenerated patches in multiple sclerosis. In the latter affection the ophthalmoscopic appearances may also be negative or be entirely disproportionate to the severity of the disorder of vision, even though the sclerotic focus is situated immediately behind the entrance of the nerve into the eye. In hysteria, however, the peripheral bundles of the optic nerve are mainly affected, in multiple sclerosis the axial bundles.

We moreover find symptoms of pressure on the nerve at the point of passage through the optic foramen. The importance of the bony foramina through which nerves pass has long been known. If it happens in the optic nerve, why may not peripheral pressure symptoms in other nerves be produced during their passage through narrow bony canals? Macroscopically these changes might be invisible, but under the microscope they would appear as partial absence of the medullary substance and finally slight atrophy. During life these changes might be much more marked than after death, when the pressure, especially that exercised by the vessels, becomes much less. It is hardly the result of pure accident that those nerves which do not pass through bony canals, viz., the motor nerves of the eye, are so very rarely or never affected by hysterical paralysis.

As each side of the body has its own vasomotor nerves, the possibility of unilateral disorders is thoroughly assured.

1. If we assume that the peripheral nerves at the point where they pass through bony canals are by very slight pressure made incapable of conducting peripheral stimuli to the central nervous system or of conveying motor impulses in the opposite direction, this constitutes a predisposition to hysteria. On account of the pressure the fluid

medullary substance will be pressed away in the unyielding bony canal. On account of the consequent impairment of conduction the transmission of stimuli will be interfered with to a greater or less extent or even abolished. It is also possible that a centripetal stimulus may be conveyed to the primary ganglia but is too weak to reach the cortex (consciousness), for example, in unilateral hysterical amaurosis with normal reaction of the pupil to light. The favorable result of rest cure in hysteria supports, to a certain extent, the theory of an insufficient amount of medullary substance in the nerves, although, as a matter of course, this does not constitute a proof.

Under such circumstances the sudden appearance and disappearance of symptoms, with complete restoration of function, would be easily explained; and the phenomenon of transfert would also be understood more readily. It would be equally plain that, after prolonged duration, nutritive disturbances and changes may develop, particularly in the nerves themselves.

It is well known that the function of the peripheral nerves, as well as that of the fibres of the central nervous system, does not begin until they acquire a medullary sheath. In different nerves and systems of fibres this takes place at different periods of fœtal life or only after birth. The investigation of such conditions has been very important in regard to our knowledge of the anatomy of the central nervous system (Flechsig). It is also well known that when the axis cylinders of the nerves are intact but the medullary sheath is wanting in places (sclerotic patches in the optic nerve, Uhthoff), the function of the nerve may be notably impaired or even abolished, without corresponding ophthalmoscopic findings. This disturbance of function can only be due to imperfect isolation of nerve conduction, and in fact the oils, among which the medullary substance must be included, have been found to be the best insulators for the electrical current.

This assumption of disturbed function by imperfect isolation presupposes conduction in the nerves in two directions, and this really appears to take place. In sensory nerves the centripetal conduction, in motor nerves the centrifugal conduction, takes place in the axis cylinders, the return conduction in the neuroglia. It has been proven anatomically and experimentally that the optic nerve, which really constitutes a part of the brain, contains an approximately equal number of centrifugal and centripetal fibres (*vide* p. 13). If the insulating medullary substance is removed at any part, the same condition is created as in the grounding of a telegraph wire; the current only flows between the point of irritation (or stimulation) and the point of interruption. If the break is not complete, further conduction is always considerably weakened or takes place only when the current is unusually strong, etc. If the break takes place, for example, in the intervertebral foramina, sensory stimuli will no longer be conveyed to the cord, but they may still give rise to the external reflexes in the domain of the sympathetic nerve. Motor impulses will not only fail to reach the muscles but will also exert no influences on the vasomotor nerves.

2. "The predisposition" is sometimes supplemented by slight excitability of the vasomotor system in whole or in part,—for example, the reflex inhibitory influence of the central organs upon the vasomotor nerves may be enfeebled or abolished and the vascular reflexes thus increased. All the conditions are then furnished which are necessary to explain the larger part of hysterical symptoms. We can also understand how hysteria may suddenly develop, if the predisposition is present, after an injury, a stroke of lightning, a fright, etc. It is only necessary that comparatively trifling causes should produce contraction or dilatation in certain of the larger or smaller vessels, to a degree insufficient to interfere notably with nutrition, but sufficiently pronounced to interfere with or abolish conduction in certain nerves.

These vascular changes may affect any part of the nervous system and, as the result of increased or diminished supply, may give rise to abnormal nutrition and to consequent increased or diminished excitability to centripetal stimuli and to increased or diminished centrifugal innervation. The points of predilection are evidently the nerves which pass through canals, where an enlargement of the vessels is apt to give rise to interference with or abolition of conduction. If, in a case of unilateral disorder from dilatation of the vessels (for example, hemianæsthesia), a comparatively slight stimulus (application of a metal or magnet) is employed, this may give rise, on account of the increased vascular reflexes, to contraction of the vessels and thus to restoration of conduction. This may even give rise to dilatation of the vessels at the corresponding part of the opposite side of the body and thus produce the phenomenon of "transfert," but the conditions are not always so simple.

I have purposely spoken of contraction and dilatation of the vessels, not of spasm and paralysis. The latter will produce much more severe symptoms, but as a matter of course it is difficult, and often impossible, to draw a sharp line between them. Moreover, transitions also occur, as, for example, from hysteria to epilepsy. On the other hand, the transition of purely functional hysterical conditions into trophic disorders has also been occasionally observed.

Apart from its conduction to the central organs, every peripheral stimulus exerts a direct influence on the vessels of the irritated part. Each function may be disturbed independently of the other. In addition to other "reflex inhibitory" fibres, the central organs contain some which have a tonic action on the vessels and inhibit the direct vascular reflexes. It is well known that the exclusion of an entire cerebral hemisphere also causes symptoms of unilateral sympathetic paralysis in the face, and particularly ditatation of the vessels. If a sensory stimulus is not conveyed to the central organs, we shall find, in addition to the disturbance of sensation, increased vasomotor reflexes, a very frequent symptom of hysteria.

If a paralysis has lasted for some time, the conviction of the impossibility of performing the movement may become so firm that it can no longer be innervated spontaneously even if conduction has been restored. In such cases the movement would be possible as an involuntary reflex.

It is also conceivable that, despite the existing interference with conduction, very violent sensory impressions would excite vigorous innervation, which would overcome the obstruction and the otherwise impossible movement would be performed.

A further proof of the mainly peripheral character of hysterical paralyses and anæsthesias is the fact that the voluntary as well as the involuntary movements are paralyzed, and that peripheral stimuli do not excite involuntary motor reflexes.

The hysterical paralyses resemble central paralyses from the fact that the interference with conduction takes place very close to the exit of the nerves from the central organ, where the nerve bundles are still arranged in the same way as in the central organ itself. The distribution of the nerve fibres to the individual branches of the nerve and the muscles takes place peripherally. Hence, as in true central paralyses, it is rare to find single muscles paralyzed in hysteria. Usually a group of muscles or an entire limb is paralyzed.

The cortical nature of hysterical disorders is disproven by the fact that in general convulsions (hystero-epilepsy) consciousness is not abolished, as a rule, as it is in true epilepsy.

Such a case, for example, as that reported by Kiepert (unilateral loss of sight and hearing, myosis and dilatation of the retinal veins on the same side) is explained very readily by dilatation of the small vessels on that side. I know of no lesion in any part of the brain which could produce such a combination of symptoms.

Without denying the occurrence of central disorders in hysteria, I believe that the majority of symptoms are better explained by an injurious influence acting upon the peripheral nerves. Who knows how many cases of ovarian tenderness are really due to tenderness of the sacral nerves? It is well known that "ovarian pain" occurs even in men. So long as hysteria was regarded as occurring almost exclusively in females, it was plausible to connect it with the uterus. But now that it has been observed so often in males, and since it has been recognized that it may develop suddenly (traumatic hysteria) from violent emotional excitement, the old term should be abandoned. On account of its brevity, however, the word hysteria will hardly be replaced by a more suitable one, for example by vasomotor neuroparesis multiplex variabilis, which best corresponds with the real condition. In time anatomical changes will probably be found more frequently in purely hysterical symptoms, especially if they have lasted for a long time. These changes should be looked for, not in the central organs, but in the peripheral nerves, especially at their passage through narrow canals.

I have really dilated upon hysteria to a greater extent than accords with my purpose in this work. But I desired to show that the

logical consideration of a single symptom—here the hysterical disorder of vision—may lead to the most important deductions concerning
the nature of a hitherto obscure disease. In hysteria, accordingly,
we have to deal with abnormal processes of innervation in the sympathetic, on the one hand with insufficient unconscious action upon
the latter by the central organs, associated with voluntary motor impulses (page 120), on the other hand with exaggerated vasomotor reflexes. The fact that hysterical individuals lose less blood, when
cupped, than others, I explain by increased vascular reflexes after
cutaneous irritation, and not, as does Gilles de la Tourette, by permanent vascular spasm.

Traumatic Neurosis.

So-called traumatic neurosis is regarded by some as a special form
of disease, but its separate existence is denied by others. Its symptoms consist of the mechanical results of the injury (for example,
paralyses of the ocular muscles in fractures of the base of the skull,
unilateral and bilateral disturbances of vision and blindness with
subsequent atrophy of the optic nerve in fissures through the optic
foramen, etc.) and of symptoms which are regarded as characteristic
of hysteria or neurasthenia. Many observers, accordingly, speak of
traumatic hysteria, neurasthenia, etc., and this really corresponds
better to the actual conditions. The French writers also deny the
existence of a special traumatic "neurosis."

Among the symptoms frequently mentioned are: general tremor,
even when the patient is not watched, intensified action of the heart,
insomnia, increase and inequality of the tendon reflexes, muscular
wasting, twitching of muscular fibres, vasomotor disturbances of all
kinds, etc. Leaving out gross changes and solutions of continuity,
these symptoms undoubtedly result, in part, from material changes
in the spinal cord (capillary hemorrhages, corkscrew-shaped, *i.e.*,
torn axis cylinders, etc.). The symptoms either terminate in
recovery or in progressive atrophic processes (so-called railway
spine).

According to Oppenheim ("Die traumatischen Neurosen," Berlin, 1889), unilateral and bilateral concentric narrowing of the field

of vision with normal ophthalmoscopic appearances is the most important symptom, and this is confirmed by Benedict, Moebius, and Jolly. Schultze, Hitzig, and Rumpf oppose these views, and Mendel remains sceptical. Although the concentric narrowing of the field of vision is simulated more or less frequently, the investigations of Wilbrand, Uhthoff, Nieden, and others prove that it does occur after traumatism (*vide* Bruns, *Schmidt's Jahrb.*, 1891, p. 81; *Neurol. Centr.*, 1892, p. 118). According to Bernhard (*Deutsch. med. Woch.*, 1888, No. 13), concentric narrowing of the field may even be the sole symptom of a traumatic neurosis, but this is denied by others. In one case Fischer (*Arch. f. Aug.*, XXIV, 2, p. 171) found exaggerated reaction of the pupil to light (dilator paralysis?) together with pronounced concentric narrowing of the field of vision. [An instance of this kind was under my notice for about six weeks. The fields were about 20° in diameter and the optic nerves excessively hyperæmic.—ED.]

In the majority of cases it is evident that the traumatism causes an outbreak of hysteria in a predisposed individual. The main visual symptoms of this traumatic hysteria are concentric narrowing of the field of vision, blepharospasm (especially in injuries near the eye) and, in addition, disorders of the cutaneous sensibility.

The trauma need not be very severe or act upon the entire body. For example, Lasègue (*Jahr. f. Aug.*, 1878, p. 254) reports the case of a girl, aged fourteen years, in whom the outbreak of hysteria occurred after a handful of sand had been thrown against her right eye. It began with a blepharospasm which lasted four months, and was followed by other symptoms.

Indeed, a violent fright is sufficient to produce the disease. Falkenstein (*Deutsch. med. Woch.*, 1891, p. 1096) reports the case of a boy of thirteen years in whom unilateral amblyopia and concentric narrowing of the field of vision developed after great fright at being threatened with a bloody knife.

It is not justifiable to diagnose traumatic hysteria after an injury from mere concentric narrowing of the field of vision. Neither does it follow that total blindness will always result from the frequent fissures through the optic foramen. A slight hemorrhage at this spot

15

may compress the optic nerve temporarily and later may be completely absorbed.

In addition, we must always remember that the symptoms which are regarded as characteristic of traumatic neurosis may be occasioned by other causes (for example, muscular tremor and accelerated action of the heart by drinking strong black coffee, smoking strong cigars, etc.). Rumpf's symptom, viz., acceleration of the pulse on pressure upon painful spots, can only decide the question when pains are really present or are simulated.

Hypnosis and Sleep.

All the symptoms of hysteria may be produced by suggestion, and intentional hypnosis may produce hysterical attacks. These circumstances render a relationship between these conditions probable, although they are entirely distinct.

Careful observation of the eye symptoms also throws a certain amount of light upon the hypnotic state.

Heidenhain mentions spasm of accommodation (especially on fixing shining objects) as the first sign of beginning hypnosis; after a while the pupil dilates but reacts vigorously to light. Foerster found the fundus normal at this stage. According to Luys and Bacchi, redness of the papilla and increased fulness of the retinal vessels (vasomotor sympathetic paresis) are present in later stages.

Opinions differ in regard to the conditions of the pupil in more advanced stages, for example, in catalepsy. According to Tamburini and Sepilli the pupil is constantly dilated and does not react to light; according to Féré it dilates and contracts during cataleptic sleep, according as the individual is directed to look at distant or near objects. In "waxen flexibility" Rumpf found wide separation of the lids, mydriasis and moderate exophthalmus, i.e., irritation of the sympathetic. During catalepsy Struebing found the eyes rotated upward, the pupils dilated and sluggish, etc.

It is an interesting fact that Berger obtained dilatation of the pupil after strong sensory irritation even in the most profound hypnotic sleep. Hence the sensory-sympathetic reflex arc (vide p. 120) is not disturbed, and conduction outside of the cranium and spinal canal is intact.

Among the frequently observed hyperæsthesias is increased visual power.

Among suggested symptoms the most interesting one to us is blindness. The pupils are then moderately dilated and the reaction to light is almost abolished. Blindness with abolition of the reaction of the pupil to light is undoubtedly a peripheral symptom. The causal condition must be located on the peripheral side of the chiasm. We are at once reminded of the optic foramen, and are inclined to place the visual disturbance, as regards its mode of development, parallel to that which occurs spontaneously in hysteria.

In every spontaneous activity of the will there is also an unconscious cerebral action upon the vessels of the nerves of the limb or organ employed, and this evidently has a tonic vaso-constrictor effect. When the function of an entire cerebral hemisphere is excluded the signs of sympathetic paresis are also present. Function of the hemisphere accordingly exerts a tonic action on the sympathetic. If one organ of special sense is chiefly employed in conscious acts, the tonic action of the brain is exerted chiefly on this organ, and the reverse is true of the other organs of special sense, whose impressions are neglected. For example, if the eyes are mainly employed in attentive reading or writing, a tonic action is exerted upon the vessels of the optic nerve (probably also upon those of the eye as a whole), while the opposite condition obtains, for example, with regard to the vessels of the auditory nerve. If the latter dilate in the bony canal, conduction in the acoustic nerve will be interfered with to a greater or less extent by the more or less complete displacement of the medullary substance. The auditory impressions reach the central organ in a much less intense degree and are consequently perceived by the brain much less distinctly. As a matter of course conduction is not entirely interrupted. As soon as attention is again paid to auditory impressions, the tonic influence of the brain upon the vessels of the acoustic nerve again comes into play, and perhaps less intense visual impressions may be neglected.

The phenomena produced by suggestion in hypnosis develop spon-

taneously in hysteria. In the latter there is evidently an entire or
partial loss of the influence of the central organs on the sympathetic,
which constricts the vessels, and of the vascular reflexes after sensory
stimuli. For this reason hysterical individuals, despite the similarity of the phenomena, are not especially well adapted for suggestion,
although easily put into a condition of purely passive hypnosis passing into catalepsy and somnambulism (*grande hystérie*, Charcot).

The color-blindness of hypnosis corresponds in the main to the
color disturbances observed in hysteria.

If, for example, any one is hypnotized by attentive fixation of an
object, the attention is concentrated upon the organ of vision; the
other impressions of special sense are entirely neglected and finally
hardly enter consciousness. I assume that now, in addition to the
functional activity of the entire visual apparatus, a certain degree of
vascular contraction develops in the latter while the opposite condition obtains in the other sense organs. As a result of this, interference with conduction is produced at the places where the sensory
nerves pass through unyielding bony canals, by the displacement of
the medullary substance. Finally, the sensory impressions are no
longer conveyed to the central organ. If the eyes are now closed,
the brain no longer receives any impressions, and sleep occurs in the
same way that a completely anæsthetic individual, who is also deaf,
falls asleep when the eyes are closed (Struempell).

This leads us to consider the question of the nature of ordinary
sleep. The theories that it is due to exhaustion, to the accumulation
of products of disassimilation, to the necessity of a stage of rest during which new energies are gathered, are no longer tenable. Mauthner believes that sleep results from interference with conduction in
the central gray matter, so that the cortex is separated centripetally
and centrifugally from the periphery. He bases his theory mainly
on the fact that sleep begins with ptosis, which he regards as a nuclear paralysis of the levator. In fact, there is more or less complete
interference during sleep with conduction between the cerebral cortex and the periphery, but, as in hypnosis, this may take place in the
peripheral nerves.

While falling asleep we first notice subjectively that the pulse is much fuller; this is especially noticeable in the vessels of the head on account of the more or less horizontal position. In sensitive individuals the pulsation of the carotids in the petrous portion of the temporal bone may be so annoying that it even interferes with falling asleep. There seems to be a general relaxation in the tonus of the sympathetic fibres to the vessels. The drooping of the lids, which Mauthner interprets as a nuclear paralysis of the levators, is an evidence of the insufficient action of the sympathetic, as is also proven by the coincident contraction of the pupils (sympathetic ptosis and myosis). Pilcz (*Wien. med. Woch.*, 1891, p. 835) also calls attention to the latter phenomenon. On simply closing the lids the pupils dilate, but in sleep they are contracted.

The general dilatation of the vessels can only exercise an effect in closed spaces from which something can be displaced. This will be mainly noticed, in nerves which pass through canals, as circumscribed displacement of the medullary substance. This will be the more marked, the finer the fibres, *i.e.*, in the sensory nerves and nerves of special sense which possess finer fibres than the motor nerves. Hence stimuli which affect the peripheral sense organs will first be conveyed feebly to the brain, finally not at all, and hence no motor impulse will result. This is not disproved by the fact that certain motor functions, for example, respiration, continue during sleep. Inspiration is not due to peripheral stimuli but to the irritation of the excess of carbonic acid in the blood of the medulla oblongata, and this is naturally the same in the sleeping and waking condition.

Awaking occurs either spontaneously or as the result of a vigorous peripheral stimulus (sensory impression from the outside, fulness of the bladder, etc.) which is able to overcome the resistance to conduction, and is converted in the brain into a motor impulse. This causes a contraction of the vessels, by means of which the conduction in the nerves is again restored. The opening of the eyes which occurs on awaking must be regarded as the result of the action of the sympathetic.

Hence, sleep will be mainly prevented, on the one hand, by external irritants, pain, etc., which are continuously conveyed to the brain and do not allow general dilatation of the vessels to occur, and on the other hand by irritative conditions in the cortex, as, for example, at the onset of so many forms of insanity which induce constant innervation impulses and thus cause contraction of the vessels. An obstacle may also be furnished by rigidity of the walls of the vessels, because the effect of the increased blood pressure is notably weakened by the impaired distensibility of the vessel. This probably occurs quite often in old age.

Hypnotic sleep differs from ordinary sleep by the existence of the possibility of "suggestion," *i.e.*, centripetal conduction is evidently not diminished as much as in natural sleep. Sensory impressions are still conveyed to the cortex, but are so enfeebled that they are regarded not as impressions from the external world but as developing in the brain itself. The speech of the operator is mistaken for the individual's own "inward" speech, and is treated as if the thought were a spontaneous one. The fact that any one who has made up his mind to rise at four o'clock in the morning, and awakes at this time without any outside assistance, is as astonishing as if he does the same thing because it was suggested during hypnosis. This is also true of one who resolves to perform a certain action after a definite interval, and does so at the proper time. It is well known that such actions may also be suggested (post-hypnotic effects).

These brief hints do not explain the whole of hypnotism. But it is important to note again that careful study of an eye symptom may be decisive as regards the interpretation of hitherto mysterious conditions. I will make one more suggestion. Attention has already been called to the fact that the nerves of sensation and special sense contain finer fibres than the motor nerves. Hence, in equal dilatation of the vessels, the centripetal paths will be more interfered with than the centrifugal tracts. The motor impulses from the cortex may therefore still be manifested in an approximately normal manner, while the sensory impressions from the outer world are conveyed to the cortex in a very imperfect manner or not at all. Hence they will have little or no effect upon actions except upon those to which the attention has been directed. Such are the words of the operator. On account of the interference with conduction they are very much weakened on reaching consciousness and are then treated as if they had developed in the individual's own brain. Under such assumptions the explanation of the phenomena of suggestion offers no difficulties, and the explanation of somnambulistic conditions is also made easier.

Neurasthenia.

In addition to a few other symptoms such as, for example, fibrillary tremor of the muscles, eye symptoms are among the most important objective signs of neurasthenia. In a large number of cases we find the often-mentioned bilateral concentric narrowing of the field of vision with or without color disturbance and with or without impairment of central vision. At the same time there is rapid exhaustion of the visual apparatus, swimming before the eyes, sometimes cloudy vision,—symptoms which are due in part to the rapid exhaustion of the nervous visual apparatus (anæsthesia retinæ associated with hyperæsthesia), in part to weakness of the muscle of accommodation and of the internal recti during convergence (asthenopia accommodativa and muscularis). According to some writers (Schweigger, Foerster, Wilbrand) there are frequent changes in the size of the field of vision, or the latter narrows in a spiral direction during the examination, in accordance with the increasing exhaustion.

The pupils are said to be often dilated although the reaction to light is intact or even exaggerated. The size of the pupils is also said to change frequently, independently of the entrance of light, of convergence or accommodation. According to Loewenfeld, Pelizæus, Beard and others, the pupils vary in size although no organic disease is present. This symptom is usually a temporary one; the pupil of one eye may always be larger or both pupils may alternate in this regard. But if the inequality of the pupils is constant and lasts a long time, a serious disease of the brain, particularly general paresis, is usually impending. We should then inquire whether paralysis of an ocular muscle has not been noticed, or whether the patient has not suffered from some disease in which such symptoms are common (albuminuria, syphilis, etc.).

A frequent symptom is inability to close the lids entirely, when standing with the legs together (Romberg's experiment). Although the lids cannot be closed completely, fibrillary twitchings appear very quickly in the orbicularis. According to Loewenfeld (*Muench. med. Woch.*, 1891, No. 50) this also occurs on standing with the legs separated and even in the sitting position. Apart from weakness of the orbicularis this symptom is evidently due to spasm of the sympathetic. Voluntary motor innervation is enfeebled, the sympathetic reflexes are increased. The former corresponds to the weakness, the latter to the irritability; both together to "irritable weakness."

The pulsation of the retinal arteries, which was seen occasionally by Raehlmann, is also probably due to the influence of the sympathetic.

From these remarks it is evident that the diagnosis between neurasthenia and mild grades of hysteria may be difficult. According to Loewenfeld concentric narrowing of the field of vision and circumscribed anæsthetic patches of skin, which are so frequent in traumatic hysteria, are not observed in typical traumatic neurasthenia. Due weight must also be attached to the absence of hysterogenous points in pure neurasthenia.

Like all other neurasthenic phenomena the eye symptoms are due either to general weakness and rapid exhaustion of muscular innervation (orbicularis, internal recti, ciliary muscle) or to increased sympathetic reflexes (contraction with or without subsequent relaxation). The latter cause probably explains the imperfect closure of the lids, the pupillary symptoms, and probably the disturbance of vision. The coincident occurrence of both kinds of symptoms, the irritable weakness, is the characteristic feature.

Perhaps the sympathetic symptoms also include the conjunctival hyperæmia and dry catarrh, which is often present. Its characteristic features are: heaviness and impaired mobility of the lids, sensation as if sand or some foreign body were lodged beneath the lids, a feeling as if the lids were lightly glued together. This condition is such a frequent one, however, that it cannot be utilized in making a diagnosis.

Hypochondria.

Like other parts of the body the eye may also be the source of complaints in hypochondria, although, in my experience, this does not happen very often. Hyperæmia of the conjunctiva, beginning presbyopia and cataract are probably the most frequent tangible causes. Muscæ volitantes and, much more rarely, vitreous opacities visible with the ophthalmoscope, may give rise to the fear of threatening blindness.

Thorough examination of the eye is imperative in every case, especially in view of the fact that the complaints concerning eye symptoms usually have some material though often trifling substratum. Even in those individual whose complaints have been declared unfounded by other physicians, I would recommend careful examination of the periphery of the fundus with feeble illumination and in the erect image.

Epilepsy.

Very many ophthalmoscopic examinations have been made in epileptics. At the beginning of the disease, and when the attacks are rare, the appearances are normal, but during its further course venous hyperæmia develops in the fundus, particularly in the retina. It is an interesting fact that Aldridge found the hyperæmia diminish under the use of potassium bromide and iodide, and again increase after their discontinuance. Koestl and Niemetschek regard the venous pulse as a constant symptom of epilepsy, but this is so frequent in healthy individuals that no weight can be attached to it. Those who have examined large numbers of epileptics also mention other ophthalmoscopic findings, such as neuritis, neuro-retinitis and atrophy of the optic nerve. But these conditions are complications which stand in no relation to pure epilepsy.

The results of ophthalmoscopic examinations during the attacks are more interesting. Immediately before an attack, while the patient was complaining of dazzling, Horstmann (*Jahr. f. Aug.*, 1874, p. 427) found considerable hyperæmia of the papilla and dilatation of the retinal veins. Raehlmann (Ber. d. Heidelberg ophth.

Ges., 1887) noticed vigorous pulsation of all the venous trunks on the papilla immediately before the attacks. The latter could be foretold by the ophthalmoscopic examination.

During the attack Koestl and Niemetschek found in one case slight dilatation of the arteries and contraction of the veins. Allbutt found hyperæmia of the fundus three times and anæmia three times during and soon after the seizure.

During the convulsions, in one case, Aldridge found the papilla markedly injected and the arteries dilated; immediately afterward there was striking pallor and narrowing of the arteries. The normal condition was restored with returning consciousness.

Horner (*Jahrb. f. Aug.*, 1874, p. 426) found enormous venous congestion during the height of the convulsive stage. Bovel (*ibid.*, 1877, p. 229) found the papillæ very hyperæmic, not infrequently to an unequal extent, during and after the attacks.

In three cases Tebaldi noticed striking contraction of the vessels immediately after the attacks.

During the status epilepticus, immediately before the beginning of the seizure, I found (Ber. d. Heidelberg. ophth. Ges., 1877) that the arteries were narrow and regained the normal calibre when the attack ceased. In addition, the size of the ophthalmoscopic image varied by fits and starts (spasms of accommodation).

In a similar case Leber (*ibid.*) found the fundus entirely normal.

The condition of the retinal vessels varies, accordingly, during the seizures. They either show no change, or their condition is the same as that which must be assumed to exist in the cerebral vessels (arterial spasm), or the opposite condition obtains, viz., hyperæmi ı during the attack, which slowly subsides and must be regarded as collateral. The latter condition is the most common one.

The hemorrhages into the lids and beneath the conjunctiva, which develop not infrequently during an epileptic convulsion, may sometimes be an important proof of the actual occurrence of a convulsion.

The eye may also be affected in other ways. A visual aura may appear prior to the beginning of the attack. This may consist of an elementary subjective impression of light, such as seeing colors or flames, or it may form a complicated hallucination. According to

Hughlings Jackson all colors may appear, from red to violet. Hilbert (*Arch. f. Aug.*, XV, p. 419) reports intense yellow vision lasting twenty-four hours before the convulsion. The visual hallucinations which appear as auræ may be repeated regularly in the same manner. For example, I observed a young man who, prior to each attack, saw French soldiers who always appeared in the same direction. While on sentry duty during the war he suddenly found himself in the immediate vicinity of a detachment of French soldiers and was compelled to beat a hasty retreat. After his safety was assured he had the first attack, and since then they have always been preceded by the aura described above.

Spasms of the ocular muscles in the form of associated movements of both eyes, rolling movements and conjugate deviation are frequent symptoms of an epileptic attack. I have been able to demonstrate with the ophthalmoscope clonic spasms of accommodation during a convulsion. During the clonic stage Beevor saw rotation of the head toward the right and conjugate deviation of the eyes to the left; later the opposite condition was noted. Nystagmus and variations in the size of the pupils (hippus) are sometimes observed during and after the attack. Simple conjugate deviation of the head and eyes is so frequent that Witkowski (*Arch. f. Psych. u. Nerv.*, IX, 3, p. 443) even regards it as a constant initial symptom of every epileptic attack.

Rudimentary attacks may consist either of part symptoms of a real attack or of symptoms which constitute the aura of real attacks; for example, blue vision with obscuration of sight (Hughlings Jackson), double blindness (Heinemann, Christensen). I have seen unilateral concentric narrowing of the field of vision advancing to complete blindness and lasting several minutes (spasm of the central artery of the retina), and unilateral micropsia (spasm of accommodation); also peripheral ocular symptoms, alternating with convulsive attacks. Nieden (*Jahr. f. Aug.*, 1888, p. 265) regarded sudden obscuration and concentric narrowing of the field in both eyes, with inability to move the eyes upward, as a rudimentary attack, because an epileptic convulsion had occurred a year before. Hughlings Jackson applies the term "epileptiform amaurosis" to attacks of blindness lasting a few minutes. But as the patient suffered from optic neuritis and later became entirely blind, no relations to epilepsy need be inferred.

After the attack, occasionally prior to the attack, concentric narrowing of the field of vision is regularly present on both sides, often with impaired central vision and not infrequently with concentric narrowing of the color boundaries or color-blindness (Westphal, Thomsen, etc.). This anomaly gradually subsides. It is also found after almost all the epileptic equivalents. The symptom may become important in suspected simulation. It is more frequent than the symptom recommended for this purpose by Echeverria· (*Jahr. f. Aug.*, 1881, p. 301), viz., post-epileptic variations in the size of the pupil without visible cause. It is said that this symptom is also observed not infrequently in the intervals between the convulsions.

The condition of the pupils during the attacks varies so greatly, even in the same attack, that no general rules can be formulated. The reaction to light may be intact or lost.

In two cases of epilepsy Féré noticed infrequency of the winking movements (Stellwag's symptom) and retardation of the upper lid on looking downward (v. Graefe's symptom), although there were no other evidences of Basedow's disease.

Siemerling (*Charité-Annal.*, XI, p. 389) calls attention to the frequent occurrence (twenty per cent) of congenital anomalies of the eye in epileptics. He includes astigmatism, pronounced hypermetropia and myopia, nystagmus, abnormal sinuosity of the vessels and poor definition of the borders of the papilla. It is questionable whether all these conditions are congenital.

Lundy observed an epileptic attack immediately after a cataract operation. On the other hand d'Abundo claims to have cured epilepsy by correction of astigmatism, Elliot Colburn and Frothingham by suitable convex glasses, Stevens by the operation for squint, Pechdo, Fumagalli, and Galezowski by enucleation of an injured eye.

All true epileptic symptoms are evidently due to spasmodic conditions in the domain of the sympathetic.

In unilateral (Jacksonian or cortical) epilepsy, Mueller (*Jahr. f. Aug.*, 1891, p. 315) calls attention to the frequent occurrence of oculo-pupillary sympathetic symptoms (ptosis, myosis, enophthalmus) in the eye of the same side.[1] In this affection we can often see very clearly that one part of the cortex is attacked after another. The symptoms observed on the part of the eye are also cortical, such as conjugate deviation, scintillating scotoma, homonymous hemianopsia or defects of the field of vision, etc. These symptoms are easily demonstrated because consciousness is intact.

The term one-sided epilepsy is preferable to cortical epilepsy because, in the ordinary form of the disease, the principal symptoms (convulsions, loss of sensibility and consciousness, conjugate deviation, bilateral visual disorders, optical aura, visual hallucinations) are due to the implication of the cortex, although the causal arterial spasm is perhaps located at the base of the brain. Nor has it been proven that the cause of cortical epilepsy is located in the cerebral cortex itself. The unilateral non-central eye symptoms all take place within the smooth muscular fibres of the eye and its vessels.

In agoraphobia or " fear of places," which is undoubtedly epileptic in character, Nieden (*Deutsch. med. Woch.*, 1891, No. 13) demonstrated considerable concentric narrowing of the field of vision for white and colors during the attack. The field was narrowed to a third of its normal dimensions in all directions. Considerable improvement occurred in a few months under the use of bromides. It is very possible that the nature of agoraphobia really consists of the paroxysmal, pronounced, bilateral narrowing of the field of vision of cortical origin, resulting from bilateral spasm of the corresponding arteries. In bilateral cortical hemianopsia from disease of the vessels a very small central field of vision is often left (*vide* p. 61). In agoraphobia which is due to temporary spasm, not to permanent destruction in the visual centres, the remaining part of the field of vision will be correspondingly greater.

Migraine and Scintillating Scotoma.

Migraine and scintillating scotoma are discussed immediately after epilepsy because the origin of both must be attributed to the

---

[1] It seems to me more probable that this should read : Enlargement of the palpebral fissure and pupil, and exophthalmus, *i.e.*, spasm of the sympathetic, in the eye of the opposite side.

sympathetic system. Moreover, they exhibit undeniable relations to epilepsy. With a certain degree of justice we may regard migraine as a rudimentary epileptic attack, and scintillating scotoma as a rudimentary migraine, inasmuch as every possible transition between them is occasionally observed.

In hemicrania or migraine there are very frequent attacks of unilateral, paroxysmal headache, which may last from a few hours to a day and may terminate in vomiting. Other symptoms on the part of the sympathetic are often seen, especially oculo-pupillary symptoms.

Apart from so-called myopathic migraine we distinguish, primarily, a paralytic and a spastic or tonic form; the former exhibiting signs of paralysis, the latter signs of spasm of the sympathetic. In the paralytic form we find enophthalmus, ptosis and myosis (with retained but correspondingly diminished reaction of the pupil to light), injection of the conjunctiva, epiphora, photophobia, etc. The fundus usually appears normal, although hyperæmia of the retinal vessels has been found in this form.

In spastic or tonic migraine the palpebral fissure and pupil are moderately enlarged, but the latter reacts to light. Distinct exophthalmus is absent or very rare.

The presence of oculo-pupillary symptoms forms an important means of distinguishing migraine from supraorbital neuralgia. They are absent in the latter, but pressure upon the nerve causes pain. Injection of the conjunctiva, epiphora and photophobia may be present in both. Gerhardt claims to have seen retinal congestion and arterial pulsation in neuralgia. Herpes zoster is distinguished from both by the sensory disorders of the skin and later by the eruption.

Among other symptoms occurring in migraine, especially the spastic form, Nicati and Robiolis (*Gaz. des Hôp.*, 1884, No. 27) mention tinnitus aurium and subjective noises, sour smell, salty taste, formication, coldness and other paræsthesiæ, anæsthesia, muscular tremor and spasms, paralyses, loss of memory, insomnia, delirium passing into temporary insanity. In the large majority of cases such symptoms must be attributed to more or less diffuse nutritive disturbances in the cerebral cortex. Temporary aphasia, hemiplegia, conjugate deviation of the eyes and head (Allbutt, Berry), spasm of convergence, etc., are also observed occasionally. There are likewise cortical brain symptoms, although the spasm of convergence may be nuclear in origin.

Scintillating scotoma (migraine ophthalmique of the French writers) is a very frequent phenomenon in migraine. It is an homonymous disturbance of vision, which begins with scintillations, either in the centre or at the periphery, and often advances to complete homonymous hemianopsia. The reactions of the pupil are retained and the ophthalmoscopic appearances are normal. The attack may last from a few minutes to a whole day, and is observed particularly in the spastic form of the disease.

In a much larger number of cases, however, scintillating scotoma occurs independently as a rudimentary attack of migraine, often without any known cause, sometimes from fasting, etc. It may disappear in a few minutes or in an hour without further symptoms, or it may be attended with one-sided pressure or pain in the head. Then it usually lasts half a day and not infrequently terminates in vomiting.

Pure scintillating scotoma is very frequent, especially in those who do much brain work, and the temporary disturbance of vision is absolutely innocuous. It may always appear on the same side or may change from one to the other. A notable disturbance occurs only in those very rare cases in which it develops on both sides at the same time, as happened to my former teacher, Professor Horner, on one occasion. In such cases, however, it may be easily recognized by the characteristic scintillations. [A surgeon of the United States army, æt. thirty-four, who frequently suffered from migraine of a severe type, was entirely relieved by cylindric glasses to correct compound myopic astigmatism and the accompanying asthenopia.—ED.]

Hardly any other explanation is possible than that of spasm of the arteries of one visual sphere, but the cause of such a condition is entirely obscure.

With advancing years migraine and scintillating scotoma become less frequent or disappear, probably on account of the diminished elasticity of the walls of the vessels.

So-called ophthalmoplegic migraine (Charcot) is known in Germany as relapsing or periodical paralysis of the ocular muscles (*vide* p. 70). This disease is improperly included under the heading of migraine because, in the majority of cases, there are positive anatomical lesions, usually basilar neuritis and perineuritis.

Chorea.

This disease is one of the few in which anatomical lesions are present with comparative frequency. Nevertheless, there are great differences of opinion concerning its nature. According to Germain Sée and others it has a rheumatic basis; according to Joffroy it is a cerebro-spinal developmental disease and is unconnected with rheumatism. According to Jackson it is due to capillary emboli in the region of the corpus striatum, and indeed emboli are found not infrequently in this region. But they are also found in other parts of the brain, and even in the spinal cord. Kahler and Pick call attention to the fact that these capillary emboli and hemorrhages are always situated in such a position that they may interfere with the pyramidal tracts. The frequent coincidence of heart disease and chorea is well known.

Two forms of chorea are recognized: *a*, ordinary chorea of children, rare in adults, and *b*, the chronic progressive form in adults, which finally leads to severe mental disturbance. The latter also includes the hereditary type (Huntingdon's chorea). Chorea (post-hemiplegic hemichorea) may also occur after spots of softening in the brain and in the course of progressive cerebral processes. In Bernhardt's case (*Arch. f. Psych. u. Nerv.*, XII, p. 405) right hemiplegia first developed, then hemichorea, hemiathetosis, aphasia, agraphia, left ptosis and sympathetic myosis (on the side of the cerebral disease!) and right hemianopsia.

Despite the fact that twitchings are almost always observed in the face, the ocular muscles are rarely affected. Lifting the eyebrows and rolling the eyes, which increase on excitement and cease during sleep, are the most frequent. Nystagmus is very rare and does not form part of chorea. Mendel (*Arch. f. Psych. u. Nerv.*, XX, 2, p. 602) describes two cases in children of twelve and thirteen years.

They suffered from chorea, nystagmus and atrophy of the optic nerve. It is possible, however, that the symptoms were due to ataxia.

Warner (*Lancet*, 1883, I, p. 273) reports two cases in which the ocular muscles were involved. Bernhardt (*Neurol. Centralbl.*, 1891, p. 377) mentions, in his report on a case of chronic chorea, that both eyes deviated to the inside and could not be easily rotated to the outside (spasm of convergence).

With the ophthalmoscope Arlidge found pallor of the papilla, while Bouchut often noted pronounced hyperæmia. The ophthalmoscopic appearances are usually normal.

An important case is that of Swanzy (Ophth. Hosp. Rep., VIII, p. 181) in which embolism of the central artery of the retina occurred at the same time with the chorea. In Sym's case (*Jahr. f. Aug.*, 1888, p. 569) sudden blindness of the right eye developed during chorea in a child æt. seven years. Ten years later mitral stenosis and atrophy of the right optic nerve were discovered. The blindness was probably due to embolism of the central artery. These two cases demonstrate the development of chorea by multiple emboli, although this is not the sole cause of the disease.

Gould claims to have cured a case of chorea by suitable glasses. According to Stevens, chorea is intimately connected with errors of refraction, but this view is strongly opposed by Bull (*Jahr. f. Aug.*, 1877, p. 366).

Athetosis.

Chorea is allied to athetosis, in which there are usually unilateral, rarely bilateral, tonic twitchings, especially in the limbs. It generally forms a part of certain forms of spastic hemiplegia.

Nothnagel mentions contraction of both superior recti in right-sided athetosis. Greidenberg (*Petersb. med. Woch.*, 1882, No. 23) describes left-sided amblyopia and mydriasis in left-sided athetosis. Goldstein (*Jahr. f. Aug.*, 1878, p. 252) reports permanent nystagmus with apparent movement of objects in athetosis, and in another case of right hemichorea, which terminated in hemiathetosis, there was right hemianopsia with ptosis and myosis. Gairdner (*Lancet*,

16

June, 1877) observed a pale, slightly swollen papilla with normal vision; Bjoernstroem (*Jahr. f. Aug.*, 1877, p. 228) found double amaurosis as the result of neuro-retinitis which terminated in atrophy of the optic nerve. These findings in the eye, which are probably accidental in great part, permit no conclusion in regard to the nature of athetosis.

Tetanus.

Traumatic tetanus produces no eye symptoms apart from the redness of the conjunctiva, prominence of the eyes, etc., which are due to the violent spasms.

In so-called tetanus hydrophobicus, however, a characteristic symptom is facial paralysis, but the so-called ocular facial (frontalis and orbicularis palpebrarum) is not always involved. This form of the disease occurs only after injuries in the distribution of the cerebral nerves. The facial paralysis is usually on the same side as the injury, or, if the latter is in the median line (bridge of the nose), it is usually bilateral.

The eye may constitute the door of entrance of the infection. It is probably not a mere coincidence that the two cases which I found in literature (Schultze, *Neurol. Centralbl.*, 1882, No. 6, and Ramiro-Guedes, *Jahr. f. Aug.*, 1886, p. 565) were due to a blow with a whip. It is well known that tetanus also occurs in horses and with comparative frequency in those who take care of horses. In Ramiro-Guedes' case recovery occurred. Schultze's case terminated fatally, but meningitis did not develop despite the fact that neuritis optica ascendens had extended to the optic foramen.

Robinson (*Lancet*, March 3d, 1883) observed blindness of three days' duration during the treatment of tetanus with calabar bean.

Tetany.

In tetany there are peculiar clonic, usually symmetrical spasms of the voluntary muscles. In mild cases and at the beginning of the disease they are often painless. In more severe cases and in the further course of the disease they are attended with pains, paræsthesiæ, and often with other nervous symptoms. The spasms begin usually in the fingers, although twitchings in the distribution of the ocular facial occur occasionally at the very start. The ocular muscles proper are very rarely affected (Hoffmann, *Deutsch. Arch. f. kl. Med.*, XXXIII, Cases 5 and 12).

Apart from the spasms themselves the main symptoms of this disease are: the production of the spasms by compression of the large arteries (Trousseau's phenomenon), increased mechanical excitability of all the motor nerves, and the increased faradic and galvanic excitability of all the motor nerves, with the exception of the facial.

Pupillary changes in tetany are not very rare, particularly spastic mydriasis (Kunn, *Wien. kl. Woch.*, 1889, p. 234). Kunn observed mydriasis associated with neuro-retinitis and terminating in atrophy of the optic nerve. Kussmaul (*Berl. kl. Woch.*, 1872, 37) observed slight retinitis at the height of an attack of tetany. Jaksch (*Zeitschr. f. kl. Med.*, XVII, Suppl., p. 171) reports double choked disc, but the case was not a typical one. It was complicated by epileptic attacks and was evidently due to an anatomical lesion in the brain. Bouchut (*Gaz. des Hôp.*, 1873, p. 202) found neuritis in one case and hyperæmia of the optic nerve in another.

Tetany may be an independent disease, occurring in small epidemics which usually attack males of the working classes, especially shoemakers, between the ages of sixteen and eighteen years; it is usually benign in character. It may also be a symptomatic affection in certain diseases of the brain and stomach, a sequela of infectious diseases (cholera, typhoid fever, varioloid), or it may follow poisoning with various substances (ergotin, chloroform). The development of tetany after extirpation of goitre is also well known. The above-mentioned complications on the part of the eye were found almost exclusively in the second form, which is a much more serious disease than the first form.

The scanty eye symptoms add nothing to our knowledge of the etiology of this peculiar disease, which is evidently a general symptom due to various causes (insufficient amount of water in the muscles and nerves? Kussmaul).

Thomsen's Disease.

This affection, which is often hereditary, is probably a peripheral disease of the muscles (partial hypertrophy and proliferation of the connective tissue): The groups of muscles which are set in action pass into temporary, painless, spastic rigidity, which subsides after a few seconds. The mechanical excitability of the nerves is unchanged, that of the muscles is increased; pains and sensory disturbances are wanting.

All the muscles or only single groups may be affected, but a preference is shown for the lower limbs. The muscles of the face and eyes are often involved.

Raymond (*Gaz. Méd. de Paris*, June 27th, 1891) published two interesting cases which furnish a good illustration of the occasional implication of the eyes in Thomsen's disease. In the first, the visual disturbances accompanied the muscular spasms of the body or followed them. Exophthalmus was simulated by blepharospasm extending behind the equator of the eye. Another symptom was "Graefe's sign," *i.e.*, on looking downward the upper lid did not follow the eye, so that the sclerotic became visible above the cornea. In addition contractures developed in the external ocular muscles; these kept the globe rigid and could be felt subjectively. They could be produced by sudden noises. Temporary amblyopia passing into complete blindness appeared at the same time. All of these symptoms were not constantly present. At a late stage hypertrophy of the recti interni, as well as of other muscles, developed in connection with insufficiency of their opponents. Spasms of accommodation did not take place.

In the second case eye movements were normal. But when the lids were shut it required some seconds to open them, and a longer time if the closure had been forcible; then the opening had to be done at successive efforts. Graefe's symptom took place. Movements made at command ended in contracture, especially upward; the opposite movement required some seconds. Spasm of accommodation could not be elicited. Pupils reacted normally but soon returned from contraction to relaxation. Occasionally there was diplopia.

Vision, color sense, visual field, refraction and fundus were normal. If the head were moved, temporary amblyopia or total blindness for some minutes would ensue. This would be sometimes attended by phosphenes, which Raymond ascribes to pressure of the extrinsic muscles on the globe or of the muscles of the neck upon the aorta.

The unstriped muscles are not implicated in Thomsen's disease; very seldom and only perhaps by accident do vasomotor disturbances occur. Therefore Graefe's symptom is not to be attributed to lesion of the sympatheticus but to contraction of the levator palpebræ superioris. Spasms of the extrinsic eye muscles as well as of other muscles can be brought about by spontaneous movements or in obedience to commands of another. The intrinsic eye muscles, the nonstriped, escape, *i.e.*, sphincter pupillæ and musculus ciliaris. In only one case (Engel) was a sluggish pupil noted.

In myoclony (paramyoclonus multiplex), which some set apart from hysteria as a peculiar convulsive type, there are lightning-like irregular contractions on both sides of the body. Others regard the condition as a type of hysteria—and Unverricht ("Myoclony," Vienna, 1891) says that sometimes all other muscles except the ocular may be concerned. This circumstance is strongly in favor of a coincidence with, or relationship to, hysteria in which the same symptoms occur.

CHAPTER II.

DISEASES OF THE SKIN.

THE relationship of skin diseases to affections of the eye is comparatively simple. We have to deal either with direct extension from the skin to the eye and *vice versa*, or with a coincident or successive appearance. The identity of diseases of the skin with those of the lids is easily demonstrable. This is more difficult as regards affections of the conjunctiva and cornea.

The external integument of the lids is distinguished from that of the remainder of the body by its great tenuity; the very loose subcutaneous cellular tissue is prone to œdematous swelling and is absolutely free from adipose cells. Even in general obesity the lids remain free from fat and often exhibit a striking contrast to the remainder of the integument of the face. A similar condition is presented only by the integument of the scrotum.

The surface of the lid is covered with fine downy hairs. Into the so-called intermarginal portion of the lid empty the so-called Meibomian glands, large acinous glands which are partly imbedded in the palpebral cartilage (tarsus). In front of their openings stand the multiple rows of eyelashes, into whose follicles unusually large sebaceous glands empty their secretion. In addition the free surface of the lids possesses large sweat glands.

All these glandular structures—the Meibomian, sebaceous, and sweat glands—take part in the corresponding diseases of the skin, and attention must again be drawn to the fact that these parts are larger than in other parts of the body.

The course of diseases of the conjunctiva and cornea often differs greatly from that of analogous affections of the skin, especially in those diseases which are attended with an eruption of vesicles. Very rarely do we find true vesicles and only for a brief period. The soft and thin epithelium soon desquamates, and we usually find only its

246

loss or an ulcer which discloses its vesicular origin merely by its rounded shape. If this evidence is wanting, the diagnosis may be very difficult (pemphigus conjunctivæ). The diseases observed during acute exanthemata will be discussed under the heading of infectious diseases; hard and soft chancre will be considered, for the sake of convenience, under the head of syphilis.

Extension by continuity from the integument of the lids and surrounding parts to the eye, or *vice versa*, is found in cancroid and in lupus, whose various forms, such as lupus erythematodes, hypertrophicus, exfoliativus, etc., occur in the face and upon the lids, where they may give rise to extensive destruction, ectropium, adhesion of the lids to the globe, secondary disease of the conjunctiva and cornea, and finally to destruction of the eye. Inasmuch as the integument of the face is affected in a characteristic manner over a large area, there is no difficulty in diagnosis except in the rare cases in which lupus first appears upon the conjunctiva.

We then notice very chronic ulcers with jagged edges, often filled with proliferating granulations, and whose neighborhood is more or less infiltrated; the corresponding lymphatic glands are also swollen. On careful examination it is often possible to find at the periphery and base of the ulcers small, whitish-yellow infiltrations as large as the head of a pin or smaller, whose presence is characteristic and corresponds to the individual lupus nodules.

As is well known, the opinion is steadily gaining ground that lupus is a tubercular affection, a local tuberculosis of the skin. As the resemblance to tubercular diseases was seen more readily in lupus of the mucous membranes, the cases under consideration were formerly often, and are now constantly, described as tuberculosis of the conjunctiva. The demonstration of tubercle bacilli in the secretion or in the granulations of the base of the ulcer now assures the diagnosis. It is to be noted that in a series of cases the lupus of the conjunctiva was the only tubercular affection, and it is also well known that lupus patients may attain an advanced age.

The eyelids are the favorite site of lupus erythematodes. Its outline is in the shape of a butterfly and leads gradually to considerable shrinking of the skin, perhaps to eversion of all the lids.

Erysipelas is observed very frequently about the eye, inasmuch as facial erysipelas in the majority of cases passes over one or both lids. The disease may also take its origin directly from slight wounds and excoriations of the lids, particularly at the inner angle, more rarely at the outer angle. Manifold variations in the appearance of erysipelas are presented, corresponding to the varying structure of the skin and subcutaneous tissues. The œdema of the lids is usually very pronounced, so that the palpebral fissure cannot be opened spontaneously, and is opened artificially only with difficulty. With the exception of more or less injection and œdema of the conjunctiva the eye is usually normal; there is often catarrh of the conjunctiva, very rarely ulceration of the cornea.

Erysipelas of the lids exhibits a marked tendency to appear in the bullous and vesicular form, even when this is not true of the remainder of the integument. Secondary abscesses of the lids occur not infrequently after the erysipelas has run its course, and more frequently than in other parts of the integument. They vary greatly in size, but they almost always heal without material injury to the eye.

Acute dacryocystitis is sometimes observed when the erysipelas, as frequently happens, passes across the region of the lachrymal sac. I have recently seen a case of this kind in which previously not the slightest sign of an affection of the lachrymal sac had been present. In other cases we merely have to deal with an acute exacerbation of a chronic affection.

In not a few cases the general practitioner mistakes an acute dacryocystitis for erysipelas. In the former the symptoms are concentrated upon the lachrymal region ; the elevated inflammatory wall at the edges and the migratory character are absent. It is also mistaken occasionally (though this is always avoidable) for herpes zoster, which stops abruptly at the median line and is characterized by paræsthesiæ, neuralgic pains, typical herpes eruption and, later, by typical radiating cicatrices.

Although the eye usually remains intact, Coursserant (*Jahrb. f. Aug.*, 1876) has observed bullous keratitis (associated with albuminuria), and Cornwell (*ibid.*, 1882, p. 380) has seen cyclitis in phlegmonous erysipelas. Orbital abscess is observed in rare cases. A more frequent finding is simple infiltration, perhaps a mere serous soaking of the orbital contents, whereby the eye is protruded to a greater or less degree. In such cases optic neuritis or ischæmia from pressure of the orbital infiltration is observed, which terminate in atrophy and usually in complete blindness. More rarely a certain degree of vision remains, with narrowing of the field of vision, scotomata and color disturbances. Pagenstecher, Hutchinson, Schenkl, Parinaud, Nettleship, Carl and others have described such cases. Paresis of the ocular muscles, for example, the levator palpebræ superioris, or partial paralysis of the motor oculi, such as mydriasis (Pagenstecher), may also persist.

Knapp (*Arch. f. Ophth.*, XIV, 3) noted thrombosis of the retinal vessels after erysipelas, probably as the result of the inflammatory infiltration in the orbit. Hutchinson (*Jahrb. f. Aug.*, 1883, p. 299) observed an elephantiasis-like condition of the lids after facial erysipelas. It is to be remembered, however, that elephantiasis Arabum runs its course upon its favorite sites (the external genitalia and lower limbs) in erysipelatoid or true erysipelatous attacks. On the other hand, a more or less doughy or firm œdema not infrequently remains for a long time after erysipelas of the lids. This resolves very slowly, but in the end it almost always disappears without leaving a trace.

Embolism of the central artery of the right eye, which was found by Emrys Jones (*Brit. Med. Journ.*, 1884, I, p. 312) in erysipelas of the left side of the face, was probably accidental, unless the conditions noted were due to compression of the optic nerve. Between them and embolism the resemblance is sometimes striking.

Carré (*Gaz. d'Ophth.*, 1882, 5) found acute inflammation of the lachrymal gland in erysipelas of the upper lid, in a man of thirty-five years. It is probable that this combination occurs more frequently.

Two cases of spontaneous recovery of uveal disease after erysipelas are reported by Nieden (*Central. f. Aug.*, 1885, March); striking improvement of a "trachoma" after erysipelas was observed by Cocci (*Jahr. f. Aug.*, 1884, p. 429), and recovery of an iridochoroiditis by Walb (*Central. f. Aug.*, 1877, June).

If erysipelas is complicated with meningitis the latter, as a matter of course, may give rise to corresponding ocular symptoms (page 150).

Eczematous disease of the integument and eye is observed coincidently or successively with equal frequency, and in the latter event the external integument as well as the eye may be the starting-point.

In severe eczema of the face, scalp and integument of the lids we often find the eye completely free or only slightly reddened; in other cases there is simple catarrh of the conjunctiva (Horner's eczematous catarrh), which is evidently due to the entrance of irritating and decomposed secretion into the conjunctival sac, and which usually disappears spontaneously with the recovery of the facial eczema. But almost purely purulent catarrhal affections of the conjunctiva may also occur, and these require independent treatment.

In many cases an outbreak of eczema occurs at the same time upon the conjunctiva and the cornea. As we have already remarked, real vesicles are rarely seen. The earliest form which I have found showed merely a club-shaped projection of infiltrated epithelium, evidently the former covering of the already ruptured vesicle. This is rapidly desquamated, and we generally find at the start a rounded ulcer which has an injected, thickened and infiltrated base upon the conjunctiva, and is surrounded by a slightly infiltrated zone when situated on the cornea.

In pure cases the more marked congestion of the vessels is confined to the immediate neighborhood of the ulcer. It is only when the latter is infected secondarily from the conjunctival or lachrymal sac, that extensive purulent processes occur. These progress in area and depth and, upon the cornea, are especially apt to bring danger to the eye (progressive ulcers with hypopyum, fascicular keratitis).

The individual ulcers vary in size from that of a pin-point to 4 or 5 mm. in diameter; the latter may deserve the term eczema impetiginodes.

Eczema of the lids, conjunctiva and cornea hardly ever occurs in the new-born, is very frequent during childhood, diminishes after the period of puberty, and is rare at an advanced age. It is then generally observed during convalescence from severe, especially infectious diseases, or in general conditions of weakness, for example, during pregnancy or the puerperal condition. [Eczema of the face may occur in subjects of advanced age having a gouty diathesis, under an acute attack with severe conjunctivitis both palpebral and ocular. It may cause extreme distress both in pain and photophobia, and may be very obstinate. —ED.]

According to my notes, eczematous disease of the nasal mucous membrane is the starting-point in three-fourths of all cases of eczema of the eyes and face, or is present at the same time. This is much more common than primary eczema of the face. The eye itself may also be the starting-point. The overflowing conjunctival secretion macerates the integument of the lids and face, and typical eczematous eruptions make their appearance. Even in many cases of so-called atropine catarrh, the first thing is extensive erythema of the face and lids, which is converted into eczema at the end of a few days.

The latter cases are evidently the result of the spread of an infectious process which produces the typical eruption after maceration and softening of the epidermis. In an entirely analogous manner we find not infrequently that eczema develops in little children upon the dorsal surface of the hands, if they press them for a long time against the spasmodically closed eyelids, on account of severe photophobia. On the other hand it is very doubtful whether the micro-organisms found by Burchardt, Gifford and others are the real cause of the disease. It is probable that they enter the ulcers secondarily from the conjunctival sac, inasmuch as inoculations proved futile. The demonstration is rendered very difficult by the fact that we are unable to examine the unruptured vesicles. We always find ulcers already contaminated by the numerous microbes of the conjunctival sac. It is also possible that several microbes possess the power of producing rounded vesicles and ulcers, while on the other hand it cannot be denied that the cause may reside in purely chemical injuries, for example, in the eczema due to the ingestion of certain articles of food, and which develops like urticaria under similar conditions. It is an interesting fact that I have repeatedly observed a typical eczema ab ingestis after the administration of cod-liver oil. As this is a frequent remedy in eczematous affections, the observation is calculated to lead us to exercise caution in its use.

One of the favorite sites of eczema squamosum of adults is the surface of the lids, where it is often sharply defined.

The propriety of calling these diseases of the conjunctiva and cornea eczematous is undoubted when an evidently eczematous disease can be recognized. If, for example, with every acute exacerbation of a nasal eczema fresh eruptions appear upon the conjunctiva and cornea and then disappear with the subsidence of the primary disease, it is difficult to regard the histologically analogous processes as not identical. Less decisive is the coincidence with so-called scrofulous symptoms in other places, and which will again be referred to in discussing scrofula. On account of the frequency of their co-existence, Arlt applied the term scrofulous conjunctivitis to the affection of the conjunctiva in question. It is remarkable that he did not apply this prefix to the analogous inflammation of the cornea, but used the term scrofulous keratitis for those forms which are known to others as interstitial or parenchymatous keratitis. Eczematous disease of the conjunctiva and cornea is known most frequently under the indifferent term "phlyctenulæ," although vesicles proper are hardly ever seen.

All diseases of the cornea and conjunctiva with roundish infiltrations leading to ulceration need not be regarded as eczematous. For example, very similar infiltrations and ulcerations, passing into vascularized patches, occur in the pannus of follicular blennorrhœa (trachoma). At the head of a growing pterygium, for example, we very often find punctate and larger rounded infiltrations and ulcers, which are probably traumatic in character. In the absence of eczema of the nose or skin, a single rounded ulcer of the conjunctiva surrounded by an injected zone, or a single analogous ulcer of the cornea, whose etiology is unknown, cannot be called eczema, although the possibility that such an assumption is correct cannot be excluded. Perhaps bacteriology may here prove successful, although the conditions in the conjunctival sac are very unfavorable to investigations.

I would suggest that the term eczema be applied to those diseases of the conjunctiva and cornea which are associated with eczema in other parts of the body and whose prognosis and course depend upon the latter. When this is not the case, the neutral term "phlyctenulæ" may be retained.

Richey (*Chicago Med. Journ. and Exam.*, April, 1888) claims to have seen two cases of eczema of the lids from "ciliary tension" in hypermetropia, and to have cured them by correcting the error of refraction. Although this is probably an extremely accidental coincidence, we may recall the fact that a connection between hypermetropia and seborrhœa of the lids and edges of the lids, and also recovery of the latter by the use of proper convex glasses, has been several times maintained with great positiveness.

Herpes zoster and vulgaris also occur upon the integument of the lids. The disease of the cornea which corresponds to herpes zoster (herpes zoster ophthalmicus) has long been known, and its course corresponds more or less to that of so-called neuro-paralytic keratitis (p. 207). The herpetic ulcers are infected from the conjunctival sac, and this is followed by progressive destruction of the more or less insensitive cornea. This may heal at any stage, but not infrequently leads to destruction of the eye.

Herpes vulgaris or febrilis of the cornea is much more common than herpes zoster. To Horner we owe the credit of its discovery. Groups of clear vesicles appear spontaneously or under conditions in which eruptions of herpes appear in other places, and not infrequently at the same time with the latter. After a certain length of time, perhaps a few hours, the epithelial covering ruptures, and this gives rise to very irregular, often branching, losses of substance. These show distinctly, however, that they are formed out of individual circular erosions. At the same time the sensibility of the cornea is moderately diminished.

The disease may pass rapidly, but its course is usually protracted, inasmuch as the base of the erosion becomes infiltrated by infection from the conjunctival sac. The disease often remains stationary for weeks, but finally heals with a very characteristic branching cicatrix. Progressive suppuration (ulcus serpens) gives rise, in rare cases, to loss of the eye.

In recent cases the disease can hardly escape recognition. At
the most it can be mistaken for certain traumatic losses of substance,
but this will be prevented by the previous history. In the later
stages the characteristic appearance is often effaced, especially after
extensive infiltration and vascularization of the cornea. The opacity
which is finally left hardly ever fails of recognition.   •

Herpes vulgaris corneæ has been described under various names,
such as furrow keratitis (Makrocki), keratitis dendritica (Emmert),
keratitis ramiformis (Hansen-Grut), etc.

It seems to me more or less doubtful whether the keratitis described by Stellwag (*Wien. kl. Woch.*, 1889, No. 31), the keratitis
maculosa of Reuss (*ib.*, No. 34), the keratitis puncta superficialis of
Fuchs (*ib.*, No. 44), Adler's keratitis superficialis centralis (*ib.*, No.
37), etc., are identical with herpes, although they appear, at least, to
be allied in character.

Seborrhœa in its different forms is very often observed coincidently on the lids and the scalp (scaly form), or in the face and upon
the nose (oily form, seborrhœa fluida). In the face as well as upon
the lids it often provokes an outbreak of acne. The acne vulgaris of
the lid is usually known as hordeoleum, and is distinguished from
acne of the face by the large size of the individual eruptions, corresponding to the much larger size of the glandular structures of the
lids, the suppuration of whose periglandular connective tissue produces the disease. It attacks chiefly the large sebaceous glands in
the vicinity of the free border of the lids, but the sweat glands of
their surface or, still more rarely, the Meibomian glands may be
affected.

As in facial acne, periods of eruption often alternate with free intervals; this also happens in styes. An outbreak may occur upon all
four lids. Both affections take place very often at the same time,
and particularly at the period of puberty.

The acarus or demodex folliculorum is found in both (*vide* Wecker,
" Traité d'Ophth.," I, p. 78).

When the affection of the lids is extensive, it gives rise to more
violent pain, œdema of the lids and of the conjunctiva. These
symptoms are especially striking when the stye is situated at the

outer angle, on account of the disturbance of circulation and of the outflow of lymph. As in acne, the most effective treatment, apart from the individual eruptions, consists in the cure of the chronic seborrhœa whose secretion gives rise to the acne pustules by occluding the excretory ducts of the glands.

While the connection between hordeolum and acne vulgaris is evident at once, the relationship between chalazion and acne rosacea is less easily understood. They both appear at the same period of life, but are found with comparative infrequency at the same time. Their mutual relationship is decided rather by the histological conditions and the connection of both with outbreaks of acne vulgaris. Both may arise from ordinary acne pustules. Acute exacerbations are often seen, and are usually followed by the growth of little tumors. The anatomical substratum common to both is a granulation-like tissue, often containing giant cells, and which, in view of the frequent recurrence of chalazion treated by operation (insufficiently!) may even lead to a suspicion of a relationship to sarcoma. The interpretation of chalazion as a local tuberculosis (Tangl), on account of the (no doubt accidental) discovery of tubercle bacilli, has been sufficiently disproven.

Among the differential features between chalazion and acne rosacea may be mentioned:

a. In chalazion the individual tumor is usually larger, corresponding to the notable size of the Meibomian glands, which are generally involved. The tumor consequently is situated firmly on the palpebral cartilage, which usually bounds its posterior surface. In acne rosacea, however, tumors are sometimes seen which do not yield in size to the largest chalazion.

b. The contents of the chalazion is often a central drop of pus, a transition to acne vulgaris, whose nodules in later life not infrequently "harden" into chalazia. A conversion into cysts is very rare in acne rosacea.

c. The more frequent solitary development of chalazion, as compared with the multiple eruptions of acne rosacea, corresponds to the smaller number, but more considerable size, of the affected glandular structures in the former. But chalazion is quite often multiple; indeed very many are often present, and the individual nodules are sometimes evidently composed of several smaller ones.

d. The mobility of the integument over the chalazion is explained by the fact that the affected glands are not situated in the cutis and the subcutaneous tissue, as in ordinary acne rosacea, but, at least at the start, within the palpebral cartilage.

On the whole we are justified for the present in regarding chalazion, despite the various possible objections, as the affection of the lids which corresponds to acne rosacea. The existing differences are readily explained by the different local conditions. Even acne rosacea of the nose often differs greatly from the similar affection of the remainder of the face.

The effective treatment of both affections is also analogous, viz., careful removal of the individual granulation foci, because relapses are apt to occur from proliferation of remaining tissue after closure of the opening made by operation. Sulphur treatment, which is often very effective in acne of the face, cannot well be applied in that of the lids. But in several cases the treatment of coincident acne rosacea of the face with sulphur caused a striking subsidence of the previously existing and recurring acute exacerbations of chalazion. No influence upon the completely developed tumor was ever observed. But the occasional spontaneous disappearance of a long-existing chalazion is not a very rare event, although the disappearance is not often absolutely complete.

[The parallel between acne vulgaris and hordeolum and between acne rosacea and chalazion was first drawn by Horner. The two latter are not yet very generally held to be cognate.]

The comparison with affections of the skin has also shed light upon so-called spring catarrh, an interesting disease of the conjunctiva, which has not long been recognized and whose relations were quite obscure.

Saemisch, who first described this affection as a special disease in his handbook, laid the chief stress upon the most striking symptom, viz., the peculiar proliferations at the rim of the cornea. These had been occasionally seen and described before, for example, by Graefe as phlyctæna pallida, but this very term shows that it was falsely regarded as a variety of conjunctival eczema. Horner and his pupil Vetsch (Diss. Zurich, 1879) then called attention to the frequently independent but often coincident change in the conjunctiva palpebrarum and the fornix.

The disease is not a catarrh. With the beginning of the warm season, generally among boys at the period of puberty, the signs of conjunctival hyperæmia make their appearance, viz., heaviness of the lids, a feeling of dryness and as if sand were in the eye, exhaustion, sensitiveness to smoke, dust and artificial illumination, a sensation " as if the eye were not properly lubricated," etc. There is no real secretion to agglutinate the lids on waking or accumulate in the angles of the eyes. Often we find merely a few shreds of mucus in the fornix of the conjunctiva. As a rule, also, there is no trace of follicles, although an accidental coincidence is not excluded.

A characteristic sleepy appearance of the patient, resulting from slight drooping of the upper lid, generally strikes us even at a distance. This feature alone often makes the diagnosis. The conjunctiva palpebrarum and the fornix show, in simple cases, merely a more or less pronounced milky opacity of the epithelium. This is not so marked or so diffuse as in croup of the conjunctiva and cannot be removed with the forceps.

In the majority of cases these are the sole findings. In others we find, especially upon the inner surface of the upper lid, more or less numerous, flattened proliferations, pedunculated like a fungus, varying in size up to that of a pea, and also covered with the cloudy epithelium. They are sometimes very numerous and, pressed closely together, may cover the entire surface of the lid. In about one-third of the cases the characteristic limbus proliferations develop around the cornea. They are generally most developed within the palpebral fissure. They are pale, yellowish-red, flat, granular thickenings of the conjunctiva, which may extend for 1 to 1½ mm. upon the cornea and occasionally surround the latter in its entire circumference. After many attacks, occasionally not until the lapse of ten years or more, everything disappears without a trace, or there is a slight limbus opacity of the cornea and slight drooping of the upper lid. Serious complications are entirely wanting.

17

The affection occurs most frequently at the period of puberty; subsequently its frequency diminishes rapidly, and it is rarely observed after the age of thirty or forty years.

The microscope shows, in all diseased parts, that the otherwise normal epithelium is strikingly thickened, even threefold and fourfold It often sends short rounded prolongations toward the conjunctival tissue. It is only in rare cases, and then only in larger limbus proliferations, that the prolongations are longer and more or less branched. The individual cells have the appearance of ordinary epithelium cells, perhaps a little larger and often very distinctly of the type of spinous (echinate) cells. At the border of the process these pass imperceptibly into the normal epithelium. At the surface of the epithelium there are often numerous cells in a condition of mucous metamorphosis, and these evidently furnish the material for the tough mucous threads, the sole secretion, in the fornix of the conjunctival sac. The conjunctival tissue beneath the epithelial thickening is absolutely normal in recent untreated cases. If it has been cauterized repeatedly, it is often richly infiltrated with cells, and then numerous round cells are found in the epithelium.

The accidental observation of the coincidence of so-called spring catarrh with an eruption of miliary warts upon the entire integument of the forehead led me to examine more closely into the analogy with warts, and, in fact, this analogy is complete if we accept the notion of warts of the mucous membrane.

The period of development is the same, and the microscopical appearances, the uselessness of all irritant treatment, the spontaneous and complete recovery after a longer or shorter period, etc., warrant us in interpreting spring catarrh as a warty formation upon the conjunctiva. The differences are sufficiently explained by the fact that the eruption occurs upon a mucous membrane. The wider diffusion is also explained by the circumstance that, as a result of the annoyance induced by the rough surface of the conjunctiva during movements of the eye, the affection is subject to more frequent irritants by rubbing, irritating treatment, etc., than in parts of the integument where it practically causes no disturbance of function. It is well known that warts are quite common upon the free margin of the lids.

I also wish to call attention to the fact that I have known very pronounced cases, without the limbus proliferations, to be explained as " chronic blennorrhœa" by competent men only a few years ago, and which were treated as such. If the term "trachoma" is to be selected for a conjunctival affection, this surely applies to "spring catarrh," which causes disturbance only on account of the rough surface of the conjunctiva. The best appellation is conjunctival warts, verrucæ, or verrucositas conjunctivæ.

As in the case of spring catarrh the coincident observation of a cutaneous eruption may serve to explain the condition, so the same relation holds good in that disease of the conjunctiva which is supposed to correspond to pemphigus. Here, likewise, a true vesicle is hardly ever observed, but merely the infiltrated loss of epithelium, the ulcer. This development of ulcers is very extensive and has given rise, in all observed cases, to adhesions between the lid and globe, with opacity and covering of the cornea, to total symblepharon, and thus to loss of sight. Treatment has been entirely useless, except that Samelsohn (Heidelberg Congress, 1879) claims to have cured two cases. The disease is very rare and was first described by White Cooper and Wecker as pemphigus of the conjunctiva, on account of coincident pemphigus of the integument. Without the latter a positive diagnosis can hardly be made. Malcolm Morris (*Mon. f. pract. Derm.*, May 15th, 1889) has collected twenty-eight cases of pemphigus conjunctivæ. They affected all ages, from infancy to the age of seventy-six years. In eight cases the disease first attacked the eye. It is probable, however, that this category includes the majority of cases which have been described as spontaneous symblepharon posterius, *i.e.*, spontaneous adhesion of the conjunctival sac. Pemphigus of the integument may have been absent for a time or may have been overlooked.

Tilly (*Jahr. f. Aug.*, 1888, p. 288) saw pemphigus of the conjuntiva develop at the age of five years, after vaccination; it is usually observed at a later age. Inoculation of the contents of pemphigus vesicles upon the conjunctiva and buccal mucous membrane of animals gave negative results (Bandler, *Prag. med. Woch.*, 1890, p. 528).

Despite the great difference in their appearance, the association of the disease of the skin and conjunctiva in pemphigus is universally acknowledged. This is by no means true of the much more frequent eczema and the forms of acne. Spring catarrh is also regarded generally as a disease *sui generis*.

In elephantiasis Arabum the lids may also take part in the process, indeed it may even be confined to the lids. Hutchinson has stated (*vide* p. 249) that elephantiasis of the lids may follow erysipelas, but the latter may be merely a symptom of the elephantiatic process, as is true of other parts of the body.

In a case of elephantiasis of the left lower limb, Michel (*Arch. f. Ophth.*, XIX, 3) found at the autopsy enormous thickening of the chiasm and right optic nerve, which was due to hyperplasia and sclerosis of the connective tissue. It is very doubtful, however, whether there is any intimate connection between the two events; at least, the relations are entirely obscure.

Xanthelasma may also be mentioned here with some justification; the microscopical findings (dilatation and thrombosis of the capillary lymph fissures and subsequent conversion into connective tissue) indicate a process which is similar to that of elephantiasis. The lids, in which xanthelasma usually begins, are in a position in which the domains of several small arteries border on one another, and hence they are in a comparatively unfavorable condition for nutrition and blood supply, and also as regards the lymph circulation. Hence, spontaneous gangrene of the lids, which, for example, I observed in one case after measles, occurs symmetrically in those parts in which the xanthelasma usually begins. However, nothing is known concerning the connection or coincidence of xanthelasma and true elephantiasis.

Elephantiasis Græcorum or lepra will be considered under the heading of chronic infectious diseases.

Ichthyosis and fibroma molluscum also occur upon the lids. In the former, apart from the ectropium due to shortening of the integument of the lids, the eye itself may be implicated, usually in the same way as in pemphigus or "trachoma" (essential retraction).

Phthiriasis of the lids, body lice and their nits upon the eyelashes, which look as if strewn with black dust, sometimes enable us, even at a distance, to suspect the presence of these parasites also on other parts of the body.

According to Goldenberg (*Berl. kl. Woch.*, 1887), lice upon the scalp are said, in a series of cases, to have given rise to relapsing conjunctival catarrh and eczema of the margin of the lids, which recovered spontaneously after removal of the lice. Herz (*Mon. f. Aug.*, 1886, October) also claims to have observed that they produce "herpes corneæ Stellwag," *i.e.*, eczema, in a reflex manner. Both are undoubtedly of a traumatic, infectious character, resulting from rubbing the eyes with fingers which had been soiled by the excretions of the lice.

Other diseases of the skin, which also occur occasionally on the lids, manifest no peculiarity.

Mooren ("Hauteinfluesse und Gesichtsstoerungen," Wiesbaden, 1884) states that during chronic cutaneous eruptions he has often seen the development of cataract. Rothmund (*Arch. f. Ophth.*, XIV, 1) also observed the development of cataract in early childhood among the children of three families with a peculiar affection of the skin. Rothmund defines the disease as reticular fatty degeneration of the stratum Malpighii and of the papillæ, with secondary atrophy of the latter and rarefaction of the epidermis. The skin disease began at the third to sixth month of life, the cataract between the ages of three and six years. Of the fourteen children of the three families, living in three adjacent villages, seven suffered from the skin disease and five of these from cataract; the other two were still under two years of age. In a girl of twenty-two years, Nieden (*Centr. f. Aug.*, 1887, p. 353) observed the formation of a cataract in combination with telangiectatic dilatation of the capillaries of the entire integument of the face. He also emphasizes the relation between skin disease and development of cataract.

Foerster observed general prurigo associated with albuminuric retinitis, but it is evident that the renal affection to which the patient finally succumbed was the cause of both conditions.

Mooren (*l. c.*) found retinal hemorrhages in extensive cutaneous burns, and regards them as a reflex neurotic disturbance of circulation. I observed two similar cases in the Zuerich Surgical Clinic, and Wagenmann (*Arch. f. Ophth.*, XXXIV, 2) has also published a case. We evidently have to deal with sepsis from absorption of products of decomposition or the destruction of red blood globules (*vide* Welti, in Ziegler and Nauwerck "Beiträge," IV, p. 251), *i.e.*, with the retinal hemorrhages of septicæmia which are found so frequently when looked for. They are much more common in extensive burns than might be inferred from the few published cases.

Mooren also reports visual disorders after suppression of perspiration of the feet, which were cured by inunctions with turpentine and spiritus formicarum which re-excited the "normal" secretion. Sudden drying of moist eczemas is said to have produced hyperæmia of the retina and hyperæsthesia of the optic nerve. In extensive eruptions on the scalp, and in favus, he claims to have repeatedly observed retinitis and optic neuritis. Rampoldi (*Jahrb. f. Aug.*, 1882, p. 334) also mentions various complications in favus: in one case bilateral iritis serosa, in two cases iridocyclitis, in one case iridocyclitis with cataract, in one case senile cataract in a former favus patient. There is no doubt that these were chiefly accidental coincidences.

Broemser (*ib.*, 1870) saw a very vascular metastatic retinal sarcoma after ligature of a pigmented mole on the cheek, but without development of a tumor in the latter locality. Hence the connection between the two is doubtful.

In a case of multiple skin tumors with the structure of fibroma lipomatodes, Hirschberg (*Arch. f. Aug. u. Ohr.*, IV, 1) found a similar tumor develop upon the cornea.

Numerous reports have been made concerning the accidental coincidence of diseases of the eye and skin, for example, of ocular affections in alopecia areata, which were collated by Froelich (*Berl. kl. Woch.*, April 6th, 1891) and present nothing characteristic. When mydriasis is found in extensive urticaria (Zunker, *Jahr. f. Aug.*, 1876, p. 296), this must be regarded as mydriasis due to cutaneous irritation. Many illustrations of this kind could be adduced.

In conclusion, a few words regarding exposure to cold. As the results of cold, Mooren mentions a melange of conjunctivitis, severe keratitis, episcleritis, retinitis, choroiditis, etc. In three cases the eyes of cataract patients, who walked on the cold floor in their bare feet a week after the operation, were destroyed by fulminant choroiditis; in one case the same effect followed nine days after putting on a cold shirt. Here the fulminant choroiditis in a latent form was undoubtedly present at an earlier period, inasmuch as it is a septic process and the carriers of infection do not enter the interior of the eye as the result of walking on a cold floor. Such an injurious agent may be the exciting cause of an acute exacerbation of an already existing process, just as similar factors may excite an acute attack of glaucoma in an already diseased eye, but not in a perfectly healthy one.

The same explanation, the acute lighting up of an already existing latent process in the affected organ, the *locus minoris resistentiæ*, must be accepted in the majority of cases of so-called colds. Very rarely is there a direct connection between an ocular affection which was not previously present and the cooling of a more or less extensive surface of the body, for example, a facial paralysis or paralysis of the iris immediately after a railway trip at the side of an open window. But if the affected individual suffers, for example, from syphilis, the case is no longer a pure one, although it is in these very cases that the "traumatic" action is undeniable. In a few cases I have cured "inflammation of the eye from a cold" by removal of a foreign body beneath the upper lid, and which the draught had blown into the conjunctival sac.

The chill or rigor from a sudden breeze, which is usually regarded by the laity as undeniable proof that they have caught cold, is a symptom of the beginning fever or the rapidly rising temperature, *i.e.*, of the already existing disease, while the real cause, the infection, may have existed a week or more beforehand, usually in an unnoticeable manner.

Articular rheumatism, which was formerly regarded as a disease *par excellence* due to cold, has proven to be a typical infectious disease. In many circles colds play the same rôle now that "the recession of the itch" did in former times. Nevertheless colds and itch do occur, but they are not at the bottom of everything with which they are charged.

In a measure, insolation may be regarded as the antithesis to a cold, although the chief part is here played not by the unusual temperature, but by the secondary loss of serum in the blood and tissues.

Hotz (*Jahrb. f. Aug.*, 1879, p. 255) mentions six cases of optic neuritis in insolation. Moore reports bilateral neuritis with subsequent pigmentation of the optic disc.

The implication of the eye in rhino-sclerosis will be considered when speaking of the respiratory organs, that of so-called myxœdema under the heading of constitutional diseases.

CHAPTER III.

DISEASES OF THE DIGESTIVE ORGANS.

WHILE there are manifold relations between diseases of the skin and eye, this is true only to a slight extent of diseases of the digestive organs, apart from a few very general occurrences.

It may first be mentioned that fluids dropped into the conjunctival sac not infrequently pass through the lachrymal canal into the nose and pharynx, are there absorbed and give rise to toxic symptoms. This is observed most frequently in the case of atropine and eserine. It should be remembered that a single drop of the ordinary solution contains half of the maximum internal dose (according to the Pharmacopœia) for an adult.

In different individuals such toxic symptoms appear with extremely varying rapidity and severity. Especially permeable lachrymal canals probably constitute the most important predisposing factor, but in other cases a personal idiosyncrasy is probably important.

The first symptom consists of a bitter taste and dryness in the throat. The further symptoms vary according to the nature of the poison.

In dogs and cats a flow of saliva appears almost immediately after instillations of atropine. The saliva is swallowed by the dog, and expectorated by the cat.

The constitutional anomalies resulting from protracted and severe gastro-intestinal diseases will be discussed in detail in the last section. Abnormal digestive processes are very frequent causes of general anæmia and hydræmia with their sequelæ: feeble accommodation, weakness of the external ocular muscles (insufficiency). In severe cases hemorrhages also take place into the retina, rarely into the vitreous body. Neuralgic processes and certain asthenic uveal diseases, such as iritis serosa, develop by preference in individuals with impaired nutrition.

Chronic gastric catarrh in poisoning by alcohol, tobacco, lead, etc., by the resulting nutritive disturbance also predisposes to disease of the optic nerve; at least, it is always present at the same time. In a measure this is also true of the retinal affections in albuminuria, the latter being often concealed by the symptoms of a chronic gastric catarrh.

The sequelæ of gastric and intestinal hemorrhages will be discussed under the heading of circulatory disturbances. Similar symptoms are occasionally observed after violent vomiting, especially with hæmatemesis. As a rule, vomiting causes dilatation of the pupils.

Chronic pharyngeal catarrh rarely extends to the eye, but smokers and drinkers very generally suffer at the same time from conjunctival catarrh, chiefly from the constant exposure to bad air. This cause is essentially traumatic in character.

Nearly everything that happens between the ages of one and seven years is attributed—and not alone by the laity—to teething. Hence the literature is quite rich in this respect.

Apart from abscesses of the gums, especially of the so-called eyeteeth, which are occasionally mistaken for dacryocystitis (the reverse is also true in many cases), a definite relation is established mainly in reflex neurotic conditions (pupillary changes, paralyses, spasms, etc.).

Conjunctivitis and phlyctenulæ are said to result from teething, but the former are so frequent that this relationship is very uncertain. During the teething period disturbances of function can only be recognized when they are very marked, and for this reason they are undoubtedly overlooked in the majority of cases.

Numerous reports have been made concerning ocular affections and toothache in adults. Keratitis, iritis, phlyctenulæ, glaucoma, paralyses, asthenopia, amblyopia without lesion, supraorbital neuralgia, exophthalmus, etc., are mentioned; restriction of accommodation is mentioned with special frequency. Schmidt (*Arch. f. Ophth.*, XIV, 1, p. 107) found it 73 times among 92 cases, either bilateral or unilateral (in the latter event, only on the side of the toothache); it was most frequent in youth and amounted to 5.0 D or more. It rarely gave rise to subjective symptoms. Schmidt believes that it is due to reflex increase of pressure in the eye, analogous to the restriction of accommodation occasionally observed as a prodromal symptom of glaucoma. Pathological anatomy, however, teaches that the latter has a much more material local cause. When limitation of accommodation appears during toothache it probably results simply from the lack of vigorous innervation on account of the distressing pain. Muscular insufficiency and diplopia, which are occasionally observed, may also be explained as paresis due to enfeebled innervation.

On the other hand spasm of accommodation has also been observed as a nervous symptom in toothache. I have seen nictitation disappear immediately after removal of a painful tooth. Gosselin makes a similar statement concerning supraorbital neuralgia. Neuralgias, particularly in the first and second branches of the trigeminus, injection of the eyes and epiphora may undoubtedly result from toothache. In view, however, of the very great frequency of toothache, the connection must at least be made probable by the fact that with the cessation of the latter, or with the removal of the diseased tooth, recovery or at least striking improvement of the neuralgia sets in. A certain degree of relief of neuralgia is probably always associated with the removal of a diseased and painful tooth.

Amblyopia and amaurosis as the result of disease of the teeth have been reported several times,--for example, by Lardier (*Rec. d'Ophth.*, 1875, p. 87), Gill (*Jahr. f. Aug.*, 1872, p. 373), Métras (*ib.*, 1873, p. 217), Keyser (*ib.*, 1872), Samelsohn (*ib.*, 1877, p. 195). When the symptoms appear without any visible lesion, the connection may be granted. But positive ophthalmoscopic findings make this extremely suspicious and lead us to think of a common cause for the pain and visual disturbance.

Hutchinson (*ib.*, 1885, p. 316) observed lagophthalmus from spasm of the levator palpebræ superioris during toothache. Blanc (*ib.*, 1871, p. 225) claims to have cured a chronic ophthalmia (?) by extraction of a tooth. Brunschwig (*ib.*, 1887, p. 302) makes the same claim with regard to two cases of iritis with hypopyon. It is easy to understand the frequent development of abscesses in the lower lid as the result of abscesses around the teeth, especially of the so-called eye-tooth. Very rare is purulent inflammation of the orbit in caries of the upper teeth. Pagenstecher (*Jahr. f. Aug.*, 1884, p. 620), Burnett (*ib.*, 1885, p. 16) and Vossius (*Arch. f. Ophth.*, XXX, 3) have observed cases of this kind. Exophthalmus may also develop from serous infiltration of the orbital tissue in the vicinity of the inflammatory focus, and it may disappear rapidly after removal of the tooth. Dimmer (*Wien. med. Woch.*, 1883, p. 299) observed metastatic choroiditis after extraction of a tooth; Sewill claims to have seen the development of cataract after this procedure. Neuschueler (*Jahr. f. Aug.*, 1889, p. 403) reports the cure of toothache by means of prisms.

On the other hand, it is to be noted that pain in the upper teeth on the same side is a frequent symptom of the so-called ciliary pains of keratitis, but particularly of iritis and cyclitis. Less frequently is there pain in the teeth of the lower jaw on the same side, and still less frequently in those of the other side; these are cases of irradiation of pain to adjacent nerve-twigs. Neuralgic toothache may be the prodromal sign of glaucoma.

Symptoms similar to those described are also attributed to the irritation of worms. A connection with diminution of accommodation and narrowing of the field of vision is conceivable, but can be explained with difficulty on account of the absence of pain. In former times a striking enlargement of the pupils was regarded as a pathognomonic sign of worms in the intestinal canal. I have not been able to find any recent data on this subject. It is, indeed, conceivable that intestinal parasites may discharge excretions which are absorbed by the intestines and cause paralysis of the sphincter pupillæ, but this is not very probable. The question merits further investigation.

Furnell (*ib.*, 1871, p. 177) claims to have cured obstinate iritis and keratitis and also nyctalopia by means of vermifuges. In this connection very fantastic accounts might be quoted. It is evident that suggestion plays a prominent part in such cures.

More intelligible symptoms are produced by tænia solium when its six-hooked cysticercus is situated in the orbit, eyelids, external ocular muscles, beneath the conjunctiva, in the interior of the eye, or in the brain. These symptoms will vary according to the locality affected. Our present knowledge of the natural history of the tapeworm compels us to discredit the statement of Montméja, that cysticerci deposit ova within the eye.

According to Reyher (*Deutsch. Arch. f. kl. Med.*, 1886, p. 43) tapeworms, especially bothryocephalus latus, produce pernicious anæmia with extreme pallor of the papilla, narrow, slightly colored vessels, and spontaneous retinal hemorrhages. The same conditions are observed in the anæmia which is connected with the presence of the anchylostomum duodenale in the intestinal canal. Slight œdema of the face, especially of the lids, is common to both conditions.

A well-known symptom is the initial œdema of the lids in trichinosis, and which may be the first objective sign of that condition. Emigration of trichinæ into the ocular muscles causes pain, interference with the movement of the eye in all directions, and drooping of the upper lid. Impairment of accommodation, which is often present, is explained by the pain caused by convergence, because the smooth muscles within the eye are secure against the emigration of trichinæ. It may also be the result of the action of ptomaines, like the mydriasis; the latter, however, is possibly sympathetic (as the result of the pains). All these symptoms occur only in severe cases and are then bilateral; they attain their height at about the fourth day and then diminish. Redness and more or less secretion from the conjunctiva are also usually present (Kittel, *Wien. med. Zschr.*, 1871, p. 254).

Profuse diarrhœas give rise, particularly in infancy, to xerosis conjunctivæ et corneæ, as in the cholera of adults, by a desiccation slough in the inter-palpebral space and subsequent infection from the conjunctival sac. In the dry cornea with a somewhat reduced temperature, however, the micro-organisms do not appear to develop with the same rapidity as in other infectious keratitides at a corresponding age.

As a general thing the appearance of xerotic keratitis in cholera infantum is a sign of impending death, and the latter generally occurs before perforation of the cornea takes place. But if the general disease tends to improve, the process may come to a standstill at any stage. With the newly awakened activity of the tissues a reactive inflammation of the cornea sets in. This checks the further spread of the desiccation necrosis and leads to exfoliation of the necrotic parts. The ultimate lesion varies according to the extent and depth of the necrosis, from a slight patch in the zone of the palpebral fissure to a dense leucoma with or without anterior synechia, prolapse of the iris, or even total staphyloma of the cornea. The milder forms are observed more frequently, because if the process extends very deeply, a fatal termination usually occurs. For reasons easily understood the disease is almost always bilateral, although it often does not exhibit the same intensity in both eyes. The intestinal disease, as such, acts as a predisposing factor, because the marked loss of fluids in profuse diarrhœa facilitates desiccation in unprotected parts. There is evidently at the same time a sort of ptomaine narcosis which gives rise to the insensibility of the cornea and the imperfect movement of the lids.

Von Graefe originally regarded this disease as a neuro-paralytic keratitis resulting from interstitial keratitis, because numerous so-called granular corpuscles were found in the brains of the little patients. These are found constantly, however, in the growing brain, and hence the diagnosis "encephalitis" is not warranted. The longer the cornea is exposed to the air, and the dryer the tissues, the greater the liability to the disease.

We may also mention three cases of corneal xerosis in adults, observed by Schoeler after strenuous treatment for obesity.

Immermann (*Jahr. f. Aug.*, 1887, p. 250) reports a case of blindness and subsequent atrophy of the optic nerve after very profuse diarrhœa.

Conjunctival hemorrhages occur not infrequently in old people with brittle vessels from straining at stool. Hemorrhages into the retina or vitreous, or into the orbit, are much rarer. Such hemorrhages possess prognostic importance because, as a rule, they are forerunners of cerebral apoplexy.

Profuse hemorrhoidal hemorrhages may cause the same visual disturbances as profuse hemorrhages from other organs. In former times a number of visual disorders were attributed to suppression of hemorrhoidal hemorrhages. In a few cases vicarious hemorrhages into the vitreous are said to have been observed, similar to the so-called vicarious menstruation. Hemorrhagic glaucoma, *i.e.*, acute glaucoma in an eye whose vessels are diseased, is also said to result from this cause, but a direct connection is doubtful. It is certain, however, that in inflammatory uveal affections of patients who suffer from piles, laxatives or leeches ad anum not alone have a general favorable effect, but also exert a favorable influence on the eye disease whatever its relations to the hemorrhoids may be.

In hepatic diseases the eye may be affected in various ways. In jaundice the yellow color of the conjunctiva is early noticeable, and also lasts longer than the icteric color of the skin. In the diagnosis of so-called hæmatogenous icterus, for example, in toxic and septic processes, the yellow color of the conjunctiva is important because the icteric color of the integument may not be pronounced.

Subjective yellow vision has been observed for a little while at the beginning of catarrhal icterus. I have again recently seen a case of this kind. We are apt to assume that this symptom results from a yellow color of the ocular media if an intense jaundice develops very rapidly. In my last case, however, the jaundice was neither very intense nor had it developed very acutely. The yellow vision reminds us exactly of that observed in santonin poisoning, where it is due to the fact that the retinal elements at first do not respond to the violet end of the spectrum and later not to the red end. It is characteristic of the peripheral nature of the symptom that, as in santonin poisoning, shadows are seen in the complementary color, *i.e.*, violet. The central perception of violet is not disturbed; the retinal elements have merely become insensitive to violet. The coincident night blindness, which also occurs without yellow vision, develops in an analogous manner (torpor retinæ, *vide* p. 34).

It is probable, therefore, that the symptom is due to the action of a ptomaine, which is not produced in every case of icterus, and may even be absent in very severe and rapidly developing forms.

Hemeralopia has been associated with hepatic affections of all kinds; compare, for example, Parinaud, Maci, and Nicati (*Jahr. f. Aug.*, 1881, p. 395 *et seq.*). Xerosis of the conjunctiva also occurs in jaundice (Leber, Parinaud, Snell), although rarely; as is well known, it is often combined with hemeralopia (*vide* p. 34).

In acute yellow atrophy of the liver, numerous retinal hemorrhages are usually found in the terminal stage, in addition to the yellow color of the conjunctiva. The hemorrhages come from the septic poisoning, as in septicæmia. The same condition is also observed, for example, in acute phosphorus poisoning.

According to Hirschberg bilateral retinitis may develop as the result of hepatic diseases. The connection is probably the same as in yellow vision, viz., direct action of an injurious substance upon the retina. Litten (*Deutsch. med. Woch.*, 1882, No. 13) emphasizes the very frequent occurrence of retinal hemorrhages in all kinds of hepatic affections which are attended with jaundice.

Xanthelasma has often been associated with diseases of the liver. Very pronounced cases are reported, for example, by Foot (*Jahr. f. Aug.*, 1876, p. 447) and Korach (*ib.*, 1881, p. 438).

Landolt (*Arch. f. Ophth.*, XVIII, 1) has observed the coincidence of retinitis pigmentosa and cirrhosis of the liver, and has assumed an intimate connection between them because the two processes are both interstitial in character. He has applied the term cirrhosis retinæ to retinitis pigmentosa. But their combination is so rare and the etiology of the two diseases differs so greatly that a connection between them cannot yet be conceded. Cirrhosis of the liver is an interstitial inflammatory process, beginning with a stage of cellular infiltration and enlargement of the organ and ending in marked retraction and diminution of volume; its cause appears to be a chronic infection or toxæmia of some sort. Under the same circumstances similar processes may appear in other parenchymatous organs, for example in the kidneys, but not a typical retinitis pigmentosa. The retinal changes in the latter are those of pure atrophy, especially of the nervous elements after destruction of the outer layers of the retina, combined with pigment emigration from the destroyed pigment epithelium. Active inflammatory processes in the retina are entirely absent or play a very subordinate part. As in other localities, the very moderate increase of connective tissue may be explained *ex vacuo*, but the destruction of the nervous elements is not initiated by cellular infiltration of the connective tissue. Retinitis pigmentosa and cirrhosis of the liver have in common only the terminal stage, the connective-tissue degeneration.

The chief site of the disease in so-called retinitis pigmentosa is in the chorio-capillaris and its nutritive territory. The affection begins at the periphery where the meshes of the chorio-capillaris are widest, and is then concentrated upon the region where the meshes are narrowest, the capillaries themselves widest, i.e., in the region of the macula lutea. As a result of the imperfect nutrition, the external layers of the retina degenerate and later its centripetal fibres. Proliferating processes, which are merely forerunners of necrosis, are found only in the pigment epithelium. The final choroidal atrophy may be more striking and intense than that of the retina. Unfortunately there are no anatomical findings in the earliest stage when there is merely peripheral torpor retinæ and visible changes are still entirely absent.

Even if typical retinitis (preferably chorio-retinitis) pigmentosa and cirrhosis of the liver occur together, the former runs a much more chronic course than the latter.

The relation of chorio-retinitis pigmentosa to cirrhosis of the liver is a very instructive illustration of the fact that, in the accidental coincidence of two diseases, an apparent anatomical similarity may lead to the assumption of an internal connection which in reality does not exist. Chorio-retinitis with retinal pigmentation and cirrhosis of the liver may both have a common cause, for example congenital syphilis, but then we have to deal, not with a typical retinitis pigmentosa, but with much more acute inflammatory processes in the eye.

18

CHAPTER IV.

DISEASES OF THE RESPIRATORY ORGANS.

THE relations between the respiratory organs proper and the organ of sight are very slight. They result chiefly from the vicinity to the eye of the beginning of the respiratory tract, viz., the nose and its auxiliary cavities, the orbit and conjunctival sac. We will therefore first consider the diseases of the nose and its cavities, with which it is best to include diseases of the ear, because the mode of connection of the latter with diseases of the eye is entirely similar. As a matter of course, tumors in these localities may penetrate the orbit and attack the eye (for example, Nieden, *Arch. f. Aug.*, XVI, p. 387), but such cases do not interest us here.

1. Diseases of the Nose

may ascend through the lachrymal duct to the conjunctival sac, or very rarely the opposite path may be followed. Tubercular (lupous) and certain follicular diseases, epithelial cancer and the like, may extend in this way from the eyelids and conjunctiva to the nose. During infectious diseases of the conjunctiva the lachrymal canals, as a rule, are impermeable on account of the swelling of their mucous membrane, and hence the disease does not reach to the lachrymal sac and nasal mucous membrane.

With far more frequency disease of the nasal mucous membrane is the starting-point for the affection of the conjunctiva and cornea.

As a rule, every acute coryza causes eye symptoms, hyperæmia, or catarrh of the conjunctiva "so that one cannot see." This is also true of so-called hay fever. Occasionally there is an outbreak of herpes febrilis upon the eyelids or cornea, which is easily recognized by its characteristic form. To this category probably belongs Decker's case (*Monatsbl. f. Aug.*, March, 1890) of herpes corneæ zoster as the result of nasal disease, unless the coincidence was merely accidental.   If herpetic disease has made its appearance on the cornea it not infrequently returns with relapses of coryza, and may severely distress the patient.

In chronic coryza we often find hyperæmic and chronic catarrhal conditions of the conjunctiva which undergo acute exacerbations from time to time, usually with the nasal disease, and require local treatment both of the eye and of the nose.   The affection is usually bilateral, but it is occasionally unilateral.   Although the lachrymal passages partake in the trouble, there is usually no obstacle to the discharge of the secretion and no noticeable pathological change in its character.   At the most—but this is very common—there is epiphora on account of the swelling of the mucous membrane, while a sound will readily pass.

Far different and more frequent is the situation when, in chronic rhinitis, the formation of ulcers and scabs in the nose occludes the opening of the lachrymo-nasal duct.   The secretion cannot escape; it decomposes, and gives rise to purulent inflammation of the lachrymal sac with all its sequelæ.

On passing a sound the obstruction is found not at the opening of the lachrymo-nasal duct into the nose, but mainly at the opening of the canaliculi into the lachrymal sac and at the transition of the latter into the lachrymo-nasal duct, about the valvular folds of the mucous membrane.

Hence we find (1) diseases of the lachrymal sac constantly relapsing despite permeability and despite the fact that the sound almost enters of itself, so long as the nasal mucous membrane is not sufficiently cured and the constant formation of crusts occludes the opening of the lachrymo-nasal duct; and (2) that a dacryo-cysto-blennorrhœa which has not lasted too long and in which there are no bony occlusions, may be cured by suitable treatment of the nasal mucous membrane and persistent emptying of the lachrymal sac.   In many cases the patient may be spared the painful treatment with sounds. The nasal treatment, however, is not perfectly simple and usually lasts a long time, because, as a rule, we have to deal with a disease that has endured for years.   Nor is it sufficient in all cases (rhinitis atrophicans).   It is important, however, under all circumstances.

Gruhn (*Muench. med. Woch.*, July 3d, 1888) found disease of the nose absent in only two cases among thirty-eight cases of disease of the lachrymal sac.

I have not been able to secure any specific statistics, but can maintain with positiveness that at least three-fourths of the cases of disease of the lachrymal sac which have come under my observation were caused by chronic rhinitis with formation of crusts.

Eczematous disease of the nasal mucous membrane very often gives rise to inflammations of the eye, particularly in children. It may be merely a more or less active catarrh (Horner's eczematous catarrh), which is very obstinate unless properly dealt with. This category also includes the large majority of diseases of the conjunctiva and cornea to which the indifferent term phlyctenular has been applied, and which have been called by Horner eczema of the conjunctiva and cornea. But not every phlyctenular disease of the eye merits this name. The isolated eruptions without a trace of eczema in any other part of the body may very well be due to some other cause. The large majority of the cases, however, are undoubtedly eczematous. In nearly 90 per cent. of the cases of phlyctenular disease of the conjuctiva and cornea in children, I can prove the coexistence of eczema nasi. In addition there is very often a striking parallelism between the relapses of the disease of the eye and nose.

Eczema of the nose may continue for a long time without corresponding diseases of the eye. But after the latter has developed, "when it has found the way," it relapses with every acute exacerbation of the nasal inflammation. Sometimes the disease is confined to one nostril and to the corresponding eye. It is a striking fact, however, that disease of the tear passages so rarely occurs, probably because eczema does not lead to the formation of crusts, as in other forms of chronic rhinitis.

As a matter of course local treatment of the eye disease, especially of severe corneal phlyctenulæ, is indispensable. But the constant relapses only cease when the nasal affection is cured.

The nasal mucous membrane is much more often the starting-point for eczematous disease of the conjunctiva and cornea, than of the skin of the lids.

In ulcerative destruction of the nasal mucous membrane vigorous blowing of the nose may cause emphysema of the lids. Such cases are reported by Newcombe (*Lancet*, 1874, II, p. 184), Jeaffreson (*ib.*, p. 221), Fano (*Jahr. f. Aug.*, 1886, p. 433) and Fieuzal (*ib.*, 1887, p. 413).

Whether suppuration in the nose may be the direct cause of iritis, as is claimed by Ziem in three cases, appears to me to be very doubtful, and, at all events, the connection is entirely obscure.

Hoadley Gabb (*Jahr. f. Aug.*, 1883, p. 549) observed in epistaxis after spasmodic cough that the blood escaped through the lachrymal duct of the corresponding side. Schmidt-Rimpler (*Monatsbl. f. Aug.*, 1877, p. 375) observed blindness of both eyes after loss of blood due to curetting nasal polypoid growths. This case is doubtful because the hemorrhage was a slight one. The ophthalmoscope showed sinuosity of the veins and slight cloudiness of the papillæ.

Rhinoscleroma will also be discussed here among diseases of the nose. It is confined almost exclusively to certain regions (Southwestern Russia, Austria and Central America). The disease consists of a tumor formation which starts from the mucous membrane of the nose or pharynx and develops, after many years, into an extremely hard, either flat or nodular, infiltration. It extends chiefly in the mucous membrane , but the muscles, cartilage and skin may also be attacked. Atrophy and cicatricial retraction do not occur until a very late period, and ulceration very rarely takes place. It seems to be due to a diplococcus which resembles Friedlaender's pneumococcus.

Among eighty-seven cases collected by Wolkowitsch (*Arch. f. kl. Chir.*, XXXVIII, p. 356), the eye, including the lachrymal sac and lids, was involved nine times. The lesions included firm nodules of the lachrymal sac with subsequent degeneration and ulceration and extension of the process to the middle of the left lower lid (Zeissl); a fluctuating tumor of the left lachrymal sac, which disappeared on pressure (Jarisch); right-sided dacryocystitis (Alvarez); right-sided epiphora (Mandelbaum); bilateral dacryocystitis (Wolkowitsch). In four cases firm, hard tumors were found upon the lachrymal sac (Weinlechner, Stukowenkoff, Schulthess, and Wolkowitsch). In

Schulthess' case (*Deutsch. Arch. f. kl. Med.*, XLI, p. 71) the entire process seems to have started from the mucous membrane of the lachrymal sac.

Gaillard (*Jahr. f. Aug.*, 1887, p. 447) reports a case of thrombosis of the orbital veins in nasal furuncle and has collected fifteen similar cases in literature.

It remains for us to consider a number of ocular reflexes in diseases of the nose. Ziem (*Berl. kl. Woch.*, 1889, No. 38; *Deutsch. med. Woch.*, 1889, No. 5, etc.) and certain American writers claim to have observed various "nervous" eye symptoms, such as asthenopia, slight pareses of accommodation, concentric narrowing of the field of vision with or without disturbance of central vision, blepharospasm, etc., in all kinds of nasal diseases, and to have cured them by suitable treatment of the nose. Even in several cases of chronic glaucoma, Ziem claims to have obtained at least temporary enlargement of the field of vision by galvano-cautery of the nasal mucous membrane. In view of the close proximity of the diseased structures, it is not astonishing that nervous ocular reflexes of the character mentioned occur occasionally, but they possess no further significance. Similar phenomena are observed in toothache and in every form of severe pain, and disappear spontaneously with the cessation of the pain.

Greater importance attaches to the visual disorders after operations on the nasal mucous membrane (galvano-cautery), and which were observed by several writers (Ziem, *Centralbl. f. Aug.*, 1887, p. 131; Berger, *Arch. f. Aug.*, XVII, p. 293, and others). In these cases the visual disturbance is not infrequently in the shape of the well-known concentric narrowing of the field of vision, with or without disturbance of central vision and the color sense. The eye on the side operated upon is attacked alone or to a more marked degree, and the condition is a temporary one in all cases. Similar symptoms result occasionally from painful procedures in the neighborhood of the eye. The reflex process is probably the same that we find in traumatic hysteria (*vide* p. 221), whose symptoms are often confined to the immediate neighborhood of the trauma. These are evidently the mildest forms of hysteria and might be termed local, temporary hysteria or hysteria minor, because the difference is merely quantitative, not qualitative, like that between the petit-mal and grand-mal of epilepsy.

After galvano-cauterization of the nasal mucous membrane in Basedow's disease, Hack (*Deutsch. med. Woch.*, 1886, No. 25) observed disappearance of the exophthalmus on the operated side.   Hopmann (*Jahr. f. Aug.*, 1885, p. 473) appears to have seen a similar case, viz., cure of a unilateral Basedow's disease by treatment of the nasal mucous membrane.   In these cases we also recognize the paresis of the sympathetic system following the injury (paralysis of the unstriped dilators of the palpebral fissure).

Trousseau (*ibid.*, 1889, p. 559) claims to have cured two cases of blepharospasm, one case of obstinate scotoma scintillans (ophthalmic migraine), one case of mydriasis and three obstinate asthenopias, by treatment of the nasal mucous membrane.   Many similar cases could be mentioned, but some of them are very doubtful.

Marckwort (*Arch. f. Aug.*, 1887, p. 452) observed the development of acute glaucoma after the protracted application of cocaine to the nasal mucous membrane.   This may result in a much shorter time, without general cocaine poisoning, by the instillation of a few drops of a cocaine solution into the conjunctival sac.

2. Diseases of the Cavities

adjacent to the nose exercise, on the whole, the same influence on the eye as do nasal diseases themselves, with which they are usually associated.   Stasis of secretion in the frontal sinuses often gives rise to supraorbital neuralgia; a similar condition in the antrum of Highmore gives rise, though much more rarely, to neuralgia of the infraorbital nerve, and these recover after removal of the obstruction to the escape of the secretion.   For this reason the nasal douche often has a favorable influence after all antineuralgics have proven useless. Ziem (*Monatschr. f. Ohrenheilk.*, Aug., 1889, p. 174) cured a case of extreme contraction of the field of vision in nasal suppuration by restoring the outflow from the antrum, and with its occlusion a relapse again occurred.

Apart from the symptoms of compression and displacement of the eye and its appendages (exophthalmus, paralyses, atrophy of the optic nerve, etc.), which result from tumors, abscesses, etc., of these cavities, inflammations of the cellular tissue and abscesses in the orbit may develop, and disease of the ethmoid may cause epiphora by pressure on the lachrymal passages (*vide* Berger and Tyrmann, "Die Krankheiten d. Keilbeinhoehle u. d. Siebbeinlabyrinthes u. ihre Beziehungen zu Erkrankungen d. Sehorgans," Wiesbaden, 1886).

Ptosis is not infrequently found in dropsy of the frontal sinuses attended by erosion of the bones and dilatation of the sinuses; this may cause pressure on the optic nerve. All symptoms may disappear after opening the cavity, but this is not true of all cases.

In diseases of the sphenoidal sinuses, caries or necrosis of the body of the sphenoid, blindness and defects of all kinds in the field of vision may be produced by implication of the adjacent optic nerve, chiasm and optic tracts.

It will be proper to consider now the diseases of the eye which are observed in

3. Ear Diseases,

because they present numerous points of resemblance to those noticed in diseases of the nose and its auxiliary cavities, to which the cavity of the tympanum also belongs.

Irritative symptoms in the external auditory canal and the cavity of the tympanum, whatever may be their origin, appear to give rise occasionally to reflexes similar to those in diseases of the nasal mucous membrane. Blepharospasm is mentioned with special frequency, for example, in cases of foreign body in the external auditory canal (Rampoldi, *Annal. di Ottalm.*, XVIII, 3), in syringing the cavity of the tympanum (Ziem, *Deutsche med. Woch.*, 1885, 49), etc. Rampoldi repeatedly checked blepharospasm of this nature by dropping cocaine into the ear.

Lucae (*Jahrb. f. Aug.*, 1881, p. 152) has several times produced "optic vertigo," with apparent motion of objects toward the opposite side, by using the air douche from the external auditory canal in cases of perforated tympanum. He also observed abduction of the

globe toward the irritated side. Evidently this was conjugate deviation of the eyes toward the side of irritation with apparent movement of objects in the opposite direction. Under certain circumstances, however, the apparent movement of objects appears to take
place in the same direction as the movement of the eyes (*vide Jahr.
f. Aug.*, 1883, p. 148). Lucae observed, at the same time, obscuration of the field of vision, probably the concentric narrowing of the
field to which attention has often been directed. He believes that
these symptoms are due to the extension of pressure of the labyrin-
thine fluid and thence to the subarachnoid cavity of the brain, possibly to irritation of the dura mater or the tympanic plexus.

Baginski (*Jahr. f. Aug.*, 1881, p. 153) found that in rabbits the
injection of water into the cavity of the tympanum produced movements of the eyes and nystagmus. These were the more marked, the
higher the pressure and the more pronounced the temperature and
chemical constitution of the injected fluid. Similar symptoms were
produced by boring into the labyrinth, if the procedure opened into
the skull at the same time. He believes that the injected fluid reaches
the brain (restiform body and ascending root of the trigeminus) and
there gives rise to these movements. Hoegyes (*ibid.*) obtained similar results, but believes they are due to irritation of the roots of the
acoustic nerve. Touching various parts of the semicircular canals
with a bristle always produced characteristic associated movements
of the eye. I also believe that we have to deal in such cases with
involuntary or voluntary reflexes due to irritation of the acoustic
nerve. Kipp (*Jahr. f. Aug.*, 1888, p. 434) observed three cases of
nystagmus in purulent inflammation of the middle ear. The six
cases (*ibid.*, 1883, p. 520) of opacity of the vitreous with cholesterin
crystals (synchysis scintillans) in disease of the middle ear were
probably accidental coincidences. In a case of polyp of the ear,
Pflueger observed nystagmus whenever the polyp was seized. Cohn
(*Berl. kl. Wschr.*, 1891, No. 43) also observed cases of nystagmus,
associated with vertigo, on irritation of the diseased ear.

Urbantschitsch (*Wien. med. Blaetter*, 1882, No. 42) often noted
the influence of ear disease (chronic catarrh of the middle ear) upon
the acuity of vision, either impairment or improvement in one or

both eyes, often an alternation between the two, in the course of the same ear disease. No definite relations could be detected, however, between the ocular and aural affections. On the other hand Rampoldi (*l.c.*) claims to have improved hearing by iridectomy in glaucoma. Davidson and Dransart claim the same result from iridectomy in leucoma of the cornea.

Much more important than these very inconstant nervous symptoms are other and material eye diseases which are observed in purulent inflammation of the middle ear, in caries and necrosis of the petrous portion of the temporal bone, etc. Apart from facial paralysis owing to destruction or lesion of the nerve in its long bony canal, there may also be sensory disturbances or paralyses of the ocular muscles, especially of the abducens. These are either temporary, and are then due to infiltration or to "remote action" upon the adjacent nerves, or they are permanent and due to destruction of the nerves. In perforation into the skull, purulent meningitis with its eye symptoms will develop (*vide* p. 150). Cerebral abscesses, especially in the temporal lobes, with their characteristic symptoms, (*vide* p. 137) also occur frequently in aural suppurations. Kipp (*Jahr. f. Aug.*, 1884, p. 324) even reports two cases of metastatic panophthalmia in aural suppuration.

Zaufal (*Prag. med. Woch.*, 1891, No. 15) recommends ophthalmoscopic examination in suppurations of the middle ear in order to determine the proper period for operative interference. So long as the ophthalmoscopic appearances are normal, there is no danger in delay. But as soon as unilateral or bilateral venous hyperæmia or an indication of neuritis or choked disc, even without disturbance of sight, is visible, then the pus must be evacuated at once on account of the danger of meningitis, even if no other threatening symptoms are present. Zaufal found that the ophthalmoscopic appearances again became normal after the operation. Others have made similar observations. In my opinion it is not advisable to wait until the development of visible ophthalmoscopic findings. In not a few of such cases, it would be too late.

Knapp (*Arch. f. Aug. und Ohr.*, II, 2, p. 191) has observed iritis serosa and keratitis parenchymatosa in combination with so-called

Menière's disease which begins suddenly with vertigo, ringing in the ears, vomiting, tottering gait and pronounced deafness, and in which no constant pathological lesion in the ear has yet been found. The real connection between these diseases of the eye and ear is very obscure. Pooley (*N. Y. Med. Rec.*, Jan. 8th, 1887) found bilateral neuritis in a man of forty-three years suffering from this disease, and who was attacked, almost at the same time, with disorder of sight and hearing; he also suffered from polyuria, without albumin in the urine. The autopsy showed nephritis and softening within the cerebrum; the cerebellum appeared normal. In the majority of ocular complications of Menière's disease, which is evidently not an anatomical unit, we probably have to deal with accidental coincidences.

Stein (*Jahr. f. Aug.*, 1887, p. 381, and *Centralbl. f. Aug.*, Jan., 1887) claims that he has produced cataract in guinea-pigs by means of tuning-forks, and all the more readily, the younger the guinea-pig and the longer the vibrations of the tuning-fork. Evidently we have to deal here in part with hypnotic conditions due to the monotonous auditory sensations which lasted hours and days; this is further supported by the drowsiness and the cessation of the pupillary reaction. There also seems to have been a development of mild nutritive disturbances in the lens, interfibrillary accumulation of fluid and the like, probably as the result of insufficient ingestion of food; this acted more rapidly in young than in old animals. Permanent opacity of the lens was not produced. What might not be expected from such experiments in man? Insanity would be the least.

4. Effects of Respiration and its Abnormalities upon the Eye.

In inspiration there is slight dilatation, in expiration, slight contraction of the pupil, as we can readily observe upon ourselves if suitable precautions are adopted. According to v. Platen the reception of oxygen and the excretion of carbonic acid during the respiration of mammals are increased by the action of light on the retina. According to Urbantschitsch (*Centralbl. f. Aug.*, 1887, p. 513) light also stimulates and darkness enfeebles the auditory function.

Reyher (*Jahrb. f. Aug.*, 1870, p. 303) observed pronounced my-driasis during sudden dyspnœa from swallowing a large piece of meat "the wrong way;" it disappeared immediately after the dislodgement of the meat by vomiting. This is easily explained by the marked suction action of inspiration upon the contents of the blood-vessels, when the air passages are occluded and the lungs contain only a moderate amount of air.

When the nostrils are closed during vigorous expiration, the air occasionally escapes through the lachrymal duct, and this produces a peculiar tickling sensation in the inner angle of the eye. Dilatation of the lachrymal sac and its distention with air may also be produced by labored breathing, for example, in pulmonary affections (Starcke, Rau).

Irritation of the conjunctiva in little children often produces sudden cessation of breathing. This is readily observed in treating blennorrhœa neonatorum with the mitigated stick. In a boy of four years suffering from atropine poisoning, Guttmann (*Deutsch. med. Woch.*, March 29th, 1888, p. 255) observed, not closure of the lids, but cessation of respiration, on touching the cornea. Hence it seems to be one of the lower reflexes, like the deep inspiration resulting from vigorous stimulation of the skin, for example, in cold douches and the like. Kuert (*Wien. med. Presse*, 1891, No. 21) noticed cessation of spasm of the glottis on touching the cornea, but this could be done more easily and certainly by tickling and irritating the nasal mucous membrane.

In dyspnœa (emphysema, etc.) venous stasis in the retinal veins is visible with the ophthalmoscope. There may even be hemorrhages into the retina, rarely into the vitreous. Hemorrhages into the conjunctiva also occur under such circumstances. It is well known that attacks of asthma occur preferably in the dark, and that the burning of a light at night decidedly diminishes their frequency.

In acute suffocation the pupil is generally, though not always, narrow, and hemorrhages, often only punctate, occur into the conjunctiva and retina. The latter are often not seen until the autopsy. The hemorrhages also often occur beneath the integument of the lids and are found occasionally in all the vascular parts of the eye and its viciuity (Schlemm's canal, Petit's canal, etc.).   In chronic suffocation, in which similar hemorrhages may finally be produced, the retinal vessels, particularly the veins, appear with the ophthalmoscope to be unusually dark, often almost black, as the result of overloading the blood with carbonic acid.

In two cases of Cheyne-Stokes breathing, Robertson (*Lancet*, 27, XI, 1886) found that the pupils grew constantly larger during the return of respiration, and constantly smaller as the respirations ceased; this was uninfluenced by illumination of the eye.   In one case, indeed, the reaction of the pupil to light was abolished (*vide* p. 171).

As is well known, sneezing is often produced on forcible opening the eyes of children suffering from photophobia.   This is particularly true of conjunctivitis and keratitis phlyctenulosa.   As a rule, nasal disease (eczema) is present in such cases, but this is not a necessary factor.   Even in adults an intense light stimulus—for example, looking at the sun—may cause one or more acts of sneezing.   In others, this effect is only produced under certain conditions,—for example, in one of my acquaintances only immediately after meals, but then with absolute certainty.   Féré (*Neurol. Centralbl.*, 1890, p. 732) could prevent this in himself by displacing his four puncta lachrymalia with serres fines; he then experienced no desire to sneeze. This reflex, he claims, necessitates irritation of the nasal mucous membrane by the abnormally secreted tears; hence it did not always occur immediately.   Whether this etiology holds good for all cases must be left undecided.

Sneezing, blowing the nose and coughing may give rise to hemorrhages in all vascular parts of the eye and its vicinity, most frequently in the conjunctiva, eyelids and retina.   In such cases there are very often coincident changes in the vessels, particularly atheroma, and these naturally constitute a predisposing factor.   If there are ulcerative processes in the lachrymal sac, or a solution of continuity has taken place in its bony walls or in those of the adjacent cavities of the nose, all these expiratory efforts may give rise to emphysema of the lids or in the orbit.   What strange things may happen is shown by Malgat's case (*Rec. d'Ophth.*, 1890, No. 4).   A lady was compelled to sneeze several times during a meal, and soon afterward felt a violent pain in the region of the inferior lachrymal duct.   Three days later a little abscess developed there and discharged a piece of salad 2 mm. long and 1 mm. wide.   Malgat had previously treated the lady for keratitis, and therefore excluded an abnormality of the lachrymal passages, but this must probably have been present.   In a young girl spontaneous luxation of the globe in front of the upper lid was observed by Dépontot (*Jahresb. f. Aug.*, 1885, p. 458) as the result of sneezing.   I have also observed this, as the result of sneezing, in Basedow's disease.

The expiratory efforts of coughing, blowing the nose and sneezing may be very painful when there is an internal inflammation of the eye.   This is also observed very often in the onset of myopia, during its acute exacerbations, and in these cases can be relieved considerably by vigorous pressure on the eyes.

Ocular affections are, on the whole, infrequent in

5. Diseases of the Respiratory Tract Proper.

We may mention the not uncommon occurrence of herpes corneæ during the course of bronchitis and acute pneumonia.   After this complication has once appeared, it may relapse—and always on the same eye—in subsequent attacks.   Bilateral herpes corneæ is very rare.

Rampoldi (*Centralbl. f. Aug.*, Nov., 1886) calls attention to the occurrence of sympathetic nerve symptoms in pulmonary diseases. It is especially in disease at the apex that the sympathetic may be implicated, and irritative as well as paralytic phenomena of vasomotor and oculo-pupillary character may be observed.   In one of his cases there was bilateral mydriasis in febrile catarrh of both apices, and it diminished with the improvement of the pulmonary affection. In the second case there was pulmonary infiltration, especially of the left apex, together with left mydriasis which passed subsequently into myosis; atrophy of the left half of the face and slight ptosis developed at the same time.

After a profuse secretion had been expectorated for six weeks in a case of purulent bronchitis, Adler (*Jahrb. f. Aug.*, 1889, p. 240) observed, in addition to chill, vomiting, fever and pain in the left shoulder, a metastatic conjunctival abscess on the left side, to the outer side of the cornea; the abscess discharged and healed in a month. Posterior synechiæ and opacities of the vitreous were present in the eye itself, but the organ was preserved. The staphylococcus pyogenes aureus was found in the pus discharged from the abscess.

Cérenville (*Rev. Méd. de la Suisse Normande*, 1888, 1 and 2) collated twenty-one cases, including six of his own, of cerebral symptoms after operation for empyema. The symptoms consisted generally of a sudden epileptic attack with wide pupils, although they were sometimes contracted at the start. The convulsions were generally unilateral, but not always on the operated side. Disorders of speech, pareses, vasomotor disturbances, amaurosis, scotoma scintillans and phosphenes may continue for a time after the seizure, usually for a few minutes only. The attacks themselves lasted from five minutes to sixteen hours. In one case the ophthalmoscope revealed a pale, cloudy papilla and narrow vessels; in two cases, hyperæmia and hemorrhages along the vessels; similar conditions may also be present in the brain. The operation wound was occasionally very sensitive; the pupils then dilated upon touching the wound or the exposed pleura, or they became unequal (like the unilateral or bilateral mydriasis upon irritation of the skin).

Two months after repeated operations for empyema, Handfort (*Brit. Med. Journ.*, Nov. 3d, 1888) observed the development of neuro-retinitis, first on the left side, then on the right, with almost complete blindness. At the end of eleven days vision again returned, but finally there was complete amaurosis. At first there was also right hemiparesis, later hemiplegia, aphasia, etc. The autopsy showed bilateral cerebral softening, especially of the angular gyri and the occipital lobes, *i.e.*, an almost symmetrical embolic process in the brain.

Of course embolic pyæmic affections may originate from purulent diseases wherever situated, but they appear to start rarely from affections of the pleura and air passages.

CHAPTER V.

DISEASES OF THE CIRCULATORY ORGANS.

WE will here consider only the abnormalities in the amount and circulation of the blood, and the direct diseases of the heart and vessels. The abnormalities in the constitution of the blood, among which the higher grades of so-called anæmia and hydræmia belong, fall in the main under the category of constitutional anomalies and infectious diseases, and will there be considered. For the affections of the vasomotor nerves, *vide* the diseases of the nervous system (p. 119).

In general anæmia and hyperæmia (plethora), the eye is not usually affected to a noticeable extent, and even if it were, it could not be regarded as a manometer. Indeed, the amount of blood in the eye may be exactly the reverse of that in other parts. In anæmic conditions, in particular, we very often find the signs and symptoms of conjunctival hyperæmia, and these are usually quite obstinate in such cases. It is only in high grades of general anæmia or venous hyperæmia that a corresponding condition is noticed in the retinal vessels. If these exhibit distinct venous hyperæmia, we will hardly ever have to deal simply with general "plethora," but complications on the part of the respiratory organs (emphysema, asthma) or the circulatory organs (valvular lesion, fatty heart) will be found at the same time. If, as the result of the stasis, hemorrhages have occurred into and around the eye (conjunctiva, retina, vitreous body, etc.), we may conclude, almost with certainty, that vascular lesions are present. It is true, however, that these are only demonstrable anatomically, and are recognized clinically merely by the appearance of the hemorrhages. Hypertrophy of the left ventricle may or may not be present.

Under normal conditions of excretion from the eye the ocular pressure is essentially a function of the blood pressure within the vessels. But increase or diminution of the blood pressure is not alone shown by a corresponding change in the ocular pressure—which may, however, be almost entirely compensated, especially in increased blood pressure, by increasing outflow—but may also produce phenomena visible with the ophthalmoscope.

Gradual diminution of the pressure and quantity of the blood, which must be assumed in many so-called anæmic conditions, must have become very pronounced before it is visible with the ophthalmoscope. We then find a pale, in the highest grades almost chalky white, papilla and narrow vessels with contents of a lighter color. But a change in the red color of the fundus itself is rarely or never seen with certainty in purely anæmic conditions. The visible changes are mainly attributable to the changed contents of the vessels. The only symptom which points directly to the diminished blood-pressure is pulsation of the retinal arteries on slight pressure upon the eye. Even spontaneous pulsation is occasionally found in intense anæmic conditions. Pressure sometimes causes entire disappearance of the veins.

After prolonged duration of such conditions we find a tendency to spontaneous retinal hemorrhages; their absorption is attended by the formation of white patches and rarely gives rise to slight pigmentation of the retina. These hemorrhages often recur if the general condition is not cured. In such cases vascular disease, either directly as the result of defective nutrition of the vessel walls or favored by it, must be assumed; it has even been demonstrated anatomically, though not in all cases. We will return to this subject in the discussion of constitutional anomalies.

In a few extremely anæmic females, between the ages of seventeen and twenty years, I saw the complete picture of retinitis albuminurica, although not a trace of albumin was discovered in the urine. The affection was very obstinate but finally recovered completely after lasting six to nine months. Such cases seem to be very rare.

If the blood pressure suddenly falls considerably, the elastic intra-ocular pressure constitutes an obstruction to the circulation of the

19

retina, which is not found in this way in other vascular tracts. The arteries pulsate and become visibly narrower; the circulation of blood becomes irregular. The papilla grows pale, everything grows dark before the patient's eyes and there is concentric narrowing of the field of vision, with which the attack of syncope begins. The same thing is observed when, after great excitement, great exhaustion ensues. The coincident subsidence of central vasomotor innervation entails imperfect supply of blood, especially to the head. On several occasions I became aware, from the suddenly developing arterial pulsation during ophthalmoscopic examination, of an impending attack of syncope, and was able to guard the patient against falling from the chair.

With the restoration of vascular tonus and filling of the vessels this usually disappears without any bad effects. In rare cases there is a more or less sudden onset of unilateral or bilateral disturbance of vision, amounting even to complete blindness. This may either be permanent or may undergo improvement. As a rule, complete restoration does not occur. Disturbances of vision and the color sense, defects of the field and, very rarely, a central scotoma are left, as in other diseases of the optic nerve.

The ophthalmoscopic appearances in such cases of "amblyopia and amaurosis after loss of blood" are often negative at the onset, or at the most we find a somewhat pale papilla and narrow vessels. More or less complete atrophy of the optic nerve develops at a later period. One finding is known as retrobulbár neuritis, and corresponds, as a rule, to a hemorrhage into the trunk of the optic nerve or its sheaths. In other cases there is neuritis and choked disc with considerable opacity of the retina and papilla, often with hemorrhages and white patches. Samelsohn observed a retinal hemorrhage develop with beginning improvement of vision after the restoration of circulation.

Such conditions are occasionally seen after hemorrhages of the most varied kinds, after hæmaturia, profuse menstruation, abortion, epistaxis, venesection, but most frequently after gastric hemorrhage and hæmatemesis. Even violent vomiting alone may produce such results. Sometimes the disorder of vision occurs only after repeated hemorrhages, or it may appear after the first hemorrhage and be absent in succeeding ones.

Three weeks after the beginning of the disease, Ziegler (" Beitr. z. path. Anat. u. Phys.," Bd. II) found fatty degeneration of the supporting elements of the optic nerve and the innermost layers of the retina, most pronounced in the scleral portion of the nerve; cell infiltration was wanting. These findings remind us of the ischæmic foci of degeneration in the brain and heart muscle. Ziegler attributes them to local contractions of the vessels owing to general anæmia.

Baqués (*Annal. di Ottalm.*, XX, 3) has studied the histological changes in the retina in temporary anæmia. If the latter lasted an hour or more, retinal hemorrhages and œdema developed at the end of eight to ten days. The ganglion cells become dropsical; vacuoles and signs of degeneration appear in the cells of the internal granular layer, with proliferation of the connective-tissue elements of the supporting fibres. These findings are not, however, directly applicable to our clinical cases.

Special attention should be paid to the following points: 1. Amblyopia and amaurosis after hemorrhage are rare and occur only in individuals who are otherwise not in a healthy condition, so that, in addition to the loss of blood, there must be some other predisposing factor. Not a single case occurred in the entire campaign of 1870–71. Fries (*Monatsbl. f. Aug.*, 1876) found only 5 cases of wounds among 106 cases—one case of hemorrhage, lasting half an hour, from a small spirting artery, three perforating wounds of the chest (hence not a simple hemorrhage), and one insufficiently described case. In all these cases recovery occurred in from three to seventy-two hours. This is otherwise not the rule, even in cases of venesection which, as a matter of course, is always done in sick individuals.

2. The typical cases generally occur after the lapse of several days (usually five to eight days, even twenty-one days or more), *i.e.*, at a time when the blood pressure has risen and the circulation has been restored. The onset is usually more or less sudden.

3. Similar conditions are sometimes observed after trifling hemorrhages, when the factor of diminished blood pressure does not come

into question. Strictly speaking, these cases do not belong in this category, because there is evidently a coincident disease of the vessels at the site of the hemorrhage and in the optic nerve.

In all cases the cause appears to be a hemorrhage into the optic nerve (which may have been absorbed at the time of examination), resulting from necrobiosis (fatty degeneration) of the previously diseased walls of the vessels. This condition is consequent on protracted disturbance of nutrition from insufficient supply of blood.

On account of the rapidly sinking blood pressure, very little or no blood enters the vessels of the retina and optic nerve, and this is true in a measure of the cranial vessels. The result is that things grow black before the eyes, and blindness and syncope set in. Under certain predisposing conditions necrobioses develop in the walls of the vessels during the depression of circulation. With returning circulation the syncope ceases and vision is restored. It is only after a certain lapse of time that solutions of continuity occur in the walls of the vessels, and hemorrhages ensue. These are comparatively harmless in the retina, but in the optic nerve they cause signs of retrobulbar neuritis with subsequent atrophy. If the hemorrhage is situated far back, the ophthalmoscopic appearances are at first negative. If it is situated near the papilla we find visible changes with opacity and stasis, because the current in the optic nerve and its sheaths normally flows toward the cranial cavity.

It is also easy to understand that a central scotoma occasionally develops. Cerebral disorders of vision, such as homonymous hemianopsia, blindness with intact reaction of the pupils, etc., are also observed at times. Such cases appear to be due to temporary disturbance of the circulation within the cranium, and the prognosis is favorable.

Early puncture of the chamber is said to have acted favorably by diminishing the ocular pressure (v. Graefe), but I find no reference to this point in later writings. If the vessels are brittle this treatment would favor the development of retinal hemorrhages, and I would advise against it.

In disorders of the circulation we not infrequently notice visible pulsation of the retinal vessels, not alone confined to the papilla, as in the physiological venous pulse, but often extending far toward the periphery of the retina.

In aortic insufficiency there is almost constantly a pulsation of the arteries, synchronous with the radial pulse, and alternating with enlargement of the veins. The conditions are very favorable to the development of pulsation in the vessels of the eye. As a result of the compensatory hypertrophy of the left ventricle the blood pressure is greatly increased during systole, and markedly diminished during diastole on account of the reflux of blood through the insufficient valves (Becker, *Arch. f. Ophth.*, XVII, 1, and *Monatsbl. f. Aug.*, 1871, p. 80; Quincke, *Berl. kl. Woch.*, 1868, No. 34). But a visible arterial pulse may also be absent in aortic insufficiency, especially if the latter is combined with a certain amount of stenosis, or it is absent during rest and appears when the heart's action is increased. It may disappear under the influence of digitalis. A similar condition is observed much more rarely in other cardiac lesions, and also in Basedow's disease and in extreme anæmia and chlorosis.

Alternate reddening and pallor of the optic papilla may also be observed in aortic insufficiency (Jaeger), and this capillary pulse is analogous to the corresponding phenomenon in the finger-nails. This symptom is, however, rarely pronounced.

Ocular symptoms may be produced either directly in the eye, or indirectly from the brain, by extensive diseases of the vessels, particularly atheroma, arterio-sclerosis, fatty degeneration and certain specific diseases, such as syphilis, albuminuria, leukæmia, chronic infections and poisons, etc. The different anatomo-pathological forms can rarely be distinguished from one another with the ophthalmoscope. Often as such vessel-changes are found post-mortem in the retina and choroid, they are seen with the ophthalmoscope with comparative infrequency, although not so rarely as was formerly supposed. For example, Raehlmann found visible changes in the retinal vessels twenty-four times among forty-four cases of general arterio-sclerosis (*Zeitschr. f. kl. Med.*, XVI, p. 606).

The walls of the arteries and veins have a white border (periarteritis and periphlebitis); in places they are thickened into a spindle shape. Often there are several thickenings upon the same vessel so that the lumen may be narrowed to complete obliteration. The walls of the vessels sometimes contain yellowish, fatty patches, especially upon the papilla. If the vessel remains tolerably transparent in endophlebitis and endarteritis, the column of blood at the diseased spot merely appears thickened. If the narrowness of the vessels, especially of the arteries, indicates considerable diminution of the blood pressure, pulsation is not infrequently observed. According to Raehlmann lateral displacements and flexions are more frequent than real changes of calibre.

In the further course of the process the vessels may be obliterated. Sinuosities and dilatations make their appearance, particularly on the veins, and often alternate with constrictions. More or less numerous miliary aneurisms may also form upon the arteries. But such cases are very rare (Michaelsen, Hirschberg, Schleich, Liebreich, Schmall, Raehlmann), and can only occasionally be demonstrated post mortem (Liouville).

The recovery of such a disease of the vessels is occasionally noted, as, for example, by Seggel, in a case of syphilis.

Such processes occur more frequently and usually much more extensively in the choroid than in the retina. Naturally their demonstration is generally possible only after death.

More striking than these diseases of the vessels, which are often discovered only after very careful examination, are their results, viz., the hemorrhages which occur either spontaneously or upon comparatively slight provocation (bending, lifting, warm or cold bath, etc.). They may occur into any vascular part of the eye, but they are usually most striking in the conjunctiva, retina and vitreous humor. They may be absorbed, but in the retina an extensive, white discoloration is often left either at the middle or at the periphery. Relapses are common. Inasmuch as disease of the vessels in the eye points to a similar affection of the vessels within the cranium,—among thirty-five cases of atheroma in the skull Raehlmann found in twenty of them a similar affection in the conjunctiva, retina or vitreous—spontaneous hemorrhages, or those which occur after trifling causes, are often forerunners of cerebral hemorrhages. If the latter are not directly fatal, they may in turn give rise, according to their location, to central disturbances of vision.   In old people we have to deal chiefly with atheroma, in younger people with other diseases of the vessels.   In both cases the prognosis is unfavorable, except in syphilis, and even the later effects may be disastrous.

According to v. Graefe arterio-sclerosis may be the cause of hemorrhagic glaucoma.   It would probably be more correct to say that this affection is glaucoma in an eye whose vessels are diseased, and that this explains its clinical peculiarities.

According to Michel and his pupils, atheroma of the carotid may give rise to cataract.   When the latter is unilateral, atheroma can almost always be demonstrated on palpating the carotid, or, if the atheroma is present on both sides, it is more pronounced on the side of the diseased eye.   Other observers have been unable to confirm these statements, and this agrees with my experience.   Indeed, I have not infrequently noticed the opposite state of affairs, viz., very slight or no atheroma upon the side of the cataract.   For this very reason I do not believe, as Michel is inclined to assume, that negative results are due mainly to insufficient examination.   Moreover, it is difficult to see how atheroma of the carotid may be the direct cause of cataract, while a similar affection in the ciliary body or processes might very well give rise to it.   But this lesion has not been proved to exist.

Hemorrhoids formerly played a great part in the etiology of diseases of the eye.   At the present time such a relation is hardly mentioned.   A profuse hemorrhoidal hemorrhage might occasionally give rise to " amblyopia or amaurosis from loss of blood," but this appears to be comparatively rare.

We have already referred to retinal pulsation in aortic insufficiency.   Schmall (Arch. f. Ophth., XXXIV, 1, p. 37) observed it twice in twenty-two cases of mitral disease, and once in obliteration of the pericardial cavity.   In such cases other factors must have given rise to a decided increase in the difference between the systolic and diastolic blood pressure.

General venous congestion or cyanosis as the result of heart disease will also be noticeable in the eye.   In the last stages of valvular lesions the cyanosis, œdema and dyspnœa will produce their effects on the eye, will give rise to hemorrhages, etc.

In a case of dilatation of the heart without valvular disease, Knapp (Jahresb. f. Aug., 1870, p. 337) observed general cyanosis, aneurismal murmurs in many parts of the body and excessive hyperæmia of the retinæ with innumerable thick and sinuous arteries and veins.

The hypertrophy of the left ventricle, which occurs quite constantly in the later stages of chronic renal disease, was formerly regarded as the cause of so-called retinitis albuminurica. It has been found, however, that the latter is due to definite disease of the vessels and occurs with or without hypertrophy of the heart.

In fatty heart there is not infrequently extensive disease of the vessels which may give rise to ocular hemorrhages. I have been unable to recognize a connection between fatty heart and the arcus senilis (fatty degeneration of the corneal tissue). As a rule, fatty heart could not be demonstrated in those very cases in which the arcus senilis appeared at a prematurely early age.

Diseases of the heart and vessels may also give rise to thrombosis and embolism in the eye. If the emboli are infectious they result in a specific new formation, an inflammation or suppuration (ulcerative endocarditis). If the embolus is not infectious, its effects may vary. Emboli of the choroid produce slight symptoms because of the large size of the choroidal vessels and the numerous anastomoses which prevent a notable interference with circulation.

A choroidal embolus looks like a chorio-retinitic patch. It forms a whitish, somewhat prominent spot with indistinct borders, over which the retinal vessels pass or in which they disappear. In the latter localities there is considerable opacity of the innermost layers of the retina. The visual disturbance is essentially a scotoma corresponding to the situation of the embolus. After a time the opacity disappears and more or less atrophy and irregular pigmentation of the choroid are left. Vision may be entirely restored. Such conditions are usually seen accidentally (because the subjective symptoms are slight), especially during convalescence from certain infectious diseases (small-pox, typhoid fever). Sometimes several foci develop at the same time or successively.

Embolism of the retinal arteries runs a much more severe course, because they are terminal arteries and are connected with the choroidal vessels merely by a few capillary anastomoses at the border of the papilla.

If the embolus lodges in the trunk of the central artery of the retina, sudden blindness will develop. With the ophthalmoscope the arteries appear empty, the veins narrow, and occasionally the column of blood in them is interrupted; the papilla is pale. Diffuse opacity of the retina soon makes its appearance. This is most dense at the entrance of the optic nerve and merely allows the region of the fovea centralis to shine through as a "cherry-red spot." As a rule the blindness remains permanent, although the retinal circulation may be restored in a few days through the anastomoses at the entrance of the optic nerve. The papilla becomes atrophic, the arteries remain narrower and occasionally degenerate into connective-tissue threads.

In embolism of a branch of the central artery the latter is merely empty beyond the site of the embolus. The opacity is confined to the corresponding vascular province, and a corresponding loss in the field of vision is demonstrable. After a while the observer may follow the development of the hemorrhagic infarction in the occluded vascular region. Numerous hemorrhages appear which may be absorbed at a later period, but as a rule leave considerable pigmentation of the retina. Patches of fatty degeneration and glistening crystals are also not infrequently seen in the degenerated retina. Later, there is partial atrophy of the optic nerve, and although vision may remain fair for a long time, there is a general tendency to impairment and to complete atrophy of the optic nerve.

The embolus itself may be visible with the ophthalmoscope, and has also been found post-mortem in a number of cases. But this is not always possible, and its origin is often clinically obscure. Suddenly developing arterial thrombi may produce the same appearances. The emboli and thrombi within the cranial cavity, which give rise to eye symptoms, have already been discussed (page 135).

In aneurism of the aorta or innominate, vasomotor and oculo-pupillary sympathetic symptoms are often found on the corresponding side.

Initial irritative symptoms give place later to paralytic symptoms (ptosis, myosis, enophthalmus, etc.). Such aneurisms may also give rise to retinal pulsation and, if the optic nerve is directly involved, occasionally lead to neuritis and choked disc. Aneurisms of the internal carotid may grow toward the orbit and produce the symptoms of "pulsating exophthalmus." Although the latter is usually the result of a traumatic aneurism of an orbital artery, ligature of the corresponding common carotid is the sole remedy. As a matter of course it is only indicated when compression of the latter artery causes disappearance of the symptoms. Recovery was effected in about half the cases of operation, but relapses may also occur. Nettleship (*Brit. Med. Journ.*, 1882, I, p. 381) observed panophthalmitis develop in the corresponding eye a few days after ligature. The patient died later of cerebral abscess.

Diseases of the thyroid gland and hypophysis cerebri may give rise to eye symptoms. In diseases and neoplasms of the thyroid, vasomotor and oculo-pupillary sympathetic symptoms may appear. As the result of pressure on the trachea, a goitre may cause symptoms of suffocation, with cyanosis and visible venous hyperæmia of the retina. The most important diseases of the thyroid, such as Basedow's disease, myxœdema, cachexia strumipriva, etc., will be considered under the head of constitutional anomalies.

Diseases of the hypophysis, particularly tumors, give rise not infrequently to local symptoms, owing to the close proximity to the chiasm and optic tract. Paralysis of the ocular muscles is not uncommon. The hypertrophy of the hypophysis in acromegaly has been mentioned on page 201.

CHAPTER VI.

DISEASES OF THE URINARY ORGANS.

AMONG the diseases of the urinary apparatus, the principal ones to be considered are those formerly known as Bright's disease, now preferably included under the term albuminuria. This unscientific name indicates a symptom, not a disease, but it is serviceable for the reason that the affections of the visual apparatus which are here described occur in the most varied anatomical and clinical diseases of the kidneys in which albumin is found in the urine.

The integument of the face and eyelids takes part in the general œdema of acute nephritis or of the terminal stages of chronic affections of the kidneys. The lids, particularly the lower one, may be œdematous at a time when nothing can be discovered in other parts. Temporary œdema occurs at a very early period, but permanent œdema is only noticed, as a rule, when other parts (ankles, etc.) are also œdematous. On the other hand, œdema of the legs and even marked ascites may have been present for a long time without noticeable œdema of the lids.

Œdema of the conjunctiva (chemosis) appears to be very rare. Brecht (*Arch. f. Ophth.*, XVIII, 2, p. 120) reported a case which was associated with detachment of the retina; later the retina again became attached.

The affections of the eye in albuminuria are either intraocular or intracranial. Among the former the most important are the affections of the optic nerve and retina. Hyperæmia of the papilla and retina, retinitis, neuritis, neuro-retinitis, choked neuritis with or without hemorrhages are found not infrequently in albuminuria, in rare cases even choked disc, which differs in no respect from the same condition when due to other causes. Such findings often lead to a recognition of the renal affection which had previously given rise only to general, vague symptoms, such as headache, malaise, digestive disturbances, etc.

The characteristic retinitis and neuro-retinitis albuminurica are distinguished by two features, viz., whitish patches and hemorrhages, which may either be combined or occur separately.

An especially characteristic feature is the presence of whitish and yellowish patches, which often look greenish-white on examination in daylight, situated mainly in the inner layers of the retina. At the very start a few are seen in the region of the macula and toward or upon the papilla; later they become more numerous and are apt to form a stellate figure around the fovea centralis. At a later period they become still more numerous and coalesce in part, so that finally the entire fundus is covered with whitish and yellow-ish-white patches which are usually sharply defined. The larger patches conceal the retinal vessels, either entirely or in places. The macula lutea and the vicinity of the entrance of the optic nerve form the point of culmination of the disease, but as a rule the fovea cen-tralis remains clearly visible as a red spot. The patches diminish rapidly in number toward the periphery of the fundus. An irregu-lar light-brown pigmentation is seen not infrequently in the larger patches.

The ophthalmoscopic appearances change slowly and often remain the same for a long time. A few patches disappear, new ones ap-pear, others grow smaller or larger. On the whole, however, a ten-dency to increase is unmistakable. Complete absorption is hardly ever observed unless the renal disease disappears.

Hemorrhages are much less characteristic. They are often scanty in numbers but sometimes very numerous. As commonly occurs, they are in streaks upon the papilla and in its vicinity, and farther toward the periphery they grow rounder. They are generally found closer to the periphery than the white patches; they rarely exceed the size of the papilla. The hemorrhages may be entirely absorbed or slight pigmentation may be left, but as a rule relapses occur. Their transition into white patches is sometimes noticeable; they then assume a whitish color, which starts either from the middle or from the periphery. This condition is also observed in other retinal hemorrhages, particularly in cachectic individuals. For a time this transformation was regarded as characteristic of the hemorrhages in pernicious anæmia.

The picture of retinitis and neuro-retinitis albuminurica consists of the patches and hemorrhages just described. If the optic nerve is implicated to a marked extent, more or less atrophic discoloration of the nerve becomes noticeable at a later period.

The affection is usually bilateral, though it does not always begin at the same time and is not equally developed in the two eyes. Unilateral cases are not very rare.

In a number of cases retinal folds develop, and there may even be detachment of the membrane. This may lead to loss of sight, but it may also disappear completely. Such cases are found particularly in the albuminuria of pregnancy, and may recover completely after delivery. Hirschberg (*Jahr. f. Aug.*, 1884, p. 387) describes double retinitis albuminurica followed by detachment of the retina in a man of twenty-two years, and Anderson (*ib.*, 1888, p. 571) reports a detachment of the retina in chronic nephritis in a child. See Liebreich's "Ophthalmoscopic Atlas."

White streaks and thickenings of the walls of the vessels, similar to those described on page 293, also appear. In extensive disease of the retina this is concealed in great part by the patches and hemorrhages. It is visible, however, before the occurrence of true retinitis, and admonishes us to examine for albuminuria.

The disturbance of vision varies greatly. It is often slight but may also be very pronounced, especially in hemorrhages into the macula. It is often strikingly disproportionate to the ophthalmoscopic appearances, being usually very slight in comparison with the extensive changes in the fundus. The impairment of vision is greater when the optic nerve is affected (retrobulbar neuritis), as shown by the subsequent atrophic discoloration of the papilla. Vision often varies without corresponding ophthalmoscopic changes. Complete blindness is very rare, and occurs almost exclusively in atrophy of the optic nerve and detachment of the retina, or in coincident "uræmic amaurosis." Disorders of the color sense and definite changes in the field of vision do not belong to the clinical history of retinitis albuminurica, and are hardly ever noticed except when the optic nerve is also affected.   In one case I observed subjective green vision which lasted several days.

Anatomical lesions are found particularly in the arteries, veins and capillaries.   The small arteries and capillaries are very much thickened, especially the intima, so that the lumen is narrowed, thrombosed and finally occluded (hyaline thickening, sclerosis).   The walls of the vessels have a waxy look but do not give the reaction of waxy degeneration.   According to Duke Charles Theodore of Bavaria, this sclerosis is due to transuding white blood globules which remain in the intima and degenerate.   The capillaries exhibit numerous dilatations, and small dissecting aneurisms are found in the arteries.

This degeneration of the vessels is found not only in the retina, but also—and often to a much greater degree—in all vascular parts of the eye, the choroid, ciliary body, iris, conjunctiva, and especially the chorio-capillaries.   The latter not infrequently contain inflammatory foci.

Infiltrations of round cells, nuclear proliferation and hyperplasia of the connective tissue are also found in parts of the retina, but are not very pronounced.

The retinal hemorrhages are found wherever the vessels extend, i.e., into the intergranular layer.   They generally take place from capillaries and veins (Weeks), and many contain numerous white blood globules.

There is almost always extensive œdema of the retina, often forming large cavities which are filled with a filamentous coagulated fluid. This gives rise not infrequently to the production of retinal folds, and later to detachment of the retina.

Granular or fatty degeneration of Mueller's supporting fibres is occasionally noticed.

The white patches, which are visible with the ophthalmoscope, consist in great part of more or less extensive fatty degeneration of the tissue.   They may also be composed of hemorrhages in the stage of absorption, accumulations of granulo-fatty cells and sclerotic ganglion cells and nerve fibres.   According to Treitel the characteristic stellate figure in the macula results from accumulations of granulo-fatty cells, and must be attributed to the absorption of tissues in a condition of fatty degeneration.

The rods and cones may long remain normal; in other cases they appear swollen. The pigment epithelium is usually unchanged, but the amount of pigment is often diminished; it sometimes appears thickened and swollen.

Treitel found amyloid granules in the chiasm and optic tract.

We have to deal accordingly with a disease of the vessels, mainly sclerosis, not alone in the retina but in all the vascular parts of the eye. All other lesions are secondary, including the hemorrhages, œdema, formation of folds and detachment of the retina, fatty degeneration and other necrobioses. These secondary changes are manifested chiefly in the retina, because its arteries are end arteries and circulatory disturbances are not compensated as readily as in other places by a collateral supply. Disease of the choroidal vessels may be much more marked without giving rise to such nutritive disturbances. With regard to the pathological anatomy of retinitis albuminurica, *vide* Leber (Graefe-Saemisch, Bd. V, p. 573), Duke Charles Theodore of Bavaria ("Beitr. z. path. Anat. des Auges bei Nierenleiden," Wiesbaden, 1887), and Weeks (*Arch. f. Aug.*, XXI, p. 54).

The albuminuric affections of the retina occur in all forms of nephritis, even in the acute forms (scarlatina). In the latter they are much rarer than in certain chronic forms. They are rare in waxy kidneys and, according to Bull, are only found in this disease when the waxy degeneration occurs in contracted kidneys. They are equally rare in the large white kidney, or parenchymatous nephritis in the stage of fatty degeneration (Leyden, *Charité-Annalen*, VI, p. 228). They occur most frequently in the terminal stage of contracted kidneys. Life is rarely prolonged longer than one year, at the most two years, after the discovery of the retinal affection.

Retinitis albuminurica may be produced by whatever gives rise to the different forms of nephritis, viz., acute and chronic infectious diseases, acute and chronic poisoning, constitutional anomalies, pregnancy, etc. After pregnancy detachment of the retina as well as recovery are comparatively frequent. After other causes, apart from acute infections and poisons, recovery is rare. The condition develops mainly at an advanced age, but it may occur even before the period of puberty. Among 103 cases collected by Bull the youngest was five years old.

Unilateral retinitis albuminurica is not extremely rare. Bull (*N. Y. Med. Journ.*, July 31st, 1886) describes ten cases. Cheatham (*Amer. Med. Assn.*, 1885, Vol. V, p. 150) found in a case of left-sided retinitis that only the left kidney was diseased. Yvert (*Rec. d'Ophth.*, 1883, p. 145) observed left retinitis in a man of forty-eight years. At the autopsy the right kidney was found entirely absent, the left was in the condition of large white kidney. But this does not permit the inference that in disease of one kidney the retinal affection will be one-sided.

In retinitis albuminurica and the previously mentioned affections of the fundus in albuminuria, the diagnosis is not infrequently made with the ophthalmoscope, because the other symptoms of the disease are often very vague. The frequency of retinal disease varies from seven to thirty per cent, according to different writers. The lower figure is probably nearer the truth.

The prognosis is the same as that of the primary disease, *i.e.*, it is bad, because the same vascular changes are found constantly in the brain. Hence, the cases with hemorrhage are more unfavorable than those with fatty degeneration alone. Temporary improvement occurs frequently, but recovery is possible only when the primary disease is capable of recovery, for example, in the acute nephritis of scarlatina and other infectious diseases, in pregnancy, when abortion or premature labor may be induced, etc. Even in such cases the acute nephritis is often followed by a chronic affection, or the former apparently recovers but returns at a later period. Even when recovery from renal and retinal affections takes place, permanent blindness may result from atrophy of the optic nerve (Graefe-Saemisch, Bd. VII, p. 83). Recoveries are most frequent in pregnant women, and may take place even after detachment of the retina. Such a complete recovery from severe disease of the kidneys and retina as is reported by Adamuek (*Centralbl. f. Aug.*, 1889, p. 98) is very rare.

The retinal affection is either the result of extensive disease of the vessels which attacks at the same time the kidneys, retina and other organs, or, as in the large majority of cases, it is the result of long existing renal disease. In the former event the noxious agent acts at the same time upon the vessels of the kidney, retina and uvea, brain, etc.; in the latter event, the insufficient excretion of harmful products of disassimilation gradually causes a sort of self-infection which, like all chronic infections and poisons, induces disease of the vessels. The consequent disturbance of circulation is easily compensated in many places, but in tissues which are supplied with end-arteries (cerebral cortex, retina, etc.) nutritive disorders and necrobiosis occur in the part supplied by the vessel. In the retina they are apt to begin in the non-vascular centre of the macula lutea, where fatty degeneration in the nerve-fibre layer gives rise to the well-known stellate figure. Œdema, hemorrhages, etc., develop at a later period. If the walls of the vessels are especially brittle as the result of the disease, hemorrhages will appear from the start. For this reason such cases have a more unfavorable prognosis.

Great importance was formerly attached to the hypertrophy of the left ventricle, which is a constant feature of the final stage of chronic renal disease, and it was regarded as a necessary link between disease of the kidneys and retina. This theory has not been confirmed. When there is disease of the vessels, the hypertrophy may favor the occurrence of hemorrhages, but it is often present when there is no affection of the retina, and the latter is not infrequently diseased when the left ventricle is normal.

Hemorrhages into the conjunctiva occur occasionally during the course of retinitis, and Talko (*Jahr. f. Aug.*, 1872, p. 353) even saw them precede the retinal affection. Samelsohn (*Virch. Arch.*, Bd. 59, p. 257) observed hemorrhages into both lower lids previous to a pure hemorrhagic retinitis albuminurica. Wharton Jones describes, as a complication, a hemorrhage into Tenon's capsule with exophthalmus and blindness.

Ophthalmoscopic appearances which are similar to those of albuminuria may be found independently of the latter,—for example, in leukæmia, diabetes, severe anæmia, after profuse hemorrhages (*vide*

20

p. 290), occasionally even without any findings. But it must be remembered that albumin may be temporarily absent from the urine in renal disease, and, in view of the comparative rarity of retinal affections from the causes just mentioned, albuminuria may be excluded only after repeated examinations. Some of the retinal affections observed in diabetes mellitus must be attributed to coincident albuminuria.

I have seen retrobulbar neuritis of one eye (*i.e.*, hemorrhage into the optic nerve) with blindness in a case of albuminuria; the other eye was attacked a year later by pure hemorrhagic retinitis which spared the blind eye. Hemorrhages into the optic nerve have also been demonstrated post mortem (Duke Charles Theodore, *l.c.*).

Anderson (*Jahr. f. Aug.*, 1888, p. 571) observed detachment of the retina in chronic nephritis in a child.

According to Deutschmann albuminuria is also the cause of cataract, but Ewetzky and others have shown that albuminuria is not more frequent among cataract patients than in other individuals of the same age.

Iritis is observed occasionally and, in the absence of any other cause, may be attributed to albuminuria, particularly as the latter is often associated with extensive disease of the vessels of the iris so that a very slight exciting cause will produce iritis. I have observed two cases of this kind, but they appear to be quite rare. Despite the extensive vessel disease in the choroid, visible choroiditic changes are rare, probably because they are concealed by the pigment epithelium. They are often found on autopsy.

Chorio-retinitis is mentioned a number of times in descriptions of the ophthalmoscopic appearances, for example, by Kepincki (*Arch. f. Aug.*, 1888, p. 388), who also reports a case of embolism of the central artery in albuminuria. Schreiber (*Jahr. f. Aug.*, 1878, p. 300) also attributes a case of disseminated choroiditis to the same cause.

A notable circumstance is the rarity of hemorrhagic glaucoma even in the purely hemorrhagic form of retinitis albuminurica. Cases have been reported by Weeks (*l.c.*), Mooren (*Jahr. f. Aug.*, 1886, p. 309), Guaita (*ibid.*, 1875), and others.

Among the intracranial ocular symptoms of albuminuria we may mention paralyses of the muscles and so-called uræmic amaurosis. Both are much rarer than changes in the fundus visible with the ophthalmoscope.

Little is found in literature concerning paralyses of the ocular muscles in albuminuria. Finlayson (*Jahr. f. Aug.*, 1877, p. 240) reports right abducens paralysis due to a hemorrhage into the pons. Foerster (*l.c.*) makes no mention of these paralyses. They are so frequent, however, that in every case of sudden or rapidly developing paralysis of the ocular muscles with the character of basilar, root, or nuclear paralysis, the urine should be examined for albumin. The cause generally appears to consist of a hemorrhage in the region of the nerve roots or nuclei, possibly even in the nerve itself. Sclerosis also occurs in the nerves of the ocular muscles, as Leber demonstrated anatomically in the abducens.

The paralyses usually recover rapidly with or without treatment, but often undergo relapses in the same or in other muscles. I have recently seen three cases of this kind in rapid succession: 1. Abducens paralysis as the sole eye symptom in albuminuria of fifteen years' standing, which had been left after typhoid fever; it relapsed twice in a few months, and then the patient died. 2. A left trochlear paralysis—on the right side hemorrhages were found in the optic nerve —in contracted kidneys of unknown duration; death in six months. 3. A complicated external ophthalmoplegia in a man of twenty-four years; the albuminuria succeeded typhoid fever two years before. On the right side the trochlearis was first attacked; then, while this was recovering, the internal rectus and the other external muscles supplied by the motor oculi; finally, attacks of unilateral and bilateral ptosis. The paralyses recovered rapidly; the right internus remained paretic longest, alternating with the inferior rectus. Soon afterward right ptosis again developed for several days, and then I lost sight of the patient. The ophthalmoscopic appearances and the internal ocular muscles were always entirely normal, and syphilis could be excluded with certainty.

The muscular paralyses also appear to be terminal symptoms of albuminuria, and are certainly indicative of changes in the cerebral vessels similar to those which are found in the retina.

Uræmic amblyopia or amaurosis is more frequent than muscular paralyses, but is much less common than the retinal affections (about one per cent). As the name indicates it is a part of uræmia, and is therefore observed particularly in those forms of nephritis in which uræmic attacks are more frequent (scarlatina, pregnancy). When the nephritis is capable of recovery, uræmic blindnses may also recover. Otherwise the attacks indicate the beginning of the end.

The disturbance of vision is bilateral, usually develops suddenly or at least rapidly, and generally passes into complete blindness. The ophthalmoscopic appearances are negative. The reaction of the pupils to light is usually retained, an indication of the cortical site of the blindness. It is commonly preceded for some time by headache. Uræmic blindness may also form part of a general uræmic attack and persist after the latter. The restoration of sight is sometimes sudden and complete, for example, on the fourth day in Monod's case (*Gaz. des Hôp.*, 1870, p. 113). Usually the restoration of sight takes twenty-four to thirty-six hours.

The secretion of urine is checked or diminished, the urine has a high specific gravity and contains a good deal of albumin. Œdema, headache, vomiting, spasm, etc., are also generally noticed; the pulse is hard and tense.

A combination with retinal disease is comparatively rare; the latter is then the prior condition. Retinal disease is more frequent during the chronic course of renal affections; uræmic disorder of sight is more frequent in acute nephritis or, at least, in acute exacerbations.

There may also be other cerebral focal symptoms, such as hemiplegia, unilateral epilepsy, aphasia, etc.

The pupil may be contracted or dilated, and may or may not react to light. In pure cases the reaction to light is unchanged; when it is absent, the optic nerve or primary optic ganglia must be implicated. In cases of this kind there are not infrequently visible evidences of stasis in the papilla of the optic nerve. Litten (*Jahr. f. Aug.*) found absence of the reaction of the pupil to light once in four cases; in this case each uræmic attack was accompanied by pronounced swelling of the disc which was visible with the ophthalmoscope.

Mydriasis is generally present during eclampsia, and Wernigk (Diss. Erlangen, 1887) found that in artificial uræmia from occlusion of the ureters or extirpation of the kidneys, the first sign of the uræmic attack was mydriasis, with pallor of the fundus (vascular spasm, probably from direct action of the poison upon the muscular coat of the walls of the vessels) and convulsions. The internal and external ocular muscles may take part in the convulsions (convergence, conjugate deviation, etc.).

After enucleation of an eye Hogg (*Lancet*, June 12th, 1875) observed uræmic blindness in the other eye which was already affected with retinitis albuminurica; partial detachment of the retina occurred later, followed by replacement.

Uræmic disorder of vision is evidently the effect of a poison, either urea or the extractive matters of the urine. The latter is more probable. Acute self-infection takes place from insufficient excretion of poisonous products of disassimilation. By long-continued vascular spasm this induces partial or total anæmia of the brain, which is manifested, in individual cases, in certain places of least resistance. The coincident increase in blood pressure causes transudation and œdema in the brain, and this may even result in visible evidences of stasis in the optic nerve.

Foerster's case (*l.c.*, p. 231) of a man suffering from albuminuria, who became hemeralopic in bed, is perhaps to be regarded as a rudimentary uræmic disorder of vision.

Temporary amblyopia and blindness in infectious diseases, especially malaria, may be the result of coincident renal disease, but may also occur independently.

Not all cases that appear to be uræmic on superficial examination prove to be so on more careful investigation. This is illustrated by Plenk's case (*Jahr. f. Aug.*, 1874, p. 400) of sudden blindness in retinitis albuminurica without uræmia, but with narrow retinal vessels (retrobulbar neuritis), and Weber's case (*ib.*, 1873, p. 376) of blindness lasting two weeks, beginning six hours after a very painful labor, without eclampsia or albumin. The latter case must be interpreted as "traumatic hysteria."

Oglesby (*Jahr. f. Aug.*, 1877, p. 241) saw double neuritis with hemorrhages in hæmaturia. Jogelson (*ib.*, 1888, p. 573) reported double optic neuritis followed by atrophic discoloration of the papilla, which caused complete blindness within a few days; six weeks later vision had improved to $\frac{2}{5}$. This occurred in a woman who suffered from retention of urine after remaining all night in a cold room.

CHAPTER VII.

DISEASES OF THE SEXUAL ORGANS.

WE have to consider not only the relations of the eye to morbid processes, but also to certain physiological conditions, such as menstruation, pregnancy, parturition, childbed, etc. We will also take up certain affections of the new-born which are connected with the process of parturition.

Chlorosis, severe anæmia and the like are included among constitutional anomalies. It is true that they develop as the result of inflammations and diseases of the uterus and its surrounding parts, but the symptoms are the same as in those cases which are due to other causes.

The affections common to both sexes will first be discussed, then those peculiar to each sex.

I. Masturbation is common to both sexes, and it has been regarded as the cause of numerous affections of the eye. In view, however, of its frequency and the comparative rarity of the latter, we must be cautious in asserting a direct causal relation between the two. Foerster (*l.c.*, p. 102) mentions a large number of cases in which " pronounced hyperæmia, catarrhal inflammation and trachomatous infiltrations, which otherwise recover rapidly in young people," improved very little or not at all in masturbators between the ages of twelve and twenty years. Landesberg (*Jahr. f. Aug.*, 1881, p. 327) also attributes cases of obstinate, often relapsing conjunctivitis to masturbation, and likewise a case of impairment of accommodation and one of unilateral central scotoma. Power (*Jahr. f. Aug.*, 1887, p. 304) maintains that masturbation in males at the period of puberty leads to functional disorders such as photopsia, muscular asthenopia, blepharospasm, sometimes even to impairment of vision and pallor of the optic nerve. Hutchinson (Ophth. Hosp. Rep., VIII, 1) reports vitreous opacities and even relapsing retinal hemorrhages, leading to blindness, in young people in whom no other cause but masturbation or excessive ejaculation could be found.   Dieu (*Journ. d'Ophth.*, I, p. 188) cured pronounced amblyopia by operation for phimosis.   It appears to me, however, that a direct connection between masturbation and the eye trouble is very doubtful in these cases.

Excessive masturbation very often leads to neurasthenia and allied conditions, and in these the above-mentioned functional disorders of the eye and hyperæmic conditions, especially of the conjunctiva, are not uncommon.   As regards other conditions, it may be said that the orgasm occurring in masturbation may act as an exciting cause of a hemorrhagic process (hemorrhages into the optic nerve, retina, vitreous, detachment of the retina), especially in an already diseased organ.   When the vessels are brittle every vascular excitement is injurious, such as lifting, bending over, a tight collar, etc.

Coitus may also act as an exciting cause of hemorrhages in an already diseased organ, especially in males, but for evident reasons this cause is often concealed.   In one of these cases I observed large retinal hemorrhages, which afterward resulted in hemorrhagic glaucoma and complete blindness.   It is evident that the walls of the ocular vessels had been previously diseased.

For lack of a more suitable place we may here discuss the gonorrhœal diseases of the genital mucous membrane.   As is well known, these are often the source of infection of the conjunctiva (blennorrhœa neonatorum and gonorrhœal conjunctivitis of adults).   An interesting circumstance is the development of the latter from washing the eyes with urine, which has been used as a household remedy (Armaignac, *Jahr. f. Aug.*, 1880, p. 291).   Gonorrhœa may also be the cause of metastatic diseases of the eye, especially of different forms of iritis.   These exhibit marked clinical similarity with so-called rheumatic iritis in articular rheumatism.   They are often associated with spontaneous coagulation of the fluid in the chamber (so-called lens-shaped or gelatinous exudation, spongy iritis), but exhibit no other peculiarities.   This form of iritis, which is observed almost exclusively in men, is usually bilateral and often returns with relapses of the gonorrhœa or alternates with inflammation of the joints.

As a rule, the iritis is associated with inflammation of the joints, especially of the knee and ankle, and the clinical similarity to rheumatic iritis is thus increased. Foerster (*l.c.*, p. 86) mentions a number of cases in which the inflammation of the joints and the iritis relapsed, but not the gonorrhœa.

The pus in the knee-joint may contain gonococci as well as other inflammatory excitants, and is often found to be the product of a mixed infection. These are evidently secondary affections, probably due to the formation of ulcerations.

II. Very little can be said of diseases of the sexual organs which are peculiar to men. Hypochondria and neurasthenia are found very often in those who cannot properly perform the sexual function. Spermatorrhœa formerly played a great part as a causal factor of both affections. Now the former is geneally regarded as a result of neurasthenia. Conjunctival hyperæmia, weakness of accommodation, weakness of the interni, and the common slight narrowing of the field of vision are very frequent symptoms, while tangible anatomical changes are hardly ever present.

Dieu (*Journ. d'Ophth.*, I, p. 188) found that pronounced amblyopia, which he attributed to masturbation, was rapidly cured almost entirely by operation upon a congenital phimosis.

III. There are numerous affections of the organ of vision which have been brought into connection with the sexual life of women and with diseases of the female sexual organs (Foerster, "Handb. d. Aug.," von Graefe-Saemisch, Bd. VII; Hutchinson, Ophth. Hosp. Rep., IX, 1; Mooren, *Arch. f. Aug.*, X, supplementary vol.; Geisler, *Berl. kl. Woch.*, 1880, p. 246; Cohn, "Uterus u. Auge," Wiesbaden, 1890, etc.).

Menstruation will be first considered. The influence of normal menstruation on the normal eye is almost *nil;* that of abnormal menstruation on the normal eye is very slight, while the influence upon a diseased eye or one that has a tendency to disease may be very considerable.

The first appearance of menstruation is sometimes preceded by other hemorrhages which cease with the onset of the former. For example, in one case Dor (*Jahr. f. Aug.*, 1884, p. 389) observed bilateral hemorrhages into the vitreous.

In normal menstruation eruptions, usually herpes, sometimes appear upon the lids (Landesberg, *Centr. f. Aug.*, May, 1883). In an extremely rare case (Ransohoff, *Mon. f. Aug.*, June, 1889) the herpes appeared upon the cornea and returned with each menstrual period. Blue rings around the eyes, slight œdema of the lids, especially the lower one, are not infrequent during normal menstruation and are perfectly harmless.

According to Vance (*Jahr. f. Aug.*, 1872, p. 343), congestion of the fundus oculi is found in the majority of cases in which disorders of the central nervous system appear during menstruation.

According to Finkelstein (Diss. Petersburg, 1887) concentric narrowing of the field of vision is noticeable during normal menstruation. It begins two or three days before, reaches its maximum on the third or fourth day of menstruation, and disappears three or four days later. There is also said to be slight contraction of the field for colors, and in a few cases green was mistaken for yellow. This is probably merely a part of the general malaise during menstruation, resulting from the preceding distention of all the blood-vessels. These hysteroid disorders of sight are occasionally quite pronounced, as in Bock's case (*Wien. med. Zeit.*, 1891).

The increased congestion and the greater blood pressure prior to menstruation may cause exacerbations of already existing eye disease or of a "predisposition" to such disease (for example, to phlyctenular disease of the cornea and conjunctiva in nasal eczema), and the affection of the eye may thus assume a four weeks' type like the quotidian and tertian types of malarial regions. This occurs in the most different kinds of disease; relapse of hordeola in seborrhœa of the edges of the lids (Dianoux, Galezowsky, Pflueger), iritis after cataract operation (Mooren), herpes corneæ (Ransohoff, *l.c.*), paralysis of the motor oculi (Hasner, *Prag. med. Woch.*, 1883, No. 10), inflammations of the lachrymal duct, etc. In one case Hirschberg (*Berl. kl. Woch.*, 1872, p. 579) observed that for many years the menses were always preceded by pains in the region of the liver and small of the back, together with jaundice and yellow vision. Trousseau (*Ann. d'Ocul.*, 115, p. 242) reports a case of iritis with hypopyon which appeared two or three days before each menstrual period and also relapsed at the corresponding periods during pregnancy.

The increased congestion may cause temporary slight exophthalmus, for example, with coincident enlargement of the thyroid and acceleration of the pulse (Cohn, *Mon. f. Aug.*, 1867, p. 351), or it may increase existing exophthalmus, particularly in Basedow's disease.   Cohn's patient evidently suffered from Basedow's disease, although all the symptoms disappeared in the intervals.

Under certain circumstances, such as disease of the vessels, menstruation may also induce hemorrhages, for example, into the optic nerve (Leber, "Handb. v. Graefe-Saemisch," Bd. V, p. 819). Whether hemorrhages can be produced in perfectly healthy eyes seems to be doubtful, but hemorrhages into the anterior chamber, vitreous, and even from the conjunctiva (Perlia, *Muench. med. Woch.*, Feb. 21st, 1888) may for a long time attend menstruation. Such hemorrhages also occur as vicarious menstruation, without coincident hemorrhage from the uterus.   A neuro-retinitis which relapses during the menses (Rampoldi) may not be regarded as vicarious menstruation, but as an exacerbation of an already existing disease.

Santos Fernandez (*Jahr. f. Aug.*, 879, p. 255) claims to have observed a case in which congenital blindness suddenly recovered, at the age of twenty-two years, with the appearance of the first menstrual period.

The influence of abnormal menstruation is more pronounced. Conditions prior to the beginning of menstruation, either dysmenorrhœa or amenorrhœa, sudden suppression of the menses and the menopause give rise, on the whole, to similar symptoms in the organ of vision.

In many cases the menstrual disorder is due to a constitutional anomaly, anæmia, chlorosis, scrofula; in others it is the result of chronic uterine disease which in turn may give rise to similar constitutional anomalies.   Hence we often find menstrual disorders associated with those eye symptoms which are attributable to constitutional disorders, asthenopia in all its forms, conjunctival hyperæmia, hysterical eye symptoms, etc.

The effect of anomalies of menstruation upon the healthy is usually not considerable. Salo Cohn found that the consecutive narrowing of the field of vision was more marked than in normal menstruation. There were no differences dependent upon the different forms of disturbance of the menses. The narrowing of the visual field was all the more marked, the more pronounced the molimina; it amounted to 10 and 15°, and often varied in its extent. When menstruation and molimina were wanting, the narrowing of the visual field was also absent. It is well known that pain alone may cause concentric narrowing of the field of vision amounting to complete blindness (things growing black before the eyes); this is generally followed by syncope unless the patient is in a recumbent posture. This is favored by loss of blood, and, according to Cohn, the narrowing of the visual field was most pronounced in profuse menstruation. Metrorrhagia may also give rise to amblyopia and amaurosis from loss of blood (page 290). In a girl of twenty years Abadie (*Union Méd.*, 1874, No. 15) observed that after the onset of menstruation, which was accompanied with violent epistaxis, complete blindness developed in both eyes. At a later period the ophthalmoscope showed atrophy and pigmentation of the optic nerve, findings which are characteristic of hemorrhage into the nerve or its sheaths.

In amenorrhœa Mooren (*l.c.*) observed interstitial keratitis with monthly irritative conditions, although the menstrual flow did not appear. This influence was not felt in disseminated choroiditis and posterior sclero-choroiditis. Hence it may be inferred that the former disease was still active, while the two latter had run their course. Vicarious hemorrhages in and about the eye may also occur in amenorrhœa. This category probably includes Leber's case (*l.c.*, p. 818) of hemorrhage into the optic nerve in a girl who had not yet menstruated and suffered from malformation of the genitalia.

Such conditions are observed oftener in dysmenorrhœa. In this affection there are often inflammatory changes in the uterus and surrounding parts, such as flexions, versions, etc. Mooren ("Fünf Lustren ophth. Thaetigkeit," Wiesbaden, 1882) observed spasm of accommodation, associated with disturbed menstruation, continuing until the latter became regular. Blindness which developed before the beginning of the menstrual period was cured by Pechlinus by artificial production of the menses (Cohn, *l.c.*, p. 32). Mooren saw monthly relapses of a pannous keratitis, which ceased with the appearance of the menses. Danthon reports the improvement of iridochoroiditis after the occurrence of menstruation.

Serous iritis and disseminated choroiditis, which are so often associated with menstrual disorders at the period of puberty in anæmic and chlorotic girls, probably owe their origin to the same cause. Schiess-Gemuseus (15 Jahresb., p. 37) observed the development of serous iritis when menstruation ceased; with the recovery from the eye disease the menses returned.

Sudden suppression of the menses may result in hemorrhages. Hemorrhages into the optic nerve or its sheaths are usually bilateral and may recover almost completely. If they are situated immediately behind the papilla, the symptoms of stasis, neuritis, retinal infiltrations and the like may also be found. In addition, there may be hemorrhages into the retina, into the vitreous, hemorrhagic detachment of the retina, hemorrhages into the anterior chamber, the conjunctiva, or into any of the vascular parts of the eye. Cerebral hemorrhages into parts which are important to vision may also produce corresponding focal symptoms. McKay (*Jahr. f. Aug.*, 1882, p. 322) states that he has seen blepharospasm in sudden suppression of the menses.

Corneal infiltrations, which were seen by Daguenet and Teillais under these circumstances, are probably an accidental finding. Brierre's unilateral hemiopia (Cohn, *l.c.*, p. 106) was probably due to a small hemorrhage into one optic nerve.

In the main, hemorrhages are also found at the menopause. At this time we meet with hemorrhagic glaucoma which sometimes exhibits monthly exacerbations. Mooren states that in a case of this kind he cured the dysmenorrhœal symptoms, pains in the back, etc., by a double iridectomy.

Profuse uterine hemorrhages at the menopause may also give rise to amblyopia and amaurosis.

The remarks made concerning menstruation also hold good, in a measure, with regard to pregnancy, parturition and childbed.

In pregnancy the eyelids, like other parts of the body, are often pigmented, sometimes in a very striking manner. A certain tendency to phlyctenular affections of the conjunctiva and cornea is often present in pregnancy, likewise (especially in the latter half of pregnancy) a certain degree of general weakness of innervation, of accommodative and muscular asthenopia. Hysterical symptoms, for example polyopia, may also develop and disappear later. Bloding (Cohn, *l.c.*, p. 123) mentions a case in which "strabismus, first of one eye, then of both eyes," was a sure indication of beginning pregnancy; this was evidently a case of spasm of convergence. Nieden (*Mon. f. Aug.*, Oct., 1891) reports a case of a primipara, in whom epiphora had lasted since the third month and had been preceded by salivation and morning sickness; both lachrymal glands were somewhat swollen. Cocaine exercised a magical effect.

The development of cataract during pregnancy must always be attributed to pre-existing disease, and serous iritis may be brought to pass.

It is well known that albuminuria is frequent, especially in the second half of pregnancy, partly as the result of stasis and mechanical obstruction to the renal circulation, partly as the result of parenchymatous nephritis. Albuminuric retinitis and uræmic amaurosis have been discussed on pages 300 and 308, and likewise unilateral and bilateral detachment of the retina, which is comparatively frequent in the former condition. All of these conditions often necessitate the induction of premature labor, although this does not always lead to recovery. Schoeler (Cohn, *l.c.*, p. 138) reports the cases of two sisters, both of whom suffered during pregnancy from detachment of the retina without albuminuria or retinitis; temporary improvement occurred in one case.

The majority of formerly reported cases of temporary amaurosis during pregnancy were probably uræmic amaurosis. Albuminuria and its eye symptoms are apt to relapse in subsequent pregnancies, and generally in a more severe form.

Violent vomiting may result in conditions which resemble the amblyopia and amaurosis after loss of blood (*vide* p. 290). To this category belongs Landesberg's case (*Arch. f. Ophth.*, XXIV, 1, p. 195) : impairment of vision until the patient was merely able to discern light, with normal ophthalmoscopic appearances; recovery in four days.

As a matter of course, all possible forms of eye disease may have been present prior to the pregnancy. This is true of Lotz's case (*Mon. f. Aug.*, Sept., 1889) : temporary detachment of the retina as the result of albuminuria during pregnancy; replacement of the retina, but persistent complete blindness after premature delivery. The woman had been myopic for a long time and had suffered from chorio-retinitis.

Jaundice during pregnancy may give rise to temporary amaurosis (Lutz, Diss. Tübingen, 1882, two cases); at the autopsy on one case, "large globular structures were found to occlude some of the smaller vessels."

The not infrequent improvement of diseases of all kinds, including eye diseases, after parturition must be attributed to the increased processes of absorption.

According to Raehlmann and Witkowsky (*Jahr. f. Aug.*, 1878, p. 132), mydriasis occurs with the beginning of labor pains, and is probably due to spasm of the sympathetic. Retinitis, but especially uræmic amaurosis and eclamptic attacks, may not occur until parturition or even childbed. Very painful labors may give rise to attacks of syncope, temporary blindness, or even traumatic hysteria. To this category belongs the case of Reuling (*Jahr. f. Aug.*, 1877, p. 41) : gradual double blindness without lesion shortly after delivery, finally complete restoration; Matteson (*ib.*, 1886, p. 309) : complete blindness lasting four days; Weber (*ib.*, 1873, p. 376) : almost complete blindness coming on six hours after very painful labor, without eclampsia or albuminuria; recovery in a month.

Hemorrhages during and after labor may cause amblyopia and amaurosis from loss of blood, and this is also true of hemorrhages during abortion. In the latter condition may be found diseases of the eye which are to be attributed to some general disease, such as syphilis, which has acted as the predisposing or exciting cause of the abortion.

Symptoms of temporary hysteria also occur during childbed.   For example, Szili (*Centr. f. Aug.*, June, 1882) reports sudden blindness on the fourth day after opening a window in a darkened room; the reaction of the pupil to accommodation was retained, the ophthalmoscopic appearances were negative; recovery in six weeks.

Albuminuria and its sequelæ are also frequent during childbed. Hemorrhages into all the vascular parts of the eye may occur in consequence, or they may be the result of non-septic emboli following venous thrombosis.   Walter (*Jahr. f. Aug.*, 1881, p. 90) observed embolism of the left central artery in phlegmasia alba dolens on the fourth day of childbed.   To this category probably belong the majority of hemorrhages into the retina, optic nerve, etc., which occur during childbed, apparently without cause, in the otherwise normal eye.   The cases of neuritis (Leber, Pflueger) are possibly hemorrhages into the optic nerve immediately behind the papilla, which have led to visible changes in the latter.   As a matter of course cerebral hemorrhages may also act as the cause of visual disorders in childbed.

During parturition and childbed local and general septic infection is produced not infrequently by various inflammatory products. Metastases may thus develop in all organs, including the eye.   In very acute sepsis from the absorption of a large amount of the chemical products of decomposition, and toward the end of life in the more chronic infections, extensive retinal hemorrhages are often found. Death occurs before they have undergone any noticeable changes. In less acute infection with organic inflammatory products, embolic suppurations (pyæmia) are produced.   In the eye, the choroid and retina are chiefly attacked, particularly the retina.   The septic embolism is rapidly followed by suppuration which extends to the entire eye (panophthalmitis) and leads to its destruction.   Whether the embolism occurred in the choroid or retina can only be determined by the ophthalmoscope at the very start; later this can only be decided by anatomical examination.   Embolic panophthalmia is most frequent during the second and third weeks of childbed, and is not infrequently bilateral.   Staphylococci, streptococci and bacteria have been found.   The prognosis is usually bad both for the eye and life.

Recovery occurs occasionally, with the loss of one eye or both eyes (Hirschberg, *Centr. f. Aug.*, 1885, p. 84; Cohn, *l.c.*, p. 169).

Abscesses have also been observed in the optic nerve (Michel, *Arch. f. Ophth.*, XXIII, 2) and beneath the conjunctiva (Feuer, *Centr. f. Aug.*, 1881, Feb.).

As a matter of course the same conditions may develop when infection has taken place from an operation upon the female genitalia.

Infection may also take place with germs which are less virulent than suppurative products. In this way we can explain the cases of iritis in childbed, reported by Galezowski, unless their development at that time was a mere coincidence. This is undoubtedly true of the cases of dacryo-cysto-blennorrhœa in childbed, which are described by this writer.

Lactation is also mentioned as a cause of eye disease. It acts mainly as a debilitating factor. This is also true of repeated childbed, especially if there have been profuse losses of blood. Under such circumstances we may find the most severe forms of general anæmia with their accompanying ocular symptoms. There may even be œdema of the lids, particularly the lower one, and extensive retinal hemorrhages.

Phlyctenular diseases of the conjunctiva and cornea, particularly the severer forms of the latter, appear to me to be more frequent in nursing than in non-nursing women. Mastitis may develop from nursing infants who are suffering from blennorrhœa neonatorum (Legry, *Prog. Méd.*, 1887, No. 35).

The new-born infant is exposed during parturition to direct and indirect injuries. Forceps delivery may be attended by contusions of the eye and lids, facial paralysis, ptosis, paralysis of the superior and external rectus (Bloch, *Centr. f. Aug.*, 1891, p. 134; Berger, *Arch. f. Aug.*, XVII, p. 191). These paralyses are due in part to direct injury by the forceps, as, for example, in facial paralysis or ptosis, in part to hemorrhages from the disturbance of circulation during delivery. Such hemorrhages occur in the lids, conjunctiva, orbit and, very often, in the retina and choroid. In the two latter cases they may possibly act as the cause of "congenital" disorders of vision, but they usually recover without leaving a trace (Naunoff,

21

Arch. f. Ophth., XXXVI, 3). Hemorrhages into the orbit may give rise to exophthalmus, disturbance of vision and muscular paralyses (Philipsen, *Ann. d'Ocul.*, Dec., 1891). In congenital paralyses, however, especially when the superior rectus and levator palpebræ superioris are involved, we must always think of congenital absence of these muscles. Atrophy of the optic nerve may result from forceps delivery (Beck, *Jahr. f. Aug.*, 1889, p. 383), probably secondary to hemorrhage into the nerve.

Infection of the conjunctiva with gonorrhœal secretion during parturition gives rise to blennorrhœa neonatorum. But even in severe forms the typical gonococci are sometimes absent, and, on the other hand, the latter are sometimes found in comparatively mild cases, so that the specific nature of the germ has been doubted. Other causes may produce clinically the same disease.

Very many infections do not take place until after birth, and often the nurse, midwife, etc., are the source of infection. I have seen little epidemics of blennorrhœa neonatorum in the clientèle of certain midwives. This is true even of children who were born with unruptured membranes. In Taylor's case (*Jahr. f. Aug.*, 1871, p. 220), the disease, which began on the third day, was unilateral. In Nieden's case (*Mon. f. Aug.*, Oct., 1891), it developed at the end of twenty-four hours, was very mild, and the secretion contained no gonococci, although the four older children had suffered from blennorrhœa neonatorum with gonococci, and the mother was suffering from vaginal blennorrhœa. It appears to me, however, to be very doubtful whether we are justified in assuming that the infection is due to injurious substances which are diffused into the fœtal waters.

Magnus (*Mon. f. Aug.*, 1887, p. 389) observed a child who suffered from blennorrhœa neonatorum and secondary corneal disease at birth. Parturition had lasted three days and repeated examinations had been made, so that the infection was easily explained.

Joint diseases after blennorrhœa neonatorum have been repeatedly observed (Darier, *Jahr. f. Aug.*, 1879, p. 231; Deutschmann, *Arch. f. Ophth.*, XXXVI, 1, p. 109). Darier also observed the development of purulent otitis media from the passage of blennorrhœic pus into the ear, but this appears to be extremely rare.

The influence of diseases of the female organs upon visual disorders is usually overestimated. The real cause appears to consist of the various forms of anæmia, which have such a fruitful source in chronic female diseases. The pains, insomnia and more or less profuse hemorrhages also play a part. We will return to this question later in discussing the results of severe anæmic conditions and constitutional anomalies.

All sorts of hysterical symptoms, including Foerster's kopiopia hysterica, a form of anæsthesia dolorosa of the retina, diminished sensibility with photophobia, combined with general symptoms of weakness, may also occur in men. I have recently seen a number of excellent illustrations. The eye symptoms are so prominent and distressing because the patients cannot do any work.

I do not believe that Mooren is justified in regarding neuro-retinitis in retroflexion of the uterus and in ovarian tumors as a mere reflex conveyed through the spinal cord. It is probably due in the majority of cases to a hemorrhage into the optic nerve immediately behind the papilla. As a matter of course, the anatomical proof of this statement can be furnished in very few cases.

According to Swanzy (*Jahr. f. Aug.*, 1878, p. 264), iritis in girls of eleven to seventeen years is associated with uterine disease. It seems more probable to me that in such cases congenital syphilis is the cause of both diseases.

Collins' (*ib.*, 1886, p. 515) temporary blindness after Porro's operation is probably to be regarded as traumatic hysteria. Similar conditions may be seen occasionally after other gynecological operations.

CHAPTER VIII.

POISONS AND INFECTIOUS DISEASES.

ALTHOUGH poisoning results from some chemical substances, and infection from a living morbific germ, nevertheless this distinction cannot always be carried out. There are numerous transitions and combinations, as we shall see in discussing ptomaine poisoning and sepsis. The results of chronic poisoning and of infectious disease are very similar.

A. Poisons.

We distinguish between acute and chronic poisoning. In the former, a large amount of the poison is received at one time; in the latter, there is a constant reception of small doses, each of which may be unable to produce any symptoms. As a matter of course there are all possible transitions between these two extremes.

Acute poisoning terminates fatally, or recovers, or it may result in a more or less chronic disease, even though recovery appeared to be complete. Thus, acute chloroform poisoning may prove fatal during the narcosis, it may pass off without doing any injury, or it may induce chronic parenchymatous changes, as, for example, in the renal epithelium and the walls of the vessels.

An acute poisoning may produce direct symptoms in the eye, such as dilatation or contraction of the pupil, paralysis or spasm of accommodation, visible changes in the fundus and its vessels (quinine, aniline, nitrobenzol, carbonic acid), yellow vision, visual hallucinations and illusions, etc. Poisons which dilate or contract the pupils usually have a corresponding effect on the vessels of the fundus oculi, but in many cases this is not pronounced.

The eye symptoms often develop indirectly, as, for example, xerosis of the cornea and conjunctiva in the death agony, icterus which is often first seen on the conjunctiva, the development of cataract when there have been violent convulsions, retinal and other hemorrhages from changes in the blood-vessels, etc. The kidneys are often affected, because very many poisons are excreted through these organs, and even the most acute hemorrhagic nephritis may result. This nephritis and the secondary chronic forms may also give rise to eye disease.

Silbermann calls attention to the importance of thrombosis in many poisons, such as aniline, chlorate of potash and corrosive sublimate, but this is stoutly denied by others (Falkenberg, Marchand, *Virch. Arch.*, 123, 3, p. 567).

Acute poisoning, especially with gases, such as sulphuretted hydrogen and carbon sulphide, may induce traumatic hysteria. In poisoning with gases we often find conjunctivitis and hyperæmia of the conjunctiva, and often more or less diminution of corneal sensibility. Narcosis after all anæsthetics (ether, chloroform, nitrous oxide) may be followed by acute mania with delirium and other mental disturbances, later by dementia. According to Savage (*Brit. Med. Journ.*, Dec. 3d, 1887), such effects may follow the action of all poisons and morbific germs which are capable of producing delirium (*i.e.*, anæmia or hyperæmia of the cerebral cortex).

In chronic poisoning, direct eye symptoms are not frequent (myosis in chronic nicotine or morphine poisoning, mydriasis in chronic belladonna poisoning, etc.). Secondary eye symptoms, on the other hand, are frequent and result from vascular disease, interstitial or parenchymatous, chronic or acute hemorrhagic inflammations (polioencephalitis acuta), partly in the eye itself, partly in the peripheral or central nervous system, and very often of the kidneys. Hemorrhages and fatty degenerations in the fundus oculi, central and peripheral disorders of vision, pains, paræsthesiæ and anæsthesiæ, spasms and paralyses may develop in this way. It is evident that vessel lesions will cause visible findings or clinical symptoms when they affect end-arteries—retina, cerebral cortex, kidneys. Certain poisons often exhibit a preference as regards localization. For example, arsenical paralysis begins in the lower limbs, lead paralysis in the upper limbs.

In the final cachectic stage there may also be general symptoms of exhaustion, such as anæmic asthenopia, slight nystagmus, hemeralopia, etc.; as a matter of course, the conjunctiva and eyelids also take part in general jaundice. A scorbutic condition may ensue in the last stage of alcoholismus.

Chronic poisoning and its sequelæ may be greatly aggravated by acute infectious disease, as, for example, toxic amblyopia by influenza.

It is impossible to give a satisfactory chemical or clinical classification of poisons, because those which are chemically very different may produce very similar clinical symptoms (alcohol, tobacco, carbon sulphide, lead). I have, therefore, decided to adopt the unscientific but practical method of arranging them in alphabetical order.

This article makes no pretence to completeness, but furnishes a good general survey.

Aconitine is a local mydriatic which is sometimes given internally in certain neuralgias. Duigenam (*Jahr. f. Aug.*, 1878, p. 335), O'Brien, Stewart (*ib.*, 1879, p. 229) and Hooper (*ib.*, 1883) found mydriasis in aconite poisoning, but Glugge (*ib.*, 1881, p. 292) mentions that it is not found constantly. O'Brien also noted twitching of the lids in his case.

Æsculin.—Vide horse-chestnut.

Alcohol poisoning is a very frequent cause of eye symptoms. After the first period of excitement, acute poisoning (drunkenness) shows paralytic symptoms (diplopia) which may pass into the most profound general narcosis with insensibility of the cornea and abolition of the reaction of the pupils to light. The ocular muscles may also take part in general convulsions. On the whole, however, eye symptoms do not play a prominent part in acute alcohol poisoning. Knapp (*Arch. f. Aug. u. Ohr.*, V, p. 383) observed detachment of the retina in a myope, aged sixty-three years, who had taken a glass of brandy to relieve diarrhœa of several days' duration.

The eye symptoms are more important in chronic alcoholism. It is well known that in this condition depressing factors (acute diseases, operations, injuries, etc.) may give rise to a usually temporary mental affection in which visual hallucinations and illusions are a prominent feature (delirium tremens). At the same time there is regularly a marked concentric narrowing of the visual field which disappears in one or two weeks. This condition often develops spontaneously, however, without any known exciting cause. As a rule, the ophthalmoscopic appearances are normal, unless complications are present.

More importance attaches to the interstitial and parenchymatous changes and to diseases of the vessels as the result of chronic alcoholism. The lesions which develop in the brain, spinal cord, peripheral nerves and kidneys are especially important as regards eye symptoms. Among these the greatest prominence belongs to axial optic neuritis, incorrectly called toxic amblyopia. It begins with misty vision which gradually increases and is almost always present in both eyes, although vision is not always impaired uniformly. Central vision may be diminished to $\frac{1}{10}$ or less. Careful examination shows that there is a central amblyopic spot in the field of vision, within which perception of red and green, in severe cases also of yellow and blue, is diminished or abolished. The external boundaries of the fields for white and colors are entirely or approximately normal. The disorder of vision and of color sense in the scotoma is characteristic of interference with conduction in the optic nerve.

Ophthalmoscopic examination shows that the outer half of the papilla is pale "like dull porcelain," the inner half grayish-red, opaque, with obliterated borders and occasionally slightly swollen; the calibre of the vessels is normal. Anatomically we have to deal with an interstitial inflammation, chiefly of the axial bundles of the optic nerve, which reaches its height near the optic foramen. The proliferation of the interstitial tissue with multiplication of the vessels, which exhibit increased fulness and sclerosis of the walls, subsequently leads to destruction of the medullary sheaths of the nerve fibres. Their axis cylinders remain intact for a while, but finally they may undergo complete atrophy. Centrally and peripherally from the point of culmination the changes in the nerve appear to be purely atrophic. Uhthoff traced the ascending atrophy to the end of the optic tract, while the descending atrophy is recognized by means of the ophthalmoscope. This shows an atrophic decoloration of the outer half of the papilla where the macular fibres are collected.

At the beginning of the affection the ophthalmoscopic appearances are sometimes negative, but they may also be very distinct where there is no disturbance of vision, especially in cases of recovery.

It is often found that the patient sees better in twilight (nyctalopia), or that there are persistent colored after-images or subjective sensations of light (phosphenes). This is probably due to irritation of the nerve fibres in the inflammatory focus. The misty vision may develop quite suddenly, and there are often striking changes from day to day. Other occasional disorders of vision, such as a sudden variation in the apparent size and distance of objects, monocular diplopia and polyopia (Daguenet, *Ann. d'Ocul.*, 62, p. 136), remind us of hysterical affections and are not connected with the lesion of the optic nerve. They are probably due to irregularities of accommodation.

It is difficult to make very definite statements concerning the frequency of the disease. As a matter of course it is much more frequent in men. Among 1,000 cases of severe alcoholism Uhthoff found ophthalmoscopic changes 139 times (60 of these suffered from amblyopia); in 9 cases there was amblyopia without abnormal findings; in 53 there was opacity of the papilla and surrounding parts, and in 6 cases marked congestion of the optic nerve.

Strong and healthy individuals are rarely attacked. The majority suffer from gastric catarrh and are poorly nourished. There are also other signs of alcoholism, such as anorexia, morning vomiting, tremor of the limbs and tongue. The disease is most frequent in the later years of life when the power of resistance to injurious agencies is impaired.

The prognosis is favorable if we can improve the general nutrition and secure abstinence from alcohol. Relapses are very common and are generally attended with still greater disturbance of vision, although complete atrophy of the nerve and blindness are rare. Injections of strychnine are highly recommended in toxic amblyopias, although recovery would probably take place spontaneously.

The disease is usually chronic. Cases with rapid onset of the inflammation and hemorrhages, such as Nettleship describes in tobacco amblyopia, are extremely rare. Disorder of vision without a central scotoma is equally rare (Vossius, *Mon. f. Aug.*, 1883, p. 291).

The connection of this disease with chronic alcoholism has been denied by a number of writers, and it has been attributed to the coincident abuse of tobacco. Hutchinson (Ophth. Hosp. Rep., VIII, 1) attributes all cases partly to tobacco, partly to adulterations of the alcoholic drink (for example, oil of absinthe), and Nettleship (*Jahr. f. Aug.*, 1887, p. 257) states that he has never seen a case of pure alcohol amblyopia. On the other hand Millingen never observed toxic amblyopia among the Turks, who smoke but do not drink, and Fumagalli (*Jahr. f. Aug.*, 1874, p. 454) denies the occurrence of tobacco amblyopia. There can be no doubt that pure alcohol amblyopia does occur and is all the more apt to develop, the stronger the alcoholic drink. Among beer drinkers it is very rare, among whiskey drinkers it is quite frequent. In the majority of cases the patients abuse tobacco as well as alcohol. In my opinion the former is more injurious than the latter. There are, however, pure cases of alcohol as well as tobacco amblyopia.

There are clinical differences, although not of an absolute character, between the two forms. Hirschberg (*Deutsch. Zschr. f. pract. Med.*, 1878, 17) states that in alcohol amblyopia the scotoma is central and always includes the point of fixation, while the tobacco scotoma is situated near, but not at the point of fixation (paracentral). This may hold good of the majority of cases but is not always true. Alcohol amblyopia is said to be bilateral and to occur more suddenly; the pupil is often wide and accommodation paretic. Tobacco amblyopia is said to be often unilateral or at least very different in degree on the two sides; the pupil is narrow (because nicotine is a myotic) and spasm of accommodation is frequent. The disorder of vision develops more gradually and is often progressive despite abstinence. [Muscular asthenopia is common with tobacco amblyopia.—Ed.]

Similar inflammatory phenomena are often observed in other nerves (multiple alcoholic neuritis). The central nervous system is also attacked in many cases. Alcoholic forms of insanity are very frequent, while we rarely observe true focal symptoms, such as hemi-

anæsthesia, including the conjunctiva and cornea, central disorders
of vision, hemiplegia. Paraplegia and paranæsthesia also occur,
and occasionally every possible systemic and focal disease of the
brain and cord may be simulated (particularly tabes, by multiple
neuritis of the sensory nerves).

In such cases the ophthalmoscopic findings may be of the highest
diagnostic importance. Alcoholic pseudo-tabes is also unattended
by myosis and reflex rigidity of the pupil, the girdle sensation, and
disturbances of the bladder and rectum. Its development is more
acute than in true tabes.

Romiée (*Rec. d'Ophth.*, 1881, p. 33) states that paresis of accommodation is one of the first disorders of vision in alcoholism; the
pupil is more often dilated than narrow. Other paralyses of the ocular muscles are only observed in combination with severe disease of
the central nervous system and with the alcoholic mental disorders.
Among 1,000 cases Uhthoff found inequality of the pupils in 25 cases,
reflex rigidity of the pupil in 10 cases, and very slight reaction to
light in 25 cases; the reaction of convergence was almost always
intact. In only 3 cases was there true muscular paralysis, each time
a double abducens paralysis (possibly spasm of convergence). Two
cases exhibited nystagmus, 13 cases nystagmus-like twitchings at
the borders of the field of vision.

Acute alcoholic paralysis of the eye muscles is due to hemorrhagic
inflammation of the floor of the fourth ventricle. This begins suddenly or within a few days, and is attended by headache, vomiting,
pains in the limbs, ataxia, delirium, etc. Then there is more or less
complete ophthalmoplegia externa with or without ptosis. Occasionally the internal muscles are also attacked. All but two cases of this
" acute superior polio-encephalitis" occurred in hard drinkers, and a
fatal termination ensued very rapidly. It is probable that similar
hemorrhagic inflammatory processes in drunkards also occur in other
parts of the central nervous system.

The parenchymatous and interstitial renal lesions which develop
in chronic alcoholism are often the cause of secondary changes in
the eye. After long-continued alcoholism there may be a general
tendency to hemorrhages (alcoholic scurvy). This may lead to hem-

orrhages into the conjunctiva, retina, optic nerve, brain, etc., as happens in other poisons, infectious and constitutional diseases.

Among 500 insane alcoholics Uhthoff found 37 cases of xerosis of the conjunctiva in the shape of small, dull, rough, often frothy patches. The necrobiosis and fatty degeneration of the epithelial cells are evidently a part of the general impairment of nutrition; this is also true of the hemeralopia which is often noticed.

Deneffe (*Jahr. f. Aug.*, 1872, p. 373) observed sudden blindness (there was merely quantitative perception of light without any ophthalmoscopic findings) in a formerly temperate individual who had been on a continuous debauch for a number of weeks. Vigorous antiphlogistic treatment resulted in rapid and complete recovery. On the other hand, Bruns observed marked impairment of vision (fingers could just be recognized at four feet) without corresponding ophthalmoscopic appearances in an individual who suddenly abstained from drink. Strychnine injections produced a cure in six days.

Amyl alcohol (fusel oil) is said by some writers to be the real cause of alcoholic amblyopia.

Amyl nitrite causes dilatation of the retinal vessels and increased redness of the papilla, according to Aldridge (*Jahr. f. Aug.*, 1871, p. 322). This was not corroborated by others, despite the dilatation of the vessels of the face and violent throbbing of the carotids. Pick (*Centr. f. d. med. Wiss.*, 1873, p. 865) observed, on looking at a light wall, a round, yellow spot, surrounded by a bluish-violet zone, *i.e.*, a subjective perception of the yellow spot and the visual purple. According to Schiff ("*Jahr. f. Aug.*, 1874, p. 151) sensibility remains intact when the amyl is administered to animals, but the pupil does not react to sensory irritation. A favorable result from the inhalation of a few drops of amyl has been observed in several cases of toxic amblyopia, but especially in ischæmic conditions of the retina and optic nerve. [Such a result has been observed by myself in several instances; once in a man who for almost twenty-four hours had blindness of both eyes, in the beginning total, and at the time I saw him slightly less serious, permitting some perception of light. The retinal arteries were almost empty and the nerves pallid. Inhalation of nitrite of amyl soon brought about the usual congestion of the face and dizziness with improved sight, and in twenty minutes vision was fully restored and the ophthalmoscopic picture became normal. —ED.]

Aniline produces, according to Galezowski (*Rec. d'Ophth.*, 1876, p. 210), a bluish-green color of the hair of workers in this dye, also vertigo, headache and nausea. In two cases it caused a photophobia, misty vision and ciliary injection; in one case it gave rise to severe relapsing iritis, first of the right eye, and five or six months later of the left eye. The connection between these conditions is obscure. According to Leloir (*Gaz. Méd.*, 1879, p. 606) there is marked mydriasis in aniline poisoning. In Mueller's case, however, there was narrowing of the pupil (*Deutsch. med. Woch.*, 1887, No. 2). In a case of poisoning with nitrobenzol which was adulterated with aniline, Litten (*Berl. kl. Woch.*, 1881, Nos. 1 and 2) found the pupils very small but still reacting, the conjunctiva of a violet color and with hemorrhages in the cul-de-sac. The fundus oculi was intensely red, the vessels looked as if filled with ink; there were also a few hemorrhages. Vision was unimpaired. In an aniline worker, aged forty-four years, MacKinlay (Ophth. Soc. of the United Kingdom, 1886, p. 144) found a brownish color of the conjunctiva and cornea, and slight impairment of vision. The condition improved when the patient left the factory.

Antifebrin produces dilatation of the retinal veins, according to Mueller, but this is denied by Herczel. After the administration of seven doses of 15 grains each within two and one-half hours, Simpson observed merely narrow and immobile pupils; the other symptoms were unimportant.

Antipyrin may produce urticaria upon the lids. Berger (*Jahr. f. Aug.*, 1889, p. 507) observed hypersecretion of tears; Guttmann (*ibid.*, 1887, p. 256) observed amaurosis, lasting one minute, after a 15-grain dose in a delicate woman of twenty-five years.

Apomorphine gives rise, according to Bergmeister and Ludwig (*ibid.*, 1885, p. 248), to complete anæsthesia of the cornea and conjunctiva within ten minutes after injection, and also produces corneal cloudiness and disagreeable general symptoms.

Argentum Nitricum.—*Vide* Silver.

Arsenic poisoning, when acute, resembles an attack of cholera. When chronic, four stages may be distinguished: 1, vomiting, diarrhœa and headache; 2, bronchial, nasal, conjunctival catarrh and cutaneous eruptions; 3, sensory disorders; 4, paralyses, especially of the legs.

Prolonged administration of arsenic may give rise to herpes zoster (Hutchinson). It is also said to be followed occasionally by a brown discoloration of the skin, which may be mistaken for Addison's disease and disappear after the remedy is discontinued (Foerster, *Berl. kl. Woch.*, 1890, No. 50).

In young workers in artificial flowers, Kittel (*Jahr. f. Aug.*, 1873, p. 240) noticed redness and scales upon the lids; in a few there were small ulcerations on the conjunctiva of the lower lid.

Liebert (*Mon. f. Aug.*, 1891, p. 181) observed retro-bulbar neuritis after the use of arsenic; on both sides there was a distinct paracentral scotoma for red and green with normal visual fields. The patient had taken arsenic for three and one-half years to relieve general psoriasis. The visual disorder developed two weeks after the dose was increased.

Hoffmann (*Jahr. f. Aug.*, 1889, p. 508) noticed slight nystagmus in chronic arsenic poisoning. According to Hutchinson (*Ophth. Review*, Jan., 1889), prolonged administration of arsenic in skin disease may give rise to vitreous opacities which he attributes to a peripheral retinitis.

According to Sury-Bienz (*Viertelj. f. ger. Med.*, Neue Folge, Bd. 49, 2), a brownish-red, later icteric color of the conjunctiva occurs in poisoning with arseniuretted hydrogen.

Atropine is the most generally used mydriatic in ophthalmology, and hence the majority of poisonings follow its application to the eyes. It is said that the most acute symptoms of poisoning have followed the introduction of a single drop of the ordinary solution. The most striking symptom is the marked dilatation of the pupils, but this does not always take place to the maximum extent, even in fatal cases.

The internal administration of atropine and belladonna also produces hyperæmia of the fundus and may even excite an attack of glaucoma, while conjunctival catarrh and eczema of the lids, sometimes extending over the entire face to the neck, are found after the instillation of impure solutions in predisposed individuals.

Visual hallucinations are sometimes prominent. They are especially apt to occur in the dark, and are frequent in chronic poisoning. The latter is sometimes induced by prolonged instillation of atropine. The bitter taste, scratching in the throat and difficult deglutition which are constantly present in acute poisoning are usually wanting in the chronic form, while the hard pulse, congested face and digestive disturbances are prominent.

Injections of morphine are the best antidote, while pilocarpine cannot be recommended so highly.

Reich (*Centr. f. Aug.*, 1889, April) observed the daily occurrence of epistaxis, which lasted about a quarter of an hour, five to ten minutes after the instillation of atropine for myopia; nothing abnormal was found in the nose. A solution of duboisine had the same effect, but to a less marked degree.

Tansley (*Jahr. f. Aug.*, 1877, p. 116) observed mydriasis of the same side after the introduction of atropine into the auditory meatus and after the application of a belladonna plaster to one side of the chest. This was probably due to the accidental transmission of the drug to the eyes by the patient.

Belladonna.—*Vide* Atropine.

Botylismus.—*Vide* Ptomaine poisoning, p. 358.

Bromide of potassium given internally not infrequently produces conjunctivitis, sometimes with phlyctenular foci of inflammation, even when a bromide eruption is not present on the rest of the body. In a patient who had taken 90.0 in twenty-eight hours, Schweig (*N. Y. Med. Rec.*, Dec. 30th, 1876) observed comatose sleep of four days' duration, lowered temperature, a feeble, accelerated pulse, superficial stertorous respiration, salivation, conjunctival catarrh and slight mydriasis lasting several days. In an epileptic lunatic of twenty-three years, who took 10.0–15.0 potassium bromide daily, Ruebel (*Jahr. f. Aug.*, 1884, p. 337) observed sudden blindness with a pale papilla and narrow vessels; recovery in five weeks after discontinuing the remedy, and a relapse on its renewal.

Caffein.— Vide Coffee.

Calabar.— Vide Eserine.

Calomel.— Vide Mercury.

Cannabis Indica.—According to Ali (*Rec. d'Ophth.*, 1876, p. 258), hasheesh acts like tobacco except that it is more injurious and more often produces toxic amblyopia. This is often unilateral, as in the case of tobacco, and there is often merely a scotoma without color disturbance. It is well known that visual hallucinations are one of the chief symptoms of a hasheesh debauch.

According to Oliver (*Jahr. f. Aug.*, 1883, p. 302), sight is veiled; there is disturbance of accommodation with normal or narrow pupils. Suesskind (*Wuertemb. aerztl. Corr.-Bl.*, 1887, No. 31) and Casiccia (*Jahr. f. Aug.*, 1883, p. 302) also saw mydriasis after poisoning with cannabis.

Werner (*ibid.*, 1886, p. 255) observed violet vision (which always precedes the yellow vision of santonin poisoning!) and a mist before the eyes, after a 0.04 dose of the extract in a small and nervous woman.

Carbolic Acid.—Falkson (*Arch. f. kl. Chir.*, XXVI, p. 204) states that in poisoning with carbolic acid the pupils are usually narrow, rarely dilated, and that they react sluggishly. Silk (*Jahr. f. Aug.*, 1881, p. 291) saw narrow pupils during profound coma in a fatal case. Nieden (*Berl. kl. Woch.*, 1882, No. 49) observed amaurosis, lasting twenty hours, with normal ophthalmoscopic appearances and dilated pupils. But as this was a case of irrigation of an empyema cavity, followed at once by collapse with great weakness and nausea, it is questionable whether the eye symptoms were due to carbolic-acid poisoning.

Carbonic-oxide poisoning is the type of a severe poisoning with gases; coal gas and illuminating gas are its most frequent sources.

In the acute initial stage the eyes are not affected, and there is very little that is characteristic in the eye symptoms during recovery. In one case Ball (*Jahr. f. Aug.*, 1878, p. 253) observed a paradoxical reaction of the pupil during the poisoning, *i.e.*, the pupil dilated on the entrance of light and contracted in the dark.

In fatal cases fatty degeneration is found in the liver, spleen, kidneys, muscles, etc. Within the skull there is marked congestion, which is absent only at the very start or after the process has lasted a long time; later there are numerous capillary hemorrhages, more rarely larger hemorrhages, especially in the basal ganglia. This results later in red and yellow softening. These changes are due to fatty degeneration of the cerebral arteries. Similar changes may also be found in the spinal cord and peripheral nerves. This explains the various affections of the brain (Illing, *Wien. med. Zeitschr.*, 1874, No. 23: homonymous hemianopsia) and spinal cord, the sensory and motor paralyses, the neuroses, temporary diabetes, etc., which are observed as sequelæ of carbonic-oxide poisoning.

The ophthalmoplegias, which have been observed in a number of cases and which almost always terminate in more or less complete recovery, are due to hemorrhagic processes in the nerve nuclei or in the peripheral nerves; Knapp (*Arch. f. Aug.*, IX, 2, p. 229), paralysis of all the ocular muscles, partial recovery in two months; Emmert (*Corresp.-Blatt f. Schweiz. Aerzte*, 1890, p. 42), central paralysis of the left motor oculi and some of the branches of the trigeminus and facial; almost complete recovery.

Herpes zoster has also been noted as a sequel in a number of cases. If the first branch of the trigeminus is implicated, the eye may also be implicated (Sattler, *Jahr. f. Aug.*, 1889, p. 555). This case is not perfectly clear because the anterior part of the eye contained mycotic thrombi, evidently starting from the cornea. (Attention must be called, however, to the fact that various writers regard herpes zoster in general as infectious.) It is a noteworthy fact that only those fibres which started from the Gasserian ganglion, not those which passed through the ganglion, were degenerated. The former could be traced to the ciliary ganglion. '

Retinal hemorrhages are rarely mentioned (Becker, *Jahr. f. Aug.*, 1889, p. 511; associated with venous congestion of the fundus), although they would undoubtedly have been found in many cases if a search had been instituted.

Carbonic-acid accumulation in the blood causes mydriasis, according to Schiff, as the result of sympathetic irritation. The ophthalmoscope shows a characteristic dark, sometimes almost black, color of the columns of blood in the vessels, particularly the veins. If the cause consists of obstruction to respiration or circulation, hemorrhages are frequent. They are produced mechanically. The pupil is often contracted.

Chloral, when given in narcotic doses, produces marked myosis, during which there is abolition of the reflex dilatation of the pupil after sensory irritation; mydriasis is said to occur after very large doses or prolonged administration. Hence, myosis is generally present in chloral poisoning. The ophthalmoscopic appearances are negative.

After a variable length of time cutaneous eruptions and conjunctivitis are observed, and may necessitate discontinuance of the remedy; urticaria of the lids also occurs.

Disturbances of vision occur in rare cases. Kirkpatrick (*Jahr. f. Aug.*, 1873, p. 214) observed temporary blindness in addition to severe nervous seizures. Mandeville (*ibid.*, 1872, p. 458) reports, in addition to considerable conjunctival irritation, diplopia of several days' duration. In a woman, who suffered twice from marked impairment of vision after a dose of chloral, he claims to have observed an exudative irido-choroiditis, which was finally cured by iridectomy. Two months later a relapse developed after another dose of chloral. Was not the irido-choroiditis already present? Patruban (*Wien. med. Presse*, 1875, p. 1,046) mentions a panophthalmitis after poisoning with chloral.

Chlorate of potash may cause uræmic amaurosis through the intense hemorrhagic nephritis which large doses will produce.

Chloroform.—Opinions differ in regard to the condition of the pupils during the administration of chloroform. During narcosis the pupils are narrow as in sleep; if the narcosis is not very profound, the pupils dilate after cutaneous irritation or on calling loudly. The pupils are usually dilated during the stage of excitement prior to narcosis proper. According to Budin (*Gaz. des Hôp.*, 1874, p. 910), vomiting may also produce mydriasis, but Schiff (*Jahr. f. Aug.*, 1874, p. 150) contradicts this in part, in view of his experiments on

22

dogs. It is evident, however, that the conditions in animals are often different from those in men.

Dilatation of the pupil during narcosis indicates the necessity for caution; sudden marked mydriasis is an evidence of impending asphyxia.

Vogel (*Jahr. f. Aug.*, 1879, p. 82) assumes three different ways in which the pupil acts during chloroform narcosis: 1. Contraction during deep narcosis, dilatation on sudden waking. 2. The opposite condition. 3. The pupil is moderately narrow and does not react.

According to Warner (*ib.*, 1877, p. 217) the ocular movements become disassociated in chloroform narcosis, but do not in ether narcosis.

Repeated use of chloroform—occasionally, perhaps, a single administration—may result in fatty parenchymatous degenerations of various organs, which may not produce clinical symptoms until a much later period.

Schirmer (*Mon. f. Aug.*, 1871, p. 246) reports a detachment of the retina which, according to the statement of the myopic patient, developed during narcosis (struggles of vomiting?). Mathewson (*Jahr. f. Aug.*, 1876, p. 238) mentions an apoplectic attack, which developed fifteen minutes after an iridectomy under chloroform, in an individual with general atheroma of the vessels. As a matter of course the vomiting and choking of narcosis may prove an exciting cause for hemorrhages in any organ.

In an individual who had drunk chloroform, Niemann (*Berl. kl. Woch.* 1887, No. 1) observed myosis followed by mydriasis.

Chromic Acid.—Yellow vision has been observed in several cases in which sweating feet were treated with a five-per-cent solution of chromic acid (*Deutsch. Militaeraerztl. Zschr.*, 1890, p. 239).

Chrysarobin ointment when rubbed into the skin is said to produce conjunctivitis. According to Trousseau (*Jahr. f. Aug.*, 1886, p. 331), it develops in a few hours, is almost always bilateral, and is free from secretion, while conjunctivitis which is due to the direct entrance of chrysarobin into the conjunctival sac is usually unilateral and attended by profuse secretion. These statements are in part disputed by others, and it is by no means certain that all the conjunctival attacks do not result from direct inoculation. In rare cases, corneal ulcers have been observed.

Cocaine may produce symptoms of poisoning, even after the subcutaneous injection of 0.004, and several cases are reported in which its introduction into the conjunctival sac produced general symptoms (Meyerhausen, *Wien. med. Presse*, 1885, No. 22; Burchard, *Charité-Annal.*, XIII, p. 653).

Maximum mydriasis appears in acute poisoning, but the cornea, according to Bettelheim, is very sensitive to touch. [This statement is of dubious value. Its probable explanation is that corneal insensibility had disappeared while mydriasis remained.—ED.] Liq. ammon. anisat. is said to act favorably as an antidote.

In chronic poisoning visual hallucinations play a prominent part. Diplopia, amblyopia, dancing of objects, colored vision, micropsia, etc., are observed occasionally.

Attacks of glaucoma following instillation of cocaine solution have been observed in a number of instances (Chisolm, Maier, Javal, Manz). Marckwort (*Arch. f. Aug.*, 1887, p. 452) also observed acute glaucoma after the protracted application of cocaine to the nasal mucous membrane.

Coffee.—Caffeine is a feeble mydriatic and is also said to be anæsthetic in a measure. Hutchinson (*Centralbl. f. Aug.*, 1887, p. 240) claims to have seen a caffeine amblyopia which closely resembled quinine amblyopia.

Coniine.—Hugo Schulz (*Deutsch. med. Woch.*, 1887, No. 23) found that inhalations of coniine gave rise to violent headache, increased heart's action, profuse perspiration, epiphora, insomnia, mental confusion, inability to keep the eyes open and burning of the conjunctiva (hyperæmia).

Creosote.—Vide Iodoform (Hutchinson's case).

Curare, when given internally, causes myosis, but at a very late period.

Cyanide of Potassium.—During the stage of asphyxia of potassium-cyanide poisoning, Mueller-Warneck (*Berl. kl. Woch.*, 1878, No. 5) observed prominence of the eyeballs and enormous dilatation of the pupils, which were completely irresponsive. Souwers (*Jahr.*

f. Aug., 1878, p. 235) attributes a swelling of the upper lids and sluggishness of the pupils, which he observed in a photographer, to his frequent use of potassium cyanide.

De Tatham (*ib.*, 1884, p. 337) claims to have observed temporary hemiopia (lasting a few hours) from inhalation of the vapor of dilute prussic acid.

Cytisin (cytisus laburnum) gave rise in one case to vomiting, small pulse, cold sweat, dilated pupils, pale optic nerve and narrow retinal vessels, symptoms similar to those of ergotin poisoning (Albutt, *Jahr. f. Aug.*, 1872, p. 344).

Daturine is said, by the majority of writers, to be identical with atropine; according to Ladenburg, it is identical with hyoscyamine.

Digitalis.—In a man of fifty-seven years, who was poisoned with 90.0 tincture digitalis, Jeanton (*Gaz. des Hôp.*, 1885, No. 56) observed mydriasis and cloudy vision, on the second day yellow vision; recovery in a week. In a fatal case Hauber (*Muench. med. Woch.*, 1890, No. 42) observed marked myosis which continued after death.

Duboisine is said to produce general symptoms more readily than atropine; they are identical with those of the latter drug. Symptoms of poisoning after dropping into the eye are reported by Carl (*Mon. f. Aug.*, 1879, p. 339), Chadwick (*Centr. f. Aug.*, 1887, p. 155), Jacubowitsch (*Jahr. f. Aug.*, 1884, p. 334). Like atropine it may also excite attacks of glaucoma.

Ergotin and other preparations of ergot cause contraction of the smooth muscular fibres, especially of the blood-vessels, and may thus lead to necrosis of the tissues. The narrowing of the vessels and the pallor of the papilla may be visible with the ophthalmoscope. Temporary disorders of vision are due directly to spasm of the vessels in the eye, possibly also in the brain. The pupils are usually somewhat dilated and react sluggishly.

The general vascular spasm may also result in hemorrhages. Davidson (*Jahr. f. Aug.*, 1883, p. 303) observed hæmatemesis, hæmaturia, hemorrhages into the lids and lips, etc. It is probable that such accidents occur in individuals with diseased vessels and poor general nutrition.

Cataract, which is usually double, develops not infrequently within a few years after ergotin poisoning which is evidently in causal relation with it. The hardness or softness of the cataract, its rapid or slow advance correspond to the patient's age. Tepljaschin (*Jahr. f. Aug.*, 1891, p. 275) reported twenty-seven cases of this kind after an epidemic in the winter of 1879–80, in which the convulsive, not the gangrenous, form of ergotism had appeared. The cause is to be regarded as nutritive disturbances in the lens, owing to spasm of the vessels of the ciliary body and also of the other vessels of the interior of the eye.

Iritis after ergot poisoning in an epidemic in Upper Hessen, which was reported by Menche (*Deutsch. Arch. f. kl. Med.*, XXXIII), appears to be very doubtful in its etiology. Hulme (*Amer. Med. News*, Nov. 5th, 1887) observed swelling of the face, dilatation of the pupils and indistinct vision after an injection of an ounce of the fluid extract of ergot.

Erythrophlœin is one of a series of remedies which, when applied locally, produces anæsthesia of the cornea and conjunctiva, but is impracticable on account of the violent irritation and even inflammation to which it gives rise.

Eserine.—Although eserine is the most generally employed myotic, distinct myosis is not always present in poisoning with preparations of calabar. Indeed, pronounced dilatation and abolished reaction of the pupil have been mentioned in a number of cases (Leibholz, *Viertelj. f. ger. Med.*, 1892, p. 284). Instillation of the solution into the conjunctival sac is the most frequent cause of poisoning. Among the symptoms of the latter, Carreras Aragó (*Jahr. f. Aug.*, 1874, p. 265) observed temporary complete blindness.

Ether.—Among 1,200 ether inhalations, Jacob (*Jahrb. f. Aug.*, 1879, p. 229) observed mydriasis in six cases. Warner (*ib.*, 1877, p. 217) mentions difference in the ocular movements during ether and chloroform narcosis. If this were constant it would prove that the ether narcosis was not very profound.

Ethyl Chloride.—In dogs, Dubois (*Arch. de Physiol.*, XX, 7 8) observed long-continued dense corneal opacities, due solely to œdema.

Ethyl diamin has been found among the ptomaines. It causes mydriasis, which is often present in ptomaine poisoning.

Ethyl Dichloride.—Dubois and Roux (*Compt. Rend.*, 1869, No. 266) observed, during and after anæsthesia, an acute attack of glaucoma with pronounced opacity of the cornea, but with slight external signs of inflammation.

Ethyl Nitrite.—Hill (*Lancet*, Nov., 1878) observed dilatation and immobility of the pupil in a case of poisoning in a child of three years.

Filix Mas.—A number of cases of blindness have been observed after the use of large doses of this remedy. Eich (*Deutsch. med. Woch.*, 1891, No. 32) collated three fatal cases, one under his own observation, which ran the course of a strychnine poisoning, with narrow pupils.

Double amaurosis is most frequent (rarely there is amblyopia), without findings; the pupils are dilated and exhibit no reaction. This pupillary symptom may also be present when there is no disturbance of vision. Immermann (*Corresp.-Blatt f. Schweiz. Aerzte*, July 1st, 1887) saw double atrophy of the optic nerve, Fritz (*Cent. f. Aug.*, 1887, p. 278) unilateral atrophy with permanent blindness, develop out of the double amaurosis without findings. This proves the peripheral nature of the visual disorder which is evidently similar to that occurring in quinine poisoning. In one case, Schlier (*Muench. med. Woch.*, 1890, No. 32) found albuminuria, so that a sort of uræmic amaurosis might be suspected, but such a condition is extremely improbable.

Fish and Meat Poisoning.—*Vide* Ptomaines.

Fungus poisoning may produce various eye symptoms, according to the nature of the fungus. The active substances which they contain vary greatly, even in the same variety of fungus. Those which contain muscarine produce spasm of accommodation and myosis (*vide* Muscarine); others, especially morchella, produce mydriasis (Bayer, *Aerztl. Intell.-Bl.*, 1881, No. 1). In still others (agaricus phalloides, Handford, *Jahr. f. Aug.*, 1887, p. 250) the pupils are unaffected. In the last-mentioned case there was also amblyopia—possibly spasm or paralysis of accommodation—and at the autopsy punctate hemorrhages of the serous membranes and marked fatty degeneration of the liver were found.   Hence it is not surprising that hemorrhages and fatty degenerations have been found occasionally in the retina.   There will be visual disturbance in case of delirium and hallucinations.

Fusel Oil.—Vide Amyl alcohol.

Gelsemium when taken internally dilates the pupils, according to Fronmueller and Ott.   According to Ringer and Murell, the pupils are contracted and mydriasis is only produced when the drug is dropped into the conjunctival sac.   But the statements are also at variance with regard to the action of the remedy under the latter circumstances, so that the action upon the eye probably depends upon accidental mixture with other substances.

Hasheesh.— Vide Cannabis indica.

Homatropine may produce general symptoms similar to those of atropine, when dropped into the conjunctival sac.   Harlan (*Mon. f. Aug.*, 1891 p. 189) observed slight attacks of glaucoma after its use. Cheney (*Bost. Med. and Surg. Journ.*, Jan. 23d, 1890) reports "hysterical" mydriasis, paralysis of accommodation and blindness from the use of hydrobromate of atropine.   De Schweinitz and Hare (*Amer. Med. News*, Dec. 24th, 1887) claim to have seen diminished frequency of the pulse as one of the symptoms.

Horse Chestnut (Æsculus hippocastaneum).—In a boy of three and one-half years who was poisoned by the green rind of the horse chestnut, Salomon (*Brit. Med. Journ*, Dec. 19th, 1887) noted great dilatation of the pupils and distressing visions, in addition to congestion of the face, a full pulse and drowsiness.

Hydracetin.—Gruenthal (*Centr. f. Aug.*, 1890, p. 73) observed small retinal hemorrhages after inunctions with hydracetin ointment; the urine contained a good deal of albumin.

Hyoscine and Hyoscyamine.—In poisoning with these two mydriatics the more or less marked dilatation of the pupils may possess diagnostic importance.

Illuminating Gas.— Vide Carbonic oxide.

Iodine and Iodide of Potassium.—Epiphora and pains in the eyes, sometimes passing into fully developed catarrhal conjunctivitis, belong to the typical history of iodism.   The eyelids take part in the

not infrequent temporary œdemas of the skin, and may also be involved in iodine eruptions (acne, etc.).   In four cases of acute iodism,
Ehrmann (*Wien. med. Blaett.*, 1890, No. 44) noted the occurrence of
trigeminal neuralgia in addition to the symptoms just mentioned.
Quinine effected rapid recovery.   The fact that all the patients suffered from syphilis may not be devoid of importance, for the reason
that under such conditions neuralgias may occur independently of
iodism.   According to Ehrmann, the latter acts as an exciting cause
by inducing hyperæmic swelling of the nerve sheaths.

It is well known that when potassium iodide is taken internally,
the dusting of calomel into the conjunctival sac is apt to produce
erosions, probably from the development of iodide of mercury
(Schlaefke, *Jahr. f. Aug.*, 1879, p. 216).   I have never observed this
accident, despite repeated experiments.   It must also be remembered
that calomel alone, if not very finely powdered, may produce erosions
of the conjunctiva.

Meurer (*Arch. f. Aug.*, XXII, p. 24) even claims that he has
seen erosions of the conjunctiva due to the external application of
potassium iodide ointment and the coincident application of white
precipitate ointment to the conjunctival sac.

Iodoform.—According to Kuester (*Berl. kl. Woch.*, 1882, No.
14), the pupils are contracted in iodoform poisoning.   In a girl of
sixteen years, upon whom iodoform applications were made during
the after-treatment of a resection of the hip joint, Hirschberg observed typical toxic amblyopia (fingers at two to three metres with
central scotoma) with normal fundus and dilated pupils.   During
the protracted internal administration of iodoform (combined with
creosote in pills) Hutchinson (*Jahr. f. Aug.*, 1886, p. 254) also observed typical toxic amblyopia with corresponding ophthalmoscopic
appearances.   Recovery after injections of strychnine and discontinuance of the pills.

Trousseau (*ib.*, 1887, p. 419))reports that a few hours after dusting iodoform upon a syphilitic ulcer of the upper lid, an erysipelatoid
swelling appeared upon the entire half of the face, and was attended
by the development of vesicles; this healed as soon as the remedy was
discontinued.   It reappeared when the iodoform was again applied.

Lead poisoning, when acute, resembles an attack of cholera and has no characteristic eye symptoms. In chronic lead poisoning, on the other hand, central and peripheral affections of sight are frequent. The remarks made concerning chronic poisoning in general also hold good here. The principal part is evidently played by changes in the walls of the vessels, particularly sclerosis and periarteritis, and these are often visible with the ophthalmoscope (*vide* p. 293). Spots of softening, ependymitis, internal hydrocephalus and hemorrhages in all possible localities of the brain are secondary in character. There are manifold disturbances both visual and motor.

The ophthalmoscope often shows a unilateral or bilateral neuritis, either of the alcoholic form (p. 327) with corresponding disturbance of sight, or as a diffusely reddened and cloudy papilla without notable swelling and sometimes with hemorrhages. In the latter event the impairment of sight may proceed to complete blindness, and may terminate in more or less atrophy of the optic nerve. A white sheath around the arteries and corresponding changes in the vessels are often visible. Retrobulbar neuritis, impairment of sight or even blindness without findings, but terminating later in atrophy of the optic nerve, are also observed.

The central disorders (due mainly to periarteritic changes, particularly in the cerebral cortex) include visual hallucinations, optical and other aphasias, hemianopsia with and without reaction of the pupils to light, etc. Concentric narrowing of the field of vision, color disorders and other hysteria-like symptoms are also observed occasionally. The clinical symptoms of tabes, multiple sclerosis, etc., sometimes develop.

There may also be disorders of movement, either central (associated disorders of movement, nystagmus), or nuclear and peripheral. Galezowski (*Jahr. f. Aug.*, 1877, p. 382) describes paralysis of accommodation; Landesberg (*ibid.*, 1880, p. 245) a unilateral paralysis of the abducens and also of the entire motor oculi, probably as the result of multiple neuritis. Ptosis is also mentioned several times.

A frequent sequel of chronic lead poisoning is disease of the kidney with albuminuria, and the latter may give rise to disorders of

sight. The ophthalmoscopic appearances of albuminuric neuro-retinitis may occur although there is no albumin in the urine.

So-called lead amblyopia and amaurosis, *i.e.*, impairment of sight
without findings, which were formerly diagnosed very often, hardly
ever come into question at the present time. Either objective findings are demonstrable after a certain lapse of time, for example, atrophy after retrobulbar neuritis, or the disorder of sight is of such a
character (hemianopsia or uræmic impairment) that it permits a definite diagnosis even without visible findings. Temporary lead amaurosis is almost always uræmic.

Although recovery from every possible form of disorder of sight
and movement is quite frequent—only those cases are directly unfavorable which are attended with considerable atrophy of the optic
nerve and are the results of renal disease—still it is well to give a
cautious prognosis in the individual case, inasmuch as we usually
have to deal with advanced stages of poisoning.

In the treatment of corneal ulcers with lead wash, the cornea is
apt to become incrusted with lead. Bellouard (*Jahr. f. Aug.*, 1882,
p. 363) has seen this develop spontaneously in individuals who suffered from ulceration of the cornea and were exposed in factories to
dust containing lead.

Marsh gas is said by Reuss (*Arch. f. Aug.*, XXIII, 3, p. 252) to
be the cause of the nystagmus of miners.

Menthol, in doses of 5.0 to 7.0, killed rabbits in five to ten minutes. Opacity of the lens developed immediately after death, but
this could not be produced during life (Charrin and Roger, *Centr. f.
Aug.*, 1888, p. 60).

Mercury and its salts, despite their frequent administration,
rarely give rise to eye symptoms. The acute poisoning very closely
resembles phosphorus poisoning, and also gives rise to similar eye
symptoms, particularly retinal hemorrhages and fatty degenerations.
During the treatment of itch with inunctions of blue ointment, Dyes
(*Deutsche Klinik*, 1871, No. 11) observed temporary blindness
which was relieved in a few days by internal and external treatment
with sulphur. Kaemmerer (*Virch. Arch.*, 59, p. 467) and Alsberg
(*Jahr. f. Aug.*, 1880, p. 228) demonstrated the presence of mercury

in the urine of individuals into whose conjunctival sacs calomel had been dusted for some time. Kohn (*Rec. d'Ophth.*, 1875, p. 365) observed violent attacks of colic at night in a child of three years whenever calomel was employed in this way.

There are also no characteristic eye symptoms in acute mercurial poisoning, while the results of chronic poisoning resemble those of chronic lead poisoning.

Methyl Alcohol.—Poisoning with this substance produced blindness within twenty-four hours, according to Mengin (*Rec. d'Ophth.*, 1879, p. 663).

Methyl chloride produces myosis in the first stage of narcosis, mydriasis in the second stage. Sudden contraction indicates danger of suffocation (Panhoff, *Jahr. f. Aug.*, 1882, p. 123).

Morphine.—In chronic morphine poisoning in the dog, Laborde (*Jahr. f. Aug.*, 1877, p. 217) found the fundus congested during the stage of vascular excitement, and pale and anæmic during narcosis. At the end of two weeks the latter condition became permanent. Ocular symptoms are also mentioned in a number of cases of morphine poisoning in man. Apart from contraction of the pupils, which possesses great diagnostic importance, Wagner (*Mon. f. Aug.*, X, p. 335) observed complete double blindness in a patient who had taken 2.0 morphine subcutaneously within five days. The papillæ were slightly cloudy, the arteries very narrow. The case was then lost to observation. Schiess-Gemuseus reports the case of a man who, after taking a sleeping-powder, slept for thirty-six hours and then suffered marked impairment of sight and hearing. At the end of three weeks he found right hemianopsia, and on the left side slight concentric narrowing of the field of vision. As in toxic amblyopias, the papilla was reddened on the inner side, pale on the outer side. Among other symptoms Schreiber (*Jahr. f. Aug.*, 1888, p. 539) observed aphasia, agraphia and alexia after acute morphine poisoning.

Although morphine is a powerful myotic, Levinstein (*Berl. kl. Woch.*, 1876, No. 14) observed mydriasis as often as myosis in morphine habitues. The mydriasis which Tupper (*Jahr. f. Aug.*, 1879, p. 229) observed, associated with spasm of the eye muscles, after an injection of morphine, was either due to pressure upon the eye or must be interpreted as mydriasis following irritation of the skin.

Jaeger's case (*ib.*, 1870, p. 380) proves that a non-aseptic injection of morphine into the temples may give rise to an orbital phlegmon, followed by exophthalmus, blindness and, later, by atrophy of the optic nerve.

Muscarine, the poisonous alkaloid of certain mushrooms, first produces maximum spasm of accommodation (in some individuals this is the sole symptom) and then myosis. This is noteworthy for the reason that ptomaine poisoning often shows the symptoms of muscarine. Indeed, the latter appears to occur as a ptomaine and has been isolated by Schmiedeberg from neurine, which is an undoubted ptomaine.

Mussel Poison.—*Vide* Ptomaines.

Naphthalin.—Bouchard (*Rev. Clin. d'Ocul.*, 1886, No. 7) found that 1.0 naphthalin per kilogramme in rabbits, when given daily, produced cataract in three to twenty days. Prior to the development of the cataract the vitreous usually contains numerous floating crystals (oxalate, sulphate and carbonate of lime, according to Panas) and there are large white patches (œdema) in the retina. At a later period these patches become retracted in the centre, atrophy and adhere to the retina (*vide* Panas, Dor, Hess, Magnus, in *Centr. f. Aug.*, 1887, pp. 104, 145, 295, 300). According to Kolinski (*Arch. f. Ophth.*, XXXV, 2, p. 29), the changes in the lens and retina appear to be due to hemorrhages into the ciliary body, ciliary processes and choroid. Such conditions have not been found in man after the use of naphthalin.

Nicotine.—*Vide* Tobacco.

Nitrobenzol poisoning gives rise, among other symptoms, to marked venous congestion of the fundus; the blood is often strikingly dark. In a case of this kind in a workman in a roburite factory, Nieden (*Cent. f. Aug.*, 1888, p. 193) observed diminution of vision to $\frac{1}{10}$ with concentric narrowing of the field, which gradually disappeared. In one eye a retinal hemorrhage, as large as the disc, was found.

Nitrous oxide gives rise, according to Aldridge (*Jahr. f. Aug.*,

1871, p. 322) to dilatation of the retinal arteries and increased redness of the papilla. Bordier (*ib.*, 1876, p. 295) observed extreme contraction of the pupils in a comatose condition an hour after the extraction of a tooth; in a few hours the condition had become normal. Narcosis with nitrous oxide is said to be attended by agreeable visual hallucinations.

Opium produces, according to Loring, no visible vascular changes in the fundus. Myosis is never absent in acute poisoning and rarely absent after prolonged administration. Hammerle (*Deutsch. Med. Woch.*, 1888, p. 838) reports impairment of vision in acute poisoning. A patient suffering from lead colic took 15.0 laudanum in one night; repeated vomiting. On the following morning, great oppression, cyanosis, maximum myosis and, during the course of the morning, impairment of vision proceeding to complete blindness. Recovery in two days. Hammerle assumes that the blindness was due to arterial spasm, but the case is not a pure one, inasmuch as the patient was suffering from lead poisoning, and, in addition, similar conditions have been seen after vomiting alone.

Galezowski (*Rec. d'Ophth.*, 1876, p. 210) reports, in two cases of chronic opium poisoning, cloudy vision, metamorphopsia, rapid exhaustion of the eyes, impairment of central vision and slight color disturbance with normal boundaries of the visual field and normal ophthalmoscopic appearances. In one case 1,000.0 opium had been taken within fifteen months; in the other the patient had taken 20.0 daily for forty years for the relief of facial neuralgia.

The corneal softening observed several times by Paster (*Muench. Med. Woch.*, 1886, No. 6) in opium smokers is a part of the general marasmus and is a bad omen as regards life. [I saw suppuration and destruction of each cornea in a Jewess, aged about sixty, who had undergone the most extreme marasmus from opium-eating during many years. Her skin was a mahogany brown, shrivelled and leathery, and emaciation extreme.—ED.]

Oxalic Acid.—In experimental oxalic acid poisoning of rabbits, Koch (Diss. Dorpat., 1879) noted dilatation of the pupils, followed by contraction (symptom of pain).

Petroleum-ether poisoning gave rise in one case (*Schmidt's*

Jahrb., 1891, 1, p. 34) to great dilatation and loss of reaction of the pupils and nystagmus; recovery.

Phenol.—*Vide* Carbolic acid.

Phosphorus poisoning produces eye symptoms which are similar to those found in carbonic-oxide poisoning. Retinal hemorrhages may appear before anatomical changes can be demonstrated in the vessels (Niederhauser, Diss. Zurich, 1875). At a later period fatty degeneration develops, chiefly in the capillaries and arteries, but also in larger or smaller foci in the retinal tissue. These may be visible with the ophthalmoscope and produce a picture similar to that of retinitis albuminurica. Hemorrhages into the optic nerve, brain, etc., may also produce eye symptoms.

As a matter of course the conjunctiva takes part in the jaundice which soon develops.

Physostigmine.—*Vide* Eserine.

Picric Acid.—After a dose of 0.3 internally, Hilbert (*Centr. f. Aug.*, 1885, March) noticed yellow vision lasting about an hour, and not followed by blue or violet vision. On account of the smallness of the dose this could not have been due to a yellow coloration of the media, and hence the yellow vision must have been the result of direct central irritation (?).

Pilocarpine when dropped into the eye acts as a powerful myotic; when given internally the myosis is slight and indeed mydriasis has been often observed. After internal administration, increased secretion from the lachrymal gland is often noticed, and this may even follow simple instillation into the conjunctival sac. In five cases Landesberg (*Mon. f. Aug.*, 1882, p. 51) observed rapid progress of cataract after pilocarpine or jaborandi treatment. On account of the powerful influence of the injections upon the general organism, temporary amblyopia may occasionally result (Fuhrmann, *Wien. med. Woch.*, 1890, No. 34).

Piscidium.—Seifert (*Berl. kl. Woch.*, 1883, No. 29) occasionally observed mydriasis after the prolonged use of ext. piscidii as an hypnotic.

Podophyllin.—The workmen who are engaged in grinding the root of podophyllum suffer from traumatic conjunctivitis and cutaneous inflammations unless properly protected. Ulceration of the cornea is an occasional sequel. According to Hutchinson (*Jahr. f. Aug.*, 1872, p. 263), the dust does not produce the irritation at once, but only on the following day.

Prussic Acid.—Vide Potassium cyanide.

Quinine.—Several dozen cases of visual disorder as the result of large doses of quinine have been reported. Almost all of them occurred in individuals whose nutrition had been notably impaired. Unilateral or bilateral impairment of vision, which may even pass into blindness, occurs more or less suddenly; seldom slowly. After a time recovery takes place or permanent disorders which correspond to partial atrophy of the optic nerve remain, viz., central impairment of sight, concentric narrowing of the visual field, defects of the visual field with or without color disturbance. Hemeralopia may also be left over (Kohn, *Rec. d'Ophth.*, 1874, p. 384).

The ophthalmoscope shows a pale or perfectly white disc and narrow vessels, occasionally slight retinal exudation. When the latter affects the macula, it shows a cherry-red spot as in embolism (Gruening, Buller). Similar ophthalmoscopic appearances are obtained experimentally in animals after the administration of large doses of quinine.

Barabaschew (*Arch. f. Aug.*, XXIII, 2) has experimented on men with doses of 2.4–3.6. He produced: 1. Acute gastritis with temporary increase of central vision. 2. Pallor of the face, vertigo, syncope, somnolence, tinnitus aurium, transient myosis passing into moderate mydriasis. 3. Considerable narrowing of the retinal arteries and pallor of the papilla, acceleration followed by slowing of the pulse. 4. Concentric narrowing of the visual field. 5. Impairment of vision passing even into temporary blindness. The latter was obsered in only one case (lasting half a minute, but recurring ten times at intervals of ten to fifteen minutes). There was no corneal anæsthesia or color-blindness, which have been observed in several cases. In one case he found transient cloudiness of the retina.

The temporary dimness of vision is caused by the arterial spasm. In case of poor nutrition, ischæmia gives rise to necrobiotic changes in the optic nerve, as in amblyopia and amaurosis after loss of blood,

ending in partial or total atrophy with permanent impairment of
vision and narrow retinal arteries. According to De Schweinitz
(*Centr. f. Aug.*, 1891, p. 118) thrombosis also occurs. The changes
which he mentions in the cortical centres of vision perhaps depend
merely upon imperfect hardening. The visual disturbances in
quinine poisoning have a decidedly peripheral character.

Favre (*Jahr. f. Aug.*, 1873, p. 114) reports the case of a man,
aged thirty-seven years, who exhibited color disturbances, with no
injury to vision, after typhus fever and prolonged treatment with
quinine. Some doubt has been thrown upon this case.

Tiffany is the only one who mentions diminution of intraocular
pressure. [For a very complete summary of cases see Atkinson
(*Journal of American Med. Assoc.*, Sept. 28th, 1889). He found
cases reported as early as 1841. De Schweinitz (Trans. Am. Ophth.
Soc., 1891, p. 23) gives the appearances during life and the pathological examination in dogs to whom large doses had been given.]

Resorcin.—Hirschberg (*Cent. f. Aug.*, 1886, Dec.) observed conjunctivitis develop after the application of a ten-per-cent resorcin
ointment to the face.

Salicylate of soda gives rise, in rare cases, to symptoms similar
to those produced by quinine. After a dose of 8.0 which was taken
within nine hours, Galli (*Jahr. f. Aug.*, 1880, p. 145) observed complete blindness, lasting twenty-four hours, in a vigorous peasant girl
of sixteen years; there was marked mydriasis. The ophthalmoscopic
appearances were normal and the urine did not contain albumin.
After 2.0 doses every two hours, Gibson and Telkin (*Jahr. f. Aug.*,
1889, p.'223) found, at the end of eight hours, extreme myosis with
loss of the reaction of the pupils to light. These symptoms disappeared thirty hours after the remedy was discontinued.

After a dose of 4.0 Rosenberg (*Deutsch. med. Woch.*, 1886,
No. 33) noticed violent burning of the skin, œdema of the lids and
a bluish-red macular eruption. After another 7.0 had been taken,
large vesicles formed upon some of the maculæ and the conjunctiva
became reddened and swollen. This affection was perhaps related
to an urticaria ab ingestis.

Santonin.—The characteristic feature of santonin poisoning is yellow vision. This begins ten to fifteen minutes after the administration of the drug, and, on careful observation, is always found to be preceded by transient violet vision. The spectrum is first contracted on the violet side, and then, but to a less extent, on the red side; it is evident that the rays in question are no longer perceived by the retina. In my own case the color is an orange yellow passing into green.

The yellow vision sometimes occurs paroxysmally, especially on looking at white and brilliant objects, while it is very little noticed in the background of a moderately light room. All shadows appear in the complementary color, viz., violet. At the same time the patient sometimes sees flashes and sparks of light. Unless complications are present, central vision, the fundus and the pupils are normal. The field is not noticeably contracted, but adaptation in the dark is very much delayed (page 29).

There is evidently a peripheral insensibility of the retina to violet and a part of the red rays, preceded by temporary hypersensibility to these rays (violet vision). Corresponding to this process I observed in myself a temporary feeling of warmth and burning of the entire integument, followed by a prolonged and striking sensation of numbness, i.e., hyperæsthesia and paræsthesia following an hyperæsthesia.

The fact that, despite the insensibility to violet light, the shadows are seen in the complementary violet, proves that the central color sense is undisturbed, that the condition is solely a peripheral non-irritability to the extreme rays of the spectrum (Rose, *Virch. Arch.*, 16, 18, 19, 20, 28; Knies, *Arch. f. Aug.*, XVIII, p. 64; Koenig, *Cent. f. Aug.*, 1889, Jan.). If small doses are administered for a considerable time, there will be insensibility to violet, but yellow vision will not be produced (Henneberg, *Jahr. f. Aug*, 1889, p. 509).

Severe santonin poisoning is attended not infrequently with cerebral symptoms, such as spasms, headache, speech disorders (aphasia), conjugate deviation, etc.; mydriasis may also be produced, sometimes unequally in both eyes. (Van Rey, *Ther. Monatsh.*, Nov., 1889).

Saponin.—Keppler (*Berl. kl. Woch.*, 1878, No. 32) states that he
23

observed in his own case severe pains, strabismus and exophthalmus
of the left eye, as the symptoms of acute saponin poisoning.

Secale Cornutum.—*Vide* Ergotin.

Silver.—The eyelids and conjunctiva take part in the general
pigmentation which appears after the excessively prolonged internal
administration of silver. The conjunctiva is much less affected than
in some cases in which nitrate of silver is applied locally.

Snake Virus produces an acute hemorrhagic diathesis, but the
effect varies with the different varieties of snakes. The disorders of
sight and of ocular movements may be due to central or peripheral
hemorrhages. If they affect the optic nerve, permanent impairment
of vision with partial or total atrophy may develop. According to
Laurençao (*Jahr. f. Aug.*, 1875, p. 376), temporary and permanent
amblyopia and blindness after snake-bites are not rare.

Strychnine.—This alkaloid, which is employed so extensively in
amblyopias, produces no eye symptoms, or at least very slight ones,
in cases of poisoning. No special action on the pupil is noticeable
despite the intense irritation of the vasomotor centres.

Acute strychnine poisoning may give rise to acute nephritis.
According to Honigmann (*Deutsch. med. Woch.*, May 30th, 1889),
this is probably due to stasis from arterial spasm. It is therefore
possible that the effects of albuminuria might become evident in the
eye.

Sulfonal.—In sulfonal poisoning Dillingham (*Jahr. f. Aug.*,
1890, p. 451) noticed ptosis which lasted two weeks, and Knaggs (*ib.*)
reports anæsthesia of the conjunctiva and of other parts of the
body.

Sulphide of carbon has given rise, particularly in india-rubber
workers, to an entire series of acute and chronic poisonings with implication of the organ of sight.

In acute poisoning a stage of excitement is followed by collapse,
often attended with motor and sensory disturbances (the cornea is
often insensible), diminished reflexes, etc. These are followed not
infrequently by impairment of vision, with or without color disturbance, concentric narrowing of the field and of the color boundaries,
monocular polyopia, macropsia and micropsia, hemianæsthesia, par-

æsthesia, local anæsthesia, paralyses and spasms, occasionally hemeralopia, etc.   These symptoms must be interpreted in part as hysterical (toxic hysteria); in fact they are the result of peripheral
neuritis.

In chronic poisoning the chief eye symptom is toxic amblyopia,
axial neuritis of the optic nerve.   This is similar to alcoholic amblyopia, and is manifested by a central scotoma with color disturbance
and corresponding ophthalmoscopic appearances.   The latter may
also be normal or color disturbance may be wanting; nyctalopia occurs occasionally.   The eye symptoms are usually preceded for a
considerable period by general nervous symptoms, such as headache,
muscular weakness, etc.   With very few exceptions the affection
of the optic nerve recovers completely under proper care.   The sensibility of the cornea and conjunctiva is slightly diminished, as in
many chronic poisonings with gases (Gallemaerts, *Ann. d'Ocul.*, 90,
p. 154; Pierre Marie, *Neurol. Centr.*, Feb. 15th, 1889; Maass, Diss.
Berlin, 1889).

Sulphur Chloride.—This substance, which is used in the caout-
chouc industry, appears to produce toxic amblyopia.   Nettleship
(*Jahr. f. Aug.*, 1884, p. 334) observed amblyopia and nervous depression; Frost reports amblyopia, neuritis and atrophy of the optic
nerve in india-rubber workers who handled carbon bisulphide and
chlorine bisulphide.

Sulphur pomade (vaselin 100.0, wax 5.0, sulphur 10.0, oil of roses
one drop) is said to have produced symptoms of poisoning after having been used for eight years (Eichbaum, *Berl. kl. Woch.*, 1887,
No. 42).   Among the symptoms were large and rigid pupils which
did not react to light or cutaneous irritation (atropine?).

Sulphuretted Hydrogen.—In a case of poisoning with this gas
Brouardel (*Jahr. f. Aug.*, 1885, p. 266) observed mydriasis, exophthalmus, insensibility of the cornea and disappearance of the corneal
reflexes.

Sulphuric Acid.—One of the few cases of acute superior polio-encephalitis (page 178) was the result of sulphuric-acid poisoning; all
the rest, with one exception, occurred in hard drinkers.

Sulphuric Ether.— Vide Ether.

Tea.—Wolfe (*Jahr. f. Aug.*, 1879, p. 230) attributes a case of liquefaction of the vitreous, with numerous floating opacities, to excessive tea-drinking, but the connection between the two is very doubtful.

Tobacco.—The toxic amblyopia which occurs during the chronic abuse of tobacco has already been discussed under the heading of alcohol. It was there stated that tobacco amblyopia is more often unilateral than alcoholic amblyopia, or is at least very different in degree on the two sides, that there is usually myosis and often spasm of accommodation, that the impairment of vision develops more gradually and, despite abstinence, increases more frequently. Lawford (*Centr. f. Aug.*, 1890, Sept.) reported nine cases in which typical tobacco amblyopia terminated in complete atrophy of the optic nerve, a very rare event in alcoholic amblyopia. Mention has also been made of the difference in the position of the scotoma, in alcohol (pericentral), in tobacco (paracentral), but this difference is by no means constant. While the majority of writers consider tobacco much more dangerous than alcohol as regards the development of toxic amblopyia (many even deny the potency of alcohol as a causal factor), Fumagalli (*Jahr. f. Aug.*, 1874, p. 454) denies the occurrence of a tobacco amblyopia.

Acute inflammatory symptoms in the fundus are rare. Nettleship (Ophth. Hosp. Rep., XI, p. 370) reports two cases of retinal hemorrhages, one associated with neuritis. Ponti (*Annal. di Ott.*, III, p. 107) observed a true neuritis. Although the disease of the optic nerve from abuse of tobacco is usually severe, corresponding affections of other nerves, paralyses, etc., are rare events. Fontan (*Rec. d'Ophth.*, 1883, p. 309) reports, in a very excessive smoker, a sudden double paralysis of the motor oculi, but with narrow pupils, which recovered slowly after abstinence from tobacco. Although the possibility of tobacco paralysis of the ocular muscles cannot be denied, their great rarity should lead to caution in assuming a direct connection between the two in individual cases.

Tobacco amblyopia rarely develops in a strong, youthful individual. It develops either at an advanced age, when the resistance to external agents has diminished, or when the general nutrition has been impaired by anorexia, chronic gastric catarrh, etc.

Complete abstinence from smoking is usually secured more easily than partial abstinence. In view of the laxative action of tobacco, care should be taken during the period of abstinence to secure regular evacuations from the bowels, preferably by means of Carlsbad water or salt.

Simons and Valzah (*Jahr. f. Aug.*, 1877, p. 216) found pronounced myosis in cases of acute nicotine poisoning, corresponding to the myotic action of nicotine, while Kosminski (*ib.*, 1871, p. 345) reports dilatation of the pupil in a servant who, in order to relieve toothache, had inserted tobacco in the hollow tooth. This case hardly appears to be one of poisoning, but, in view of the visual impairment and concentric narrowing of the field of vision, reminds us decidedly of traumatic hysteria. This is probably also true of O'Neill's case (*ib.*, 1879, p. 229) of mydriasis in poisoning from the external application of tobacco. After an enema of tobacco, Wilkinson (*ib.*, 1889, p. 510) observed vomiting, diarrhœa and blindness lasting a quarter of an hour; the latter was possibly due to the vomiting. [In detecting tobacco amblyopia allowance must be made for the better appreciation of red at a distance of 3° to 5° from the macula than directly at the visual centre. This difference is slight, but is normal to many persons. As yet no measure of this normal difference has been established. That tobacco alone can cause central optic neuritis is an undeniable fact of clinical experience, yet the diagnosis founded on pallor of the infero-temporal quadrant of the optic disc must be taken with great caution; unless supplemented by scotoma for red and more or less amblyopia, judgment should be held in suspense. Furthermore, this local pallor may be wanting in genuine cases. For valuable knowledge on this topic see Uhthoff in *Graefe's Archiv*, Bd. XXXII, Abth. IV, 95-188, 1886; and Bd. XXXIII, Abth. I, 257-318, 1887. A most convincing case in clinical experience is reported by Filehne in *Graefe's Archiv*, Bd. XXXI, Abth. II, p. 27, 1885, as personal to himself. See also Groenow, *Graefe's Archiv*, Bd. XXXVIII, Abth. I, p. 1.—ED.]

It is not astonishing that workers in cigar factories, in which the air is filled with the irritating dust, should suffer from conjunctivitis and photophobia (Bierbaum, *Viertelj. f. ger. Med.*, 1889, Suppl., p. 115).

Despite its incompleteness, this *résumé* of the eye symptoms of poisonings will give an idea of the manner in which the organ of vision, in the widest sense, can be implicated. Poisoning with ptomaines, meat, fish, ice-cream, toxalbumin, etc., constitutes a transition to the infectious diseases proper.

Ptomaines and toxalbumin which are poisonous to man are to be found in living animals, for example, in certain poisonous mussels and fishes. Often these are only poisonous at certain times or under certain conditions, and indeed the toxic character may belong only to special organs (liver, sexual glands). This was true of the cases of mussel poisoning at Wilhelmshafen a few years ago, which ran the course of acute alkaloid poisoning; no micro-organisms could be discovered.

Ordinary ptomaine poisoning is due to toxic substances which are produced by microbes from articles of food, although there may not be any decided phenomena of decomposition. The virus produced in this way may alone be absorbed, or the bacteria are absorbed, increase in the body, and there produce toxic products for a longer or shorter period, or both conditions prevail. In the first event we have to deal with an acute poisoning, not infrequently with an eruption of erythema and urticaria. It develops whenever the microbes cannot develop any further in the human body, or when they have been destroyed by the process of cooking, etc. In the second event, a more or less acute infectious disease develops, sometimes resembling cholera, sometimes resembling typhoid fever. In the third event the acute poisoning is followed by the acute infectious disease.

There is not always a single definite microbe or toxic substance in these cases, and there may even be a number of microbes in the same case. The toxic products of disassimilation also vary in character, and the chemical constitution of some has been accurately determined. Certain ones have a specific effect upon the eye. Muscarine and neurine produce spasm of accommodation and myosis, tyrotoxine causes paralysis of accommodation and mydriasis, a very frequent symptom of meat poisoning. Other substances produce no special eye symptoms but are very similar to well-known poisons, such as curare, strychnia, atropia, etc. Scratching and dryness in the throat, difficulty in deglutition, gastritis and gastro-enteritis, dysuria, etc., are common symptoms of poisoning.

As a general thing vision is not impaired, but a number of cases of amblyopia without findings have been mentioned. They generally improve rapidly, but may also remain permanent. A sufficient explanation of these cases is yet wanting.

Apart from the very frequent injection of the conjunctiva, acute ptomaine poisoning is accompanied most frequently by paresis or paralysis of accommodation, generally with, more rarely without mydriasis, which is always bilateral. In very mild cases this may be the sole symptom. Thus, I observed slight paresis of accommodation, lasting twenty-four hours, in two individuals who had very probably been poisoned by an apparently innocuous fish (*Jahr. f. Aug.*, 1886, p. 258). Such cases probably occur very often, but, as a matter of course, are easily overlooked. Paresis or paralysis of accommodation, usually with mydriasis, much more rarely the latter alone, are among the typical symptoms of poisoning by bad meat, although they are often absent (Cohn, *Arch. f. Aug.*, IX, 2, p. 148; Leber, *Arch. f. Ophth.*, XXVI, 2, p. 236; Ulrich, *Mon. f. Aug.*, 1882, p. 230; Roth, Federschmidt, *Jahr. f. Aug.*, 1883, p. 303; Virchow, *Berl. kl. Woch.*, 1885, No. 48).

Among other paralytic symptoms ptosis is the most frequent (Puerkhauser, *Jahr. f. Aug.*, 1877, p. 219; Fleury, *ib.*, 1885, p. 269; Hirschfeld, *ib.*, p. 266; Kaatzer, *Deutsch. med. Woch.*, 1881, No. 7). Every other possible form of paralysis of the ocular muscles is also occasionally observed. They are generally nuclear in character (hemorrhages?), or are perhaps due to basilar neuritis, for example, complete or partial paralysis of the motor oculi (Fleury, *l.c.*), sometimes bilateral paralysis, with or without implication of other eye muscles (Eichenberg, *Jahr. f. Aug.*, 1880, p. 247; Alexander, *Centr. f. Aug.*, 1888, p. 87), double external ophthalmoplegia (Guttmann, *Berl. kl. Woch.*, 1891, No. 8), etc. Although the symptoms are usually alike when due to the same cause, and only differ quantitatively, according to the amount ingested, nevertheless a symptom may be present or absent in the same "epidemic," although we can offer no sufficient explanation.

In such poisonings there is either rapid recovery or death, or a transition into the second form, usually a typical infectious disease with a period of incubation lasting several days. Clinically and anatomically (swelling and ulceration of Peyer's patches) the condition exhibits a great resemblance to ordinary typhoid fever. This second form of ptomaine poisoning does not produce eye symptoms, but we occasionally observe all those visual symptoms which may occur in acute infectious diseases.

B. Infectious Diseases.

Infectious diseases are toxæmias which develop in a peculiar manner. The poison is produced by animal or vegetable parasites, but is rarely absorbed at the start in large quantities (septicæmia). As a general thing the organized parasite is received in small quantities, proliferates within the body, and it is only after a certain length of time that its products of disassimilation cause symptoms of poisoning, in other words, the infectious disease. This is preceded by a stage of latency or incubation, which varies in duration from a few hours to several weeks or even years. Its duration varies according to the nature of the disease, but there may be considerable variations even in the same disease. The outbreak of the characteristic symptoms is preceded by a longer or shorter prodromal period. This presents great similarity in the different infectious diseases, but may also exhibit characteristic features.

It was plausible to assume that particular excitants of inflammation always produced definite clinical infectious diseases, but experience shows that this is not always true. Very few infectious diseases (splenic fever, glanders, tetanus, recurrent and typhoid fevers) possess a single micro-organism which occurs only in that disease and always produces the corresponding clinical history. In other cases a disease which is clinically the same may be produced by different microbes, for example, pyæmia, erysipelas, purulent meningitis. Indeed, a single bacterium, for example, the diplococcus pneumoniæ, may produce, according to circumstances, groups of symptoms which are clinically distinct. Moreover, infection by a number of excitants is very common, either during the course of the disease or from the very start. This is facilitated whenever the infectious disease has given rise to losses of substance or ulcerative processes in the integument, in the digestive, respiratory, or genito-urinary tract. A number of complications and sequelæ are due to these secondary or mixed infections. It is especially the ubiquitous staphylococcus pyogenes aureus and albus, streptococcus pyogenes, diplococcus pneumoniæ, and a few others, which produce similar complications, particularly purulent processes and abscesses, in the most varied infectious diseases.

In many infectious diseases a causal micro-organism has not yet been demonstrated, although the disease can hardly be explained except on the theory of living infection (small-pox, measles, scarlatina, etc.).

At first there was a tendency to exaggerate the part played by specific microbes. The fact that many individuals are unsusceptible, even when infected, proves the great importance of the living nutrient substance. On the other hand there may also be an abnormal susceptibility to certain infectious diseases, as in one of my acquaintances who, despite the fact that he had had small-pox and had twice been vaccinated successfully, was again attacked by small-pox shortly afterward and died. In making cultures it is found that the virulence of the germs may be increased or entirely extinguished, according to the character of the nutrient. An individual may also be insusceptible from the start to the most virulent infectious substances. Probably this is due to certain chemical substances (products of disassimilation of the micro-organism?) which are present in one individual, and are either absent or exist in less quantities in another. The fact that acquired immunity is secured by passing through the infectious disease in question, points toward this explanation. On the other hand it is well known that one attack of one infectious disease (malaria, articular rheumatism, croupous pneumonia, erysipelas) may predispose to others. In such cases the exciting causes of inflammation must remain within the individual.

Vegetable parasites are by far the more frequent, particularly bacteria, while animal parasites are much rarer (as in trichinosis, probably malaria, and certain cases of dysentery).

From these statements it is evident that there will be a certain degree of uniformity in the complications and sequelæ of all infectious diseases. Whatever happens in one may be found occasionally in all the others. This was shown very distinctly in the last influenza epidemic which exhibited all the complications and sequelæ to be found in infectious diseases (la grippe or grip).

The eye may be the point of entrance of the infection, but this is comparatively rare (syphilis, tuberculosis, diphtheria, splenic fever, lyssa). Gonorrhœal conjunctivitis and blennorrhœa neonatorum have also given rise to nephritis and swelling of the joints. [Deutschmann reports finding the diplococci Neisseri in the serous effusion taken from the inflamed knee-joint of an infant having purulent conjunctivitis.—Ed.]

The eye is affected somewhat more frequently during the incubation or prodromal period, and this is not infrequently a guide when the symptoms are still entirely indistinct. Conjunctivitis and photophobia are characteristic of measles, and œdema of the lids of trichinosis, as pains in the throat are characteristic of scarlatina, and pain in the back characteristic of variola.

Diseases of the eye are still more frequently a part of the infectious disease itself, such as conjunctivitis in measles and typhoid fever, pustules on the lids, conjunctiva, cornea and lachrymal sac in variola, spasms and paralyses of the ocular muscles and optic neuritis in cerebro-spinal meningitis, diphtheria of the conjunctiva in diphtheria of the nose and throat, retinal hemorrhages in septicæmia, purulent inflammation of the choroid and retina in pyæmic processes, etc.

Chronic infectious diseases (syphilis, leprosy, tuberculosis) may take effect in the eye and the central parts of the visual apparatus, and always as the result of an embolic process.

The hemorrhages into the lids, conjunctiva and orbit during whooping-cough are purely mechanical. Similar conditions may be produced by any violent coughing spells, severe vomiting, etc.

Hæmatogenous icterus, which is a symptom, is not infrequently the first noticed in the conjunctiva and the last to disappear.

In infectious diseases, which often give rise to outbreaks of herpes, the eruption may appear upon the cornea. It is well known that this is frequent in pneumonia, rare in typhoid fever, and is practically never seen in meningitis, so that it may be important in differential diagnosis.

The most serious importance attaches to the complications and sequelæ which affect the eye either directly or secondarily.

If an acute infectious disease begins with severe symptoms of toxæmia, hemorrhages may occur very quickly into the different organs, although the microscope does not reveal any change in the vessels. An acute scorbutic condition develops and is followed rapidly by a fatal termination. Under such circumstances more or less extensive retinal hemorrhages are found, as is well known with regard to septicæmia, extensive burns, etc. They are much more frequent than the scanty reports in literature would lead us to believe; they have been found in the dead-house more often than with the ophthalmoscope. They may be present in all "hemorrhagic" infectious diseases, particularly in hemorrhagic small-pox, in which, however, an ophthalmoscopic examination has hardly ever been made.

When the infection does not begin so acutely, we often find extensive acute fatty degenerations, either in the intima of the vessels or in the parenchyma of one or more organs. The latter are degenerated diffusely or in patches, especially the liver, kidneys and nervous system (including the retina). This is almost always associated with inflammatory changes in the vessels, particularly the capillaries and small arteries (vasculitis, endovasculitis, perivasculitis). Hemorrhages may occur at the same time or secondarily. In fatty necrobiosis of the parenchyma the direct action of the virus is evidently the essential feature. This causes coagulation necrosis and subsequent fatty degeneration, terminating either in restitution by the cellular elements which still remain, or in chronic inflammation. This is especially frequent in the kidneys which are the chief channels of elimination of the poisonous products of disassimilation and, in part, of the agents of the inflammation. Hence the frequency of albuminuria, sometimes of the most acute hemorrhagic nephritis, at the height of the disease.

When the course is still less acute the action of the virus is chiefly expended, apart from the vessels, upon the interstitial connective tissue, while the parenchyma proper remains healthy or is involved only in the stage of retraction of the interstitial process.

In very chronic infections, the vessels alone are at first attacked. They are degenerated very extensively, perhaps only in certain localities, and often in a manner which is characteristic of the disease. They exhibit vasculitis, perivasculitis and endovasculitis, which give rise, on the one hand, to narrowing of the lumen and even complete obliteration, on the other hand to dilatations and aneurisms with their sequelæ, such as circulatory disturbances, stasis and œdema, hemorrhages and thromboses. In the retina these changes may be ophthalmoscopically visible.

Although the influence of vascular lesions upon the tissues appears to be relatively slight, on account of their very chronic course and the possibility of collateral nutrition, they become harmful through long duration and may give rise to injury, especially to the vulnerable nervous system, years after the original infection has been cured. Such lesions are of a simple atrophic, not of an inflammatory character. Focal and systemic diseases of a sclerotic and purely atrophic character in the brain, cord and peripheral nerves are due not infrequently to infectious diseases, even after the lapse of years.

We may distinguish four principal classes, although they shade into one another.

1. Acute hemorrhage without noticeable vascular changes (hemorrhages in septicæmia).

2. Acute hemorrhagic inflammations with acute fatty degeneration of the vessels and parenchymatous tissues.

3. Inflammatory lesions of the vessel, with diffuse inflammatory, mainly interstitial, changes; this includes multiple neuritis, the cyclitis of relapsing fever, etc.

4. Chronic inflammatory or necrobiotic lesions of vessels (fatty degeneration) without marked interstitial or parenchymatous changes,

and which rise to simple atrophy after the lapse of years (tabes and tabetic atrophy of the optic nerve after syphilis).

All four forms occur in acute and chronic infections, and may even be observed in the same infectious disease, according to the acuteness of its development (malaria, syphilis).

In the affections hitherto considered we have referred solely to the poisonous action of the infection. There is also a second factor, which constitutes the difference between poisoning and infection, viz., infectious embolism and thrombosis. As a matter of course, embolism and thrombosis may result from the vessel disease alone and then act mechanically as an obstruction to the circulation. But if they contain specific inflammation-producers or the latter colonize in a certain part of the tissues, they may give rise to a specific local disease. This will vary according to the nature of the micro-organism and its more or less violent action upon adjacent tissues (necrosis, necrobiosis, inflammation, suppuration, hyperæmia). In many cases the specific focus is characteristic and possesses the highest diagnostic value. In this sense a pyæmic abscess, a diphtheritic inflammatory focus, a variola pustule, or an eruption of measles is to be regarded as a specific focus, like a typhoid lymphatic gland, a gumma, a nodule of tubercle or leprosy (so-called specific neoplasms).

Such specific foci do not develop in all infectious diseases, particularly in those which are very acute and in which the poisonous action predominates (anthrax). But even those in which focal diseases occur constantly may occasionally prove fatal before the outbreak of the eruption, if the primary infection is very violent (scarlatina, variola).

A less acute infection leads to focal affections which either disappear without leaving a trace (measles, scarlatina) or they suppurate and, if a fatal termination does not ensue, leave cicatrices (pustules, abscesses). It is only in comparatively chronic processes that a specific neoplasm may be expected. All these cases depend upon quantitative differences in the action of qualitatively different inflammation-producers upon the tissues.

Not all the focal affections which arise in infectious diseases are the direct effects of the specific germs. Many are the results of secondary or mixed infection. This is especially true of abscesses. These may contain merely the specific germ of the infectious disease, provided it possesses pyogenic properties. More often they contain pyogenic staphylococci, streptococci, diplococci and bacilli.

The symptoms of an infectious disease also depend very materially upon the nutrient medium or the soil itself, as represented by the individual. The infection may not take at all, or its action may be intensified into acute gangrene, especially during the late stages of the disease when the power of resistance of the body has considerably diminished. It is often found that specific foci and neoplasms multiply with increasing severity the longer the disease has lasted.

An infectious disease may act favorably or unfavorably upon existing eye diseases. It has a favorable action when absorption is facilitated by the increased process of disassimilation (clearing up of pannus, vitreous opacities, etc.), and an unfavorable action when chronic inflammations are excited to acute exacerbations (relapses of iritis, cyclitis and choroiditis). Acute attacks of glaucoma in eyes which suffer from chronic peripheral anterior synechia of the iris and cornea may be provoked by an infectious disease, acute myopia may develop after measles, necroses may occur in old corneal leucomata during many infectious diseases, etc.

We may summarize the affections of the eye which may occur in connection with infectious diseases as follows:

1. Hemorrhages in all parts of the peripheral and central visual apparatus from the most varied causes in all stages of the disease, and consecutively every possible disorder of vision, motion and sensation.

2. Foci of fatty degeneration and softening in the central organs and the eye, visible in the retina with the ophthalmoscope, and often associated with hemorrhages.

3. Inflammatory changes in the vessels in all localities, with the above-mentioned consequences.

4. More or less diffuse inflammations of the tissues of the eye, especially of the uvea and retina; iritis, cyclitis, choroiditis, retinitis, diffuse interstitial keratitis, etc. Meningitis with its various eye symptoms also develops in the same way.

5. Changes (chronic and acute hemorrhagic forms) in the optic nerve, chiasm, tractus, motor and sensory nerves (multiple neuritis).

6. Pure atrophy of the nerve tissues (central organs and optic nerve), occurring after the lapse of years and probably the final outcome of the vessel lesions.

7. Focal hyperæmias and inflammations (metastases) in various degrees from a chronic to an acute hemorrhagic and purulent process, or even terminating in acute gangrene. These are found in the integument of the lids (eruptions, sometimes leading to gangrene), the sclera (scleritic foci), uvea (disseminated choroiditis and chorioretinitis, embolic suppurations), retina (benign, but usually septic emboli), orbit (metastatic suppurations), lachrymal glands (embolic abscesses, dacryo-adenitis), optic nerve and brain, etc.

8. Specific neoplasms (syphilis, tubercle, leprosy) in almost every part of the eye and surrounding structures and in the central nervous system.

The functional results of these lesions are:

1. Visual disorders of all kinds, of peripheral, intermediate and central origin.

2. Paralyses and spasms of a central, nuclear and peripheral character, and even due to direct disease of the muscles.

3. Neuralgias, anæsthesias and paræsthesias of every possible mode of origin.

4. Other affections, such as adhesion of the lids in conjunctival catarrh, disorders of lachrymal secretion and conduction in affections of the lachrymal glands and canal, etc. In infectious diseases which are attended with high fever and congestion of the meninges and cortex, visual hallucinations and illusions are also seen. These diseases may terminate in more or less severe forms of insanity.

Infectious diseases which have a typhoid stage often lead to desiccation keratitis on account of suspension of winking; protracted convalescence causes asthenopic symptoms from weakness of accommodation and convergence; the latter forming part of the general muscular weakness. We may find in them the exciting cause of internal squint, especially from measles and whooping-cough.

Phlyctenular diseases of the cornea and conjunctiva often develop during the period of convalescence.

In many cases the visual apparatus is affected secondarily by the complications of an infectious disease (uræmic amaurosis, retinitis and paralyses in renal affections, amblyopia and amaurosis from loss of blood, in meningitis, thrombosis of the cerebral sinuses, diseases of the walls of the orbit, etc.).

During the course of an epidemic, some complication or sequel of the infectious disease may be observed in an individual who has not suffered from the infectious malady or has only exhibited ill-defined symptoms. Thus, paresis of accommodation is noticed in epidemics of diphtheria, acute nephritis in epidemics of scarlatina, etc. These are undoubtedly the result, in the majority of cases, of an infection which has taken an unusual course.

Among all infectious diseases the greatest resemblance to a toxæmia is exhibited by

Septicæmia.

This is due to the rapid absorption of a large amount of putrid virus. It is entirely analogous to the first form of ptomaine poisoning, except that the absorbed ptomaines and toxalbumins are different from those there described. Hence, as a rule, mydriasis and paralysis of accommodation are absent.

Among eye diseases we need only mention foci of fatty degeneration and hemorrhages into the retina, which may be found at a very early period (second or third day) in acute septicæmia. As a rule death then occurs very quickly (although the retinal affection *per se* is benign), because there is also a tendency to hemorrhage and intense degenerations in vital organs. The retinal hemorrhages may also possess diagnostic value in so-called spontaneous septicæmia (Leube).

If the septicæmia does not prove fatal, it usually terminates in

Pyæmia.

This is the result of the absorption of the agents of inflammation. Under their influence the thrombi which have formed at the point of reception of the germs undergo softening and parts of the throm-

bus, laden with the germs, enter the circulation. They then produce
emboli, thrombi or hemorrhages in other parts, and a purulent focus
again develops. The point where the germs enter may be situated in
any part of the body; external integument (injuries, pressure gangrene, etc.), vesical mucous membrane (cystitis), peritoneum (perito-
nitis), bones (suppuration from the ears), uterus (puerperal fever),
etc. The causal disease is sometimes trivial, as, for example, a
furuncle, small abscess of the lip (Foerster, *l.c.*, p. 181), a "post-mortem tubercle," an inflamed phalangeal joint (Leber), even the infected
wound of a cataract operation (del Toro, *Jahr. f. Aug.*, 1881, p. 288).

A pyæmic focus may develop in any vascular portion of the eye,
but particularly in the choroid and retina. It appears to be most
frequent in the retina, probably on account of the small size of the
vessels and their being "end arteries." As a rule, septic embolism
leads to suppuration of the eye and external evacuation of the pus
after destruction of the cornea. An early fatal termination may be
predicted with tolerable certainty after the development of pyæmic
suppuration of the eye, but rare exceptions have been observed.
Hirschberg (*Centr. f. Aug.*, 1883, Sept.), Salo Cohn (*l.c.*, p. 169)
and Beck (*Jahr. f. Aug.*, 1877, p. 211) mention cases in which pyæmia recovered even after double pyæmic panophthalmia. Among
thirty-five cases of septic general disease Litten (*Zeitschr. f. kl.
Med.*, II, 3, p. 32) saw three unilateral and five bilateral panophthal-
mias.

In rare cases pyæmic disease of the eye runs a milder course; it
does not terminate in perforation but in simple shrivelling of the
globe (Landsberg, *Berl. kl. Woch.*, 1877, No. 38).

The pus in the eye contains pyogenic staphylococci, streptococci,
diplococci, and even bacilli.

Pyæmic abscesses in the brain, orbital abscesses and thromboses
of the cerebral sinuses give rise to corresponding eye symptoms. In
the last days of life, sometimes immediately before death, numerous
retinal hemorrhages, without suppuration, are observed in cases of
pyæmia. This is a sort of scorbutic terminal stage.

Weiss (*Mon. f. Aug.*, 1875, p. 393) found double metastatic choroiditis as the sole pyæmic metastasis in a case of compound fracture.
24

Leber (*Jahr. f. Aug.*, 1880, p. 324) reports unilateral purulent choroiditis after a " post-mortem tubercle," and bilateral choroiditis starting from inflammation of a phalangeal joint.   During an apyrexial puerperal state, Feuer (*Centr. f. Aug.*, 1881, Feb.) saw two abscesses develop in the sclera and terminate in suppuration and shrivelling of the eye.   It is doubtful whether this was a case of pyæmia.

In a particular case it is not always possible to make a sharp distinction between pyæmia and septicæmia, and both may develop together or successively.

Extensive Burns

resemble pure septicæmia because, on account of the absorption of products of decomposition from the burned surface of the skin, an acute self-intoxication is produced with a tendency to fatty degeneration and hemorrhage in all the tissues.   Retinal hemorrhages are often abundant after the third or fourth day.   Acute hemorrhagic nephritis occurs under the same circumstances, as after poisoning with chlorate of potash (Fraenkel, *Berl. kl. Woch.*, 1889, No. 2).

Death may happen from shock at the end of a few hours, or with the signs of an acute toxæmia in a few days, or after a still longer period from exhausting suppuration.

As in many other toxæmias, destruction of numerous blood-globules, hyaline thromboses, etc., may be demonstrated anatomically.

In two cases I have seen extensive retinal hemorrhages after severe burns.   There are few reports in literature (Wagenmann, *Arch. f. Ophth.*, 34, 2, p. 111), although they are undoubtedly very frequent. Mooren also mentions them.

Retinal hemorrhages are an unfavorable prognostic sign, but recovery may take place despite their occurrence.

Glanders

is substantially, in its clinical history, a pyæmia which has been produced by the pyogenic glanders bacillus.   The abscess due to the latter may also affect the eye.

In a case of subacute glanders Schoby-Buch (*Berl. kl. Woch.*, 1878, p. 74) observed purulent conjunctivitis, and also pustules from the size of a pea to that of a hazel-nut, upon a red, hard and painful base, situated on the forehead, nose and eyelids. In acute glanders, Boyd (*Jahr. f. Aug.*, 1883, p. 301) noticed an orbital abscess of the left side, after abscesses had developed in other parts; it was attended with swelling of the lids and exophthalmus and produced blindness, although the cornea remained transparent.

As a matter of course there may be an occasional localization in any of the vascular parts of the eye, and this is also true of

Ulcerative Endocarditis.

This disease is merely a variety of pyæmia. Its septic, coccus emboli may lead to hemorrhages, thromboses and abscesses. In addition to miliary foci in many other organs, Michel (*Arch. f. Ophth.*, XXIII, 2, p. 213) found numerous ecchymoses in the conjunctiva and retina, and congestion of the optic nerve. Numerous capillary emboli and miliary abscesses were found in the latter. Doepner (Diss. Berlin, 1877) mentions three cases of acute puerperal endocarditis with retinal hemorrhages. These would undoubtedly be found very often if search were made.

Splenic Fever,

anthrax, malignant pustule, malignant œdema, etc., when it does not remain localized—as is generally the case in man—produces its effects less by embolism than by the impoverishment of nutrition and toxic excretions by the bacilli which proliferate with enormous rapidity. Finally, there develops a sort of scorbutic terminal stage with a tendency to hemorrhages.

One of the favorite sites of infection is the delicate skin of the lids, where it may lead to very extensive destruction. The eyelids and the angle of the eye are also said to be favorite sites of the Aleppo bubo, which is closely allied to anthrax (Altonnyon, *Jahr. f. Aug.*, 1886, p. 312).

Tetanus and Lyssa.

In tetanus it is certain, and in lyssa it is probable, that a specific poison is produed by the morbific germs, so that both diseases resemble an acute alkaloid poisoning.   Concerning the occasional findings in the visual organ during tetanus, *vide* page 243.   In lyssa the pupils are often much dilated; during the paroxysms the eyes are wide open and glistening, and the conjunctiva is congested.   I am unacquainted with any other eye symptoms in this disease, although hemorrhages probably occur as the result of the interference with respiration and the fatal suffocation.

Penzoldt (*Berl. kl. Moch.*, 1882, No. 3) reports a case in which the conjunctiva was probably the point of entry of the infection.   In addition to a bite in the lip, the dog's snout had produced a slight wound in the conjunctiva.   There were violent pains in the eye during the prodromal stage.

Rheumatism.

Acute articular rheumatism must be regarded as an infectious disease, although a typical agent of inflammation has not been discovered or, when found, was not characteristic.   The general infection in gonorrhœa (*vide* p. 312) presents great similarity in its clinical course.   In the localizations of the former the same organisms are commonly found as in articular rheumatism; the gonorrhœal diplococcus is much less frequent.

In addition to the well-known outbreaks in the joints and serous membranes, acute articular rheumatism produces a series of inflammatory diseases of the eye, such as iritis, cylitis, scleritis and inflammation of Tenon's capsule, perhaps also glaucoma and parenchymatous keratitis.

Even in the prodromal stage there may be affections of the eye, but they are not characteristic,—for example, temporary blindness (Woinow, *Jahr. f. Aug.*, 1871, p. 342).   Michel (*Mon. f. Aug.*, X, p. 167) reported sudden and complete paralysis of the motor oculi, probably hemorrhagic in origin.   Such conditions are extremely rare.

Rheumatic iritis is in a measure characteristic.   In many cases it cannot be distinguished from an ordinary attack, but it is often signalized by great pain (spontaneous, on pressure and on moving the eye) and the appearance of a coagulating exudation in the anterior chamber.   Even hemorrhages may occur in the latter situation.

All these signs may also be present in iritis from other causes, but they are more frequent in rheumatic than in other forms. Purulent iritis is very rare (Thiry, *Jahr. f. Aug.*, 1873, p. 283); a grayish yellow, so-called cyclitic hypopyon is more frequent.

The iritis is either synchronous with the joint affection or alternates with the latter. Relapses are very common. The attacks of iritis occasionally recur for years at the same season (Higgens, Laqueur, *ib.*, 1874, p. 326).

Transitions to cyclitis occur in all forms, from the most severe to "cyclitis minima" (Bouchéron), in which the main symptom is spasm of accommodation.

Scleritis, inflammation of Tenon's capsule (exophthalmus, chemosis, pain on pressure and on moving the eye) and certain forms of glaucoma, especially very acute and painful ones, are also observed. The symptoms are not characteristic, except that anti-rheumatic treatment (salicylate of soda) often exercises a favorable influence on their course. In many of these cases, however, it remains doubtful whether they are really due to articular rheumatism.

In the last twenty years I find mention made of only two cases of rheumatic optic neuritis (Macnamara, *ib.*, 1890, p. 353).

Parenchymatous keratitis appears to be associated more frequently with articular rheumatism (Arlt, Foerster), but it does not differ from the same disease in congenital syphilis.

The diagnosis of a rheumatic affection of the eye may not be made simply because we know of no other cause, or on account of the great pain, or the patient's statement that he has caught cold or has been exposed to a draught. It is necessary to prove that the same noxious agent which produced the disease of the joints and serous membranes was also the cause of the disease of the eye, and this is rarely possible.

The large majority of so-called rheumatic muscular paralyses do not belong to this category.

Rheumatic endocarditis is one of the most frequent causes of embolic processes in the eyes and brain, particularly of benign emboli which act mechanically and excite very little or no inflammation. There is every possible transition, however, into the severe infectious emboli of ulcerative endocarditis, which may with equal propriety be called pyæmic.

An unprejudiced survey leads to the conviction that acute articular rheumatism is a typical infectious disease, but that it may be produced clinically by several micro-organisms (the gonococcus occasionally produces exactly the same effects). As the microbes found in rheumatism are also met with in other diseases, nothing remains but the assumption that the condition and quality of the patient determine the characteristic course—on the one hand, the localization mainly in the joints, the serous membranes, endocardium, occasionally the eye; on the other hand, the form of inflammation (serous, plastic, very rarely purulent) and the peculiar character and course of the symptoms.

The micro-organisms are usually not destroyed entirely or excreted in the first attack. Hence, very slight causes may excite a relapse in the *locus minoris resistentiæ*, *i.e.*, in the tissues and organs which have already been diseased. Chronic rheumatism then develops under the influence of the repeated inflammations and the consequent changes in the processes of disassimilation. Although chronic rheumatism has been regarded as the cause of numerous diseases of the eye, this assumption is only justified, in a measure, in regard to certain forms of uveitis (iritis, glaucoma) and scleritis.

Measles.

Pronounced conjunctival catarrh with more or less photophobia is usually present toward the end of the period of incubation and in the prodromal stage of measles. When the other symptoms are doubtful, this may be very important in making an early diagnosis. It continues during the eruptive stage and may be associated with phlyctenulæ. Affections of the cornea are rare in this stage. According to Galezowski (*Jahr. f. Aug.*, 1887, p. 251), measles may appear solely as a phlyctenular conjunctivitis, but it would be very difficult to prove this statement in an individual case.

Other complications are rare. They are, in part, the results of a complicating meningitis, like Nagel's three cases of double blindness after measles, which was associated with other meningeal symptoms.

In two of these cases the blindness was permanent, the third case slowly recovered. In one of the former the ophthalmoscope showed optic neuritis. During convalescence from measles, Graefe also observed bilateral, complete, but temporary blindness with slight inflammation of the optic nerve and adjacent retina (Foerster, *l.c.*, p. 161).

Other cases of optic neuritis after measles, due in the main to complicating meningitis, have been reported. Wadsworth (*Arch. f. Aug.*, X, 1, p. 100) reports three cases, one terminating in atrophy of the nerve and blindness, one in death, one complicated by abducens paralysis. Carreras Arago (*Centr. f. Aug.*, 1882, Oct.) observed one fatal case, and Galezowski (*Jahr. f. Aug.*, 1881, p. 401) also reports a case.

Measles is followed very often by suppuration from the ear. This may give rise to meningitis and then involve the eye. For example, Keller (*Mon f. Ohr.*, 1888, No. 6) reports a case of acute catarrh of the middle ear followed by double choked disc and left abducens paresis.

V. d. Stok (*Arch. f. Aug.*, 1888, p. 391) describes a case of right hemiplegia and hemianopsia, terminating in recovery, after the disappearance of the measles eruption in a man aged twenty years.

Among the sequelæ during convalescence and later, special attention should be drawn to phlyctenular diseases of the conjunctiva and cornea. They are often very severe and obstinate and occasionally lead to destruction of the cornea (Dujardin, *Jahr. f. Aug.*, 1868, p. 247, ten cases). Obstinate eczema of the lids and their edges is also frequent.

Keratomalacia after measles has been reported by Bezold (*Berl. kl. Woch.*, 1874, p. 408), Fischer and Beger (*Jahr. f. Aug.*, 1874, p. 310).

A very rare symptom is spontaneous gangrene of the lids. Noma is much more frequent and may extend to the lids. Fieuzal (*Jahr. f. Aug.*, 1887, p. 422) describes three cases of gangrene of the upper lid terminating in ectropium. In one infant I observed spontaneous gangrene of all four lids, approximately in the locality where xanthelasma usually develops. Recovery was attended by a deep scar but without deformity, inasmuch as the edges of the lids had remained intact.

Lindner (*Wien. med. Woch.*, 1891, p. 683) observed an acute inflammation of both lachrymal glands with considerable swelling of the cervical, sublingual and submaxillary glands, after measles, in a boy of eight years. Adler ("3. Ber. d. Krankenh. in Wieden") also observed dacryo-adenitis after measles in a boy of seven years.

Horner (*Mon. f. Aug.*, I, p. 11) reports retinitis albuminurica after measles. This is not surprising in view of the fact that renal affections often follow measles.

Fialkowsky (*Jahr. f. Aug.*, 1888, p. 538) claims to have seen recovery of trachoma, ectropium and pannus as an effect of the increased disassimulation during measles, while Hirschberg observed, under similar circumstances, spontaneous necrosis and ulceration of a leucoma adhærens, followed by granuloma of the iris.

Roetheln (Bastard Measles).

Among my notes I find mention made by St. Martin of a case of facial paralysis and gangrene of the upper lid after roetheln.

Scarlatina.

This gives rise to conjunctival catarrh in quite a proportion of cases, but by no means so constantly as measles.

The most important affections of the eye are due to complications. Acute nephritis, which often develops very suddenly, may produce uræmic blindness. This is generally associated with eclamptic attacks, often with œdema of the lids; the reaction of the pupils to light may be retained or absent. Less acute diseases of the kidney may cause retinitis albuminurica, and the latter is occasionally combined with the uræmic disorder of vision.

On the whole, the prognosis of these acute infectious renal diseases and their effects upon the eye is better than when the former develop independently of infection, but partial atrophy of the optic nerve may also follow scarlatina. Not infrequently chronic nephritis is the result, and this may prove fatal long afterward, for example, after the lapse of twenty years in Aufrecht's case (*Jahr. f. Aug.*, 1888, p. 571).

Scarlatinous nephritis may appear after very mild attacks of scarlatina. Hutchinson (*Jahr. f. Aug.*, 1871, p. 293) even speaks of scarlatinous renal retinitis without an eruption.

Complicating meningitis may produce optic neuritis, spasms and paralyses of the eye muscles; when it runs a chronic course it may lead to deafness and idiocy.

Despite the frequent coincidence of pharyngeal diphtheria, post-diphtheritic paralysis of accommodation is quite rare. It should not be mistaken for the simple weakness of accommodation during convalescence. Other muscular paralyses after scarlatina are also rare.

Severe phlyctenular affections of the conjunctiva and cornea, and aural suppuration with its results (facial paralysis, meningitis, etc.) are more frequent after scarlatina than after measles. Alt (*Arch. f. Aug. u. Ohr.*, 1878, p. 54) observed post-scarlatinous kerato-malacia, without disease of the lids or trigeminus paralysis.

Diseases of the lachrymal sac are also quite common. Kendall (*Brit. Med. Journ.*, 1883, I, p. 1,225) reports a case of this kind associated with suppuration of the right eye.

Atrophy of the optic nerve from orbital inflammation is reported by Nettleship (*Jahr. f. Aug.*, 1880, p. 282, Case 3), who also describes (*ib.*, 1886, p. 248) an erysipelatous orbital suppuration after scarlatina, which gave rise to optic atrophy.

Lindner (*Wien. med. Woch.*, 1891, No. 16) reports two cases of purulent inflammation of the lachrymal gland with scarlatina, while dacryo-adenitis after measles does not lead to suppuration. He claims that scarlatina is a more pyogenic affection than measles.

The embolism of the central artery of the retina, observed by Hodges (*Jahr. f. Aug.*, 1885, p. 260) in a girl of eighteen years during convalescence, was probably a mere coincidence, although a causal connection cannot be absolutely denied.

Schiess reports a case of partial albinism of the upper lid after scarlatina. A white streak was situated on the forehead and right upper lid; the brow was perfectly white in one place, the eyelashes partly white, partly black. This part was also free from freckles which covered the rest of the face. Motion and sensation were normal. Although there can be no doubt of a causal connection between infectious diseases and such trophic disturbances, no definite localization (superior cervical ganglion?) has yet been ascertained.

Small-pox.

Compare Manz (*Jahr. f. Aug.*, 1871, p. 178) and Adler (*Vierteljahr. f. Dermat. u. Syph.*, 1874). The eruption of small-pox may attack the eye. According to Adler (*ib.*) it appears upon the integument of the lids in twenty per cent of the cases. If the pustules are very numerous, œdema develops; if they are hemorrhagic, blood will be found in the lids. Erysipelas, phlegmons, abscesses and furuncles of the lids are not uncommon results. Pustules upon the free border lead to partial trichiasis, ankyloblepharon, etc. Ciliary seborrhœa and blepharitis ensue not infrequently. We then find hordeolum, chalazion, ectropium, thickening of the lids, trichiasis and distichiasis; also bending and shrinking of the lids if the process has extended more deeply. Caries and periostitis of the rim of the orbit have also been observed (Landesberg, *Jahr. f. Aug.*, 1874, p. 505; Magnus, *ib.*, 1887, p. 133).

The conjunctiva often exhibits congestion and catarrh, especially when the lids or their edges are affected. There may even be blennorrhœa, with chemosis and conjunctival hemorrhages. If pustules develop upon the conjunctiva, an ulcer alone is visible because the top of the pustule is soon exfoliated. They look exactly like large phlyctenulæ, and even large ones may heal without any bad after-effects. But if they are situated, as frequently happens, at the edge of the cornea, progressive corneal suppuration with hypopyon is apt to develop. This may heal at any stage—often with a very characteristic sickle-shaped opacity of the cornea. It often leads to loss of one eye or both, especially in variola which has not been modified by vaccination. According to Geissler the loss of both eyes after variola is rare only because such individuals usually die. Extensive conjunctival hemorrhages may occur in hemorrhagic small-pox.

Some deny (Adler, Foerster), others affirm (Horner) the development of primary pustules on the cornea, but at all events they are rare.

The fact that purulent keratitis often does not begin until one or two weeks after the outbreak of the eruption, or even during the period of convalescence, shows that the progressive suppuration is due to secondary infection of the pustules.[1] Severe corneal suppurations of this kind are found in bad cases of small-pox and in marasmic individuals. Simple marasmic keratitis (kerato-malacia) also occurs under such circumstances.

Acute and chronic inflammations of the lachrymal passages are frequent in small-pox, and often from the formation of pustules upon their mucous membrane.

Pustules upon the eye are rarer in varioloid and run a milder course than in true variola, although even mild cases sometimes cause severe affections of the eye.

Among the complications of small-pox meningitis is very rare, while renal disease is frequent. Uræmic disturbance of vision or retinitis albuminurica may be observed at an earlier or later period.

Retinal hemorrhages undoubtedly occur in hemorrhagic small-pox, although I find no references in literature. Hemorrhages into the optic nerve will also occur occasionally, and there is no doubt that such hemorrhages near the entrance of the nerve must have been the cause of some of the affections described as neuritis with and without stasis, and with or without termination in optic nerve atrophy. Adler (l.c.) describes two cases of diffuse neuro-retinitis as the pustules passed into the stage of desiccation; both recovered.

Sequelæ of small-pox are rare, and do not develop until after the third week. We may mention diffuse iritis (associated, in hemorrhagic small-pox, with hemorrhages into the anterior chamber), cyclitis, choroiditis; in many cases the latter affections are only revealed by opacities in the anterior or posterior part of the vitreous, as in relapsing fever. This sometimes terminates, at a later period, in cataract (Romiée, Hutchinson). According to Adler (l.c.) many

[1] Hence it follows that the destructive suppuration of the pustules can be prevented by strict antisepsis or asepsis. Even if this cannot be done over the entire body, excellent results would follow its employment in the face alone.

cases exhibit merely ciliary irritation, *i.e.*, ciliary injection, photophobia and epiphora, without decided inflammatory symptoms.

Localized diseases of the choroid may also occur after small-pox. I have seen several chorio-retinitic foci in the fourth week. Retinitis is undoubtedly very rare, but Manz mentions a case.

Outbreaks of glaucoma, due to small-pox, have been reported by several writers.

Parenchymatous keratitis after small-pox has been reported by Adler (*l.c.*) and Bock (*Centr. f. Aug.*, 1890, p. 361). More or less severe phlyctenular diseases are much more frequent.

Paralyses of the eye muscles and central disorders of sight are very rare. Wohlrab (*Jahr. f. Aug.*, 1872, p. 512) observed aphasia in varioloid and, during the fourth week, neuro-paralytic keratitis. Both were probably due to meningitic processes.

Geissler (*Arch. f. Heilk.*, 1872, p. 549) noticed complete and permanent clearing of an old corneal opacity during variola, while the previously healthy cornea of the other eye was destroyed.

Vaccination.

The laity generally attribute to vaccination everything that happens to a child in the first year of life. It cannot be denied that phlyctenular diseases of the conjunctiva and cornea, eczema of the face, etc., may follow vaccination, as they do many other infectious diseases. In feeble children such affections of the eye are occasionally quite severe and protracted. If these affections occur long after vaccination, as usually happens, there can be no ground for admitting the causal influence of the vaccination.

The vaccine virus is sometimes inoculated accidentally upon the lids or in the conjunctival sac. A typical vaccine pustule may be produced in this way (Hirschberg, *Arch. f. Aug.*, VIII, p. 166; Senut, *Jahr. f. Aug.*, 1886, p. 440; Berry, *Brit. Med. Journ.*, 1890, p. 1,483). Hirschberg (*Jahr. f. Aug.*, 1885, p. 464) also mentions a vaccine blepharitis from the introduction of lymph into the conjunctival sac. In the case of a physician who accidentally struck himself in the eye while vaccinating, Critchett (*ib.*, 1876, p. 268) observed violent keratitis with sero-purulent infiltration of the cornea, despite immediate irrigation of the eye. The keratitis terminated in a leucoma, *i.e.*, a vaccine pustule of the cornea. Vaccine pustules of the eye are most apt to develop on the free borders of the lids and in the canthi. [A picture of the appearance of the lids when vaccine infection attacks them is given in the editor's text-book of diseases of the eye, Plate VII. See description, p. 816.—ED.]

If syphilis is transmitted in vaccination, any of the syphilitic affections of the eye may follow, among which iritis is usually the earliest.

After vaccination at the age of five years, Tilly (*Journ. of the Amer. Med. Assn.*, Feb. 4th, 1888) observed pemphigus of both conjunctivæ (in addition to other parts). Within a year this caused loss of sight by obliteration of the conjunctival sac. The coincidence of the two affections seems to me to have been accidental.

Varicella.

Apart from eruptions upon the lids, varicella rarely produces eye symptoms. According to Comby (*Jahr. f. Aug.*, 1884, p. 323), the eruption on the lids sometimes precedes that on other parts of the body. Steffan (*ib.*, 1873, p. 284) noticed purulent iritis with hypopyon during convalescence from varicella in a child of three years; the case terminated fatally. As this case is unique it is impossible to determine whether the coincidence was accidental or not (possibly due to secondary or mixed infection).

As nephritis is sometimes found after varicella (Oppenheim, Henoch, Janssen), this might occasionally give rise to eye symptoms. As the varicella may run a very mild course and is often hardly noticed, it would be difficult to recognize sequelæ as such, especially if they do not develop until after the lapse of several weeks.

Typhoid Fever.

In typhoid fever the eye may be attacked during the disease or at a later period. Conjunctivitis in varying degrees of severity is not uncommon. During convalescence, paresis of accommodation with dilatation of the pupil is very frequent, not as a true paralysis but as a part of the general weakness. True paralyses of the eye muscles may also begin long after the disease has run its course. Considerable dilatation of the pupil is often noticed, despite normal vision and accommodation (Segal, *Arch. f. Aug.*, 1888, p. 386). During convalescence, and still more at a later period, there is a great tendency to phlyctenular affections of the conjunctiva and cornea, even passing into kerato-malacia. This should not be mistaken for marasmic xerosis corneæ, which develops in the somnolent stages of typhoid from desiccation of the conjunctiva and cornea and subsequent external infection. Next to cholera this is observed most frequently in typhoid fever.

Temporary and permanent impairment of sight may also be produced by neuritis or retrobulbar neuritis, with or without subsequent atrophy of the optic nerve. A complicating meningitis often forms the connecting link between the infectious disease and the affection of the optic nerve. Von Petershausen (*Jahr. f. Aug.*, 1873, p. 362) reports double neuro-retinitis with macular hemorrhages during typhoid fever; Munier (*ib.*, 1875, p. 345) describes neuro-retinitis and deafness, evidently as the result of meningitis. The first eye was attacked three months after the typhoid fever, the second eye four months later.

Meningitic processes, or "irritation" with more or less involvement of the cortex, are evidently quite frequent and probably the main cause of the delirium and hallucinations. At the base of the brain they give rise to optic neuritis, irritation and paralyses of the nerves; at the convexity they may cause cortical disorders of vision, hemianopsia (very rare) and double amaurosis without findings, especially in children. The cause of an hemianopsia may also be located in the tractus.

Leber and Deutschmann (*Arch. f. Ophth.*, XXVII, 1, p. 272) reported a case of double blindness with secondary atrophy of the optic nerve and pigmentation of the entrance of the nerve. Seggel (*Jahr. f. Aug.*, 1884, p. 323) reported a case of impaired vision during the second week; rapid improvement in the right eye—with the left only movement of the hand could be seen, and later there was atrophy of the optic nerve. Both these cases must be regarded as retrobulbar neuritis, probably from hemorrhage into the optic nerve.

Unilateral and bilateral iritis, cyclitis, choroiditis, chorio-retinitis, etc., are also met with, but not as often as in relapsing fever. They occur in serous and plastic forms, and are often complicated with opacities of the vitreous. Cataract may develop more or less rapidly as the result of these uveal diseases, for example, Arens (*Jahr. f. Aug.*, 1885, p. 428, ripe cataract in two young sisters a year after typhoid fever), Trélat (*Gaz. des Hôp.*, 1879, p. 417), Romiée, Fontan (*Rev. gén. d'Ophth.*, April, 1887), etc. In such cases the examination of the urine should not be neglected, because diabetes may occur as a sequel of typhoid fever.

In muscular paralyses occurring some time after typhoid fever (Runeberg, *Jahr. f. Aug.*, 1875, p. 504; left trochlear paralysis one and one-half years afterward), albuminuria should always be looked for as the intermediate cause. Chronic nephritis is a quite frequent sequel of typhoid fever, and appears to me to possess in these cases a certain tendency to the production of paralyses of the eye muscles. These recover quickly and also relapse quickly, and are nuclear in character. I cannot regard it as a mere coincidence that almost all the albuminuric paralyses which I have seen of late years could be attributed to a previous typhoid fever. During the fever the paralyses are very rare. Nothnagel reports double ptosis and right abducens paralysis, with aphonia, at the beginning of the third week (Foerster, *l.c.*, p. 167).

Noma may also extend to the eyelids and, if recovery occurs, may give rise to ectropium. It is well known, however, that the majority of such cases terminate fatally.

Profuse hemorrhage is a not infrequent cause of amblyopia and amaurosis (*vide* p. 290). It is usually intestinal, but an epistaxis (Ebert, *Berl. kl. Woch.*, 1868, p. 21) or menorrhagia (Williams) may also act as the cause.

Psychoses, focal and systemic diseases of the brain and cord are occasional sequelæ, which sometimes do not develop until after the lapse of years. They may give rise to a corresponding affection of the organ of sight. Proelss (Diss. Berlin, 1886) observed after typhoid fever a paralysis of the cervical sympathetic which terminated in facial hemiatrophy.

Typhus Fever.

Very little has been published concerning affections of the eye during and after typhus fever. They are probably similar to those following typhoid and relapsing fever.

According to Salomon (*Jahr. f. Aug.*, 1880, p. 239), conjunctival catarrh is a constant symptom and exhibits very striking injection of the bulbar conjunctiva. This is not true, however, of all epidemics.

The uveal inflammations which are so frequent after relapsing fever are also found after typhus, though not so frequently. Hersing (*Arch. f. Ophth.*, XVIII, 2, p. 69) reports an annular, equatorial chorio-retinitis, with hemeralopia and annular scotoma, after typhus fever.

In a patient aged twenty-eight years, Lindner (*Wien. Med. Woch.*, 1891, p. 283) observed an acute inflammation of the lachrymal glands without suppuration; death occurred upon the tenth day.

Relapsing Fever.

Vide *Jahr. f. Aug.*, 1870, p. 319; Logetschinkoff (*Arch. f. Ophth.*, XVI, 1, p. 353); Foerster (*l.c.*, p. 169); Brieger (*Charité-Annal.*, VI, p. 136); Haenisch (*Deutsch. Arch. f. kl. Med.*, V, p. 53); Peltzer (*Berl. kl. Woch.*, 1872, No. 37); Estlander (*Arch. f. Ophth.*, XV, 2, p. 108), etc.

Conjunctival catarrh is a frequent symptom in relapsing fever. At the height of the attacks there may be temporary unilateral or bilateral disorders of vision, even blindness (Brieger, *l.c.*). As is shown by the reaction of the pupils to light, these may be due to peripheral as well as central action, and perhaps result, in part, from the direct action of the virus.

The symptoms during convalescence resemble those after typhoid fever, viz., weakness of accommodation, temporary enlargement or inequality of the pupils, tendency to phlyctenular diseases, etc.

The most characteristic sequel is commonly called cyclitis, although all parts of the uveal tract may be affected separately or in combination. Simple plastic iritis is comparatively rare; while the serous type, with deposits upon the posterior wall of the cornea, and perhaps hypopyon, is more common. Cyclitis and choroiditis are manifested chiefly by vitreous opacities. In mild cases a ring of opacities is detached from the ciliary body and is slowly absorbed in the anterior parts of the vitreous body. In severe cases the entire vitreous is more or less filled with opacities and there is more or less impairment of sight. The deficiency may be much more pronounced than the density of the opacities would lead us to believe. If the fundus is visible, the ophthalmoscope shows merely a reddened disc. The choroiditis is not visible with the ophthalmoscope, because it is diffused and the pigment epithelium is very slightly changed; it is shown only by the vitreous opacities, but a part of these are undoubtedly derived from the retina. The implication of the latter is undeniably the cause of the disproportionate impairment of vision. After the disease has run its course, disseminated changes are found not infrequently in the anterior part of the choroid.

The disease of the eye must then be regarded as a diffuse inflammation of the uvea, which usually reaches its greatest intensity in the region of the ciliary body. It appears in all possible degrees of acuteness, from hardly visible ciliary injection to the most violent inflammatory phenomena; and the pain will be in proportion to their intensity. There are also transitions to the plastic-purulent forms, such as are peculiar to meningitis. In severe cases the tension of the eye is considerably diminished; shrinking of the vitreous, detachment of the retina and, finally, phthisis bulbi may ensue. Secondary cataract is rare and appears usually as posterior cortical cataract.

This uveitis is essentially a sequela; it rarely begins between or during the attacks, usually not until the third week, and occasionally much later.

The prognosis is generally favorable. Complete recovery occurs after the lapse of weeks and months by the gradual absorption of the vitreous opacities. On the other hand, complete loss of one eye or even of both eyes may result from pupillary occlusion, vitreous opacities and detachment of the retina. More or less dense vitreous opacities may also persist for a long time.

The frequency and severity of ocular complications vary in different epidemics. In some they are found in nearly ninety per cent of

25

the cases, in others in very few. In about twenty per cent the trouble attacks both eyes. Mild epidemics may be attended by numerous and severe diseases of the eye, severe epidemics by rare and mild affections. Men are attacked more frequently than women.

It is evident that the virus of the disease slowly produces certain tissue changes, in great part, probably, in the vessels. These are followed after a while by a more or less acute interstitial inflammation (*vide* p. 363), on account of some unknown cause (embolism or thrombosis?). In view of the great vascularity of the uvea, it is not astonishing that this should be affected so often. Moreover, diseases in this region are most apt to produce subjective and objective symptoms. A similar process in another locality would produce very vague symptoms or none at all. Even in the pia mater, which is genetically co-ordinate and equally vascular, an analogous condition would not produce notable symptoms until it had acquired considerable extent. However, symptoms of a more diffused or circumscribed meningitis are frequent in all forms of typhoid disease. Like the uveitis of relapsing fever, the meningitis of non-pyogenic infectious diseases also recovers completely, as a general thing, but sometimes leads to very annoying complications and sequelæ.

The uveitis which has just been described occurs most frequently in relapsing fever, but is also found in other typhoid affections, small-pox, influenza, etc.

Every possible transition is found between it and the typical plastic, purulent irido-choroiditis of simple and cerebro-spinal meningitis, and also purulent inflammation of the choroid and retina in pyæmic processes.

Very little has been published concerning any other complications and sequelæ of relapsing fever.

Cholera.

Vide Foerster (*l.c.*, p. 177), Graefe (*Arch. f. Ophth.*, XII, 2, p. 198).

The characteristic changes in the eye are those which depend upon the loss of the serum of the blood and tissues, and the consequent interference with circulation. Similar changes in a milder form are also seen in cholera morbus and cholera infantum. As a matter of course the face and eyelids take part in the often striking and general cyanosis.

On account of the loss of fluid in the orbital tissues, the eyeball sinks deep into its socket; the loss of fluid in the lids causes them to shrink so that they are closed imperfectly. It is true that the voluntary action of the orbicularis palpebrarum will close the lids, but during sleep the palpebral fissure is more or less open and the imperfectly covered eyes are rotated upward. The lowermost part of the cornea and the bulbar conjunctiva below it are thus exposed to external injurious influences. Hyperæmia and inflammation of the exposed conjunctiva develop, but usually not until the stage of reaction. In severe cases these parts become desiccated (xerosis). This is facilitated by the complete apathy of the patient in the typhoid stage.

Spontaneous hemorrhages into the conjunctiva sometimes occur and have a very grave prognostic import. Xerotic keratitis is also very ominous, although it may possibly heal at any stage after exfoliation of the dried slough. In favorable cases, a more or less extensive leucoma is left upon the lower half of the cornea.

In very severe cases slaty-gray patches of irregular shape occasionally appear below or alongside the cornea. As they may also develop beneath the closed lid, they cannot always be due to simple drying of the conjunctiva and sclera. According to Boehm and Graefe they are the result of the loss of fluid in the corresponding part of the sclera. It appears much more probable to me that in those cases which may not be attributed to simple desiccation, we have to deal with choroidal hemorrhages which shine slaty-gray through the thinned sclera. Such extravasations have been found in a number of autopsies, and the very rapid development of these patches also favors this theory.

It is readily understood that the lachrymal and conjunctival secretion is diminished.

At the height of the disease the retinal arteries are seen to be narrow, filled with dark blood, and are readily made to pulsate or even emptied by pressure. The more marked these appearances are, the more unfavorable is the prognosis, because the ophthalmoscopic findings simply represent the general condition of the blood and the blood pressure. The veins are also very dark, their calibre unchanged, the column of blood sometimes interrupted, and the circulation then takes place by fits and starts as happens during restoration of the retinal circulation after embolism of the central artery of the retina.

Vomiting might possibly result in the same disorder of vision as losses of blood (*vide* p. 290), but no reports are furnished in literature. Perhaps this explains Roorda Smit's case (*Arch. f. Aug.*, XVIII, 3, p. 383), in which a previously healthy man of forty years became entirely blind within two weeks after an attack of cholera and later exhibited atrophy of the optic nerve (so-called retrobulbar neuritis, probably retrobulbar hemorrhage into the optic nerve).

The condition of the pupils varies, but they are usually contracted. During the attack, Campart and St. Martin (*Jahr. f. Aug.*, 1885, p. 356) often found pronounced mydriasis and inequality of the pupils; sometimes unilateral and bilateral myosis without any change in accommodation. Myosis predominated toward the end of the attack. According to Joseph, myosis is also present in the typhoid stage and mydriasis is only found in severe collapse (Foerster, *l.c.*, p. 178). According to Corte (*Deutsch. med. Woch.*, Jan. 22d, 1891), the condition of the pupils during the algid stage decides the prognosis. If the reaction to light is preserved, the prognosis is favorable whether the pupil is narrow or wide, and however violent the other symptoms (cyanosis, collapse, pulselessness, etc.) may be. Among sixty-six patients with intact reaction of the pupils to light, all recovered; a fatal termination is, however, possible during the stage of reaction. If the pupils are immovable we may be certain of a fatal termination despite the slight severity of the symptoms and the restoration of various functions.

Little is known concerning sequelæ of cholera, so far as they affect the eye. The same conditions may occasionally develop as after other infectious diseases. Williams (*Jahr. f. Aug.*, 1885, p. 26) reports a case of iritis after cholera, and also reports another in which the iritis was cured by an attack of cholera. Delens (*ib.*, 1886, p. 475) reports a cure of leukæmic orbital tumors by an attack of cholera.

Dysentery.

Disorders of sight play a comparatively insignificant part in this disease. It often leads to nephritis, and occasionally the effects of the latter will be observed in the organ of sight. Muscular paralyses, similar to those after diphtheria, have been attributed in a few cases to dysentery. Marcisiewicz (*Wien. med. Presse*, 1888, p. 561) describes an inflammation of both lachrymal glands as a sequel of dysentery.

Diphtheria.

In this disease eye symptoms play an important part. We will not consider diphtheritic affection of the conjunctiva and lids, which may occur either independently or in combination with the same disease in the nose, pharynx and larynx. Apart from these, the most important sequelæ are paralyses of various kinds. Special importance attaches to paresis, more rarely to complete paralysis of accommodation. As a rule, it begins a few weeks after the diphtheria has run its course and usually disappears spontaneously in four to eight weeks. It is almost always bilateral, and the pupil is hardly ever affected. Hence some suspicion attaches to Jeaffreson's case (Foerster, *l.c.*, p. 172) of unilateral paralysis of accommodation after diphtheria, because there was mydriasis on the same side. The paresis of accommodation is often isolated, but there may also be other post-diphtheritic paralyses, especially in the larynx and pharynx. The large majority of cases occur in children.

This characteristic paralysis may occur after diphtheria of all localities (conjunctiva, vulva, wounds, etc.), but is most frequent after pharyngeal diphtheria. It may follow mild cases and even those which have been entirely overlooked. It is not impossible that a diphtheritic general infection and subsequent paralysis may develop without any visible localization. During epidemics of diphtheria, typical paresis of accommodation, occasionally in conjunction with other paralyses, may be observed although there has been no trace of previous local disease. In other cases, the patient suffers from pain in the throat or difficulty in swallowing, but the most careful examination fails to reveal any diphtheritic membrane. In these cases, however, typical paresis of accommodation must be regarded as evidence of previous diphtheritic infection.

During this paresis of accommodation, refractive power is sometimes moderately diminished, especially in latent hypermetropia, or slight astigmatism becomes manifest. Adams (*Lancet*, 1882, No. 4) observed violent spasm of accommodation following a diphtheritic paresis.

According to Hasner (*Wien. med. Zeit.*, 1873, p. 120) a frequent complication of the paresis is congestion of the retina with impairment of vision which recovers very slowly. In the many cases which I have seen the fundus has always been found normal.

Paralyses of other eye muscles, with or without paresis of accommodation, are also occasionally seen after diphtheria. Squint is mentioned in a number of cases. Callan (*Jahr. f. Aug.*, 1875, p. 485) reports double paralysis of accommodation with left hemiplegia and incomplete ptosis; Rumpf (*ib.*, 1877, p. 381) describes double paralysis of accommodation and of the internal recti. Henoch reported double abducens paralysis (*Deutsch. med. Woch.*, 1889, No. 44); this was possibly a spasm of convergence, like Rosenmeyer's (*l.c.*) two cases of paresis of both externi.

Complicated paralyses of the eye muscles also occur. Uhthoff (*Berl. kl. Woch.*, 1884, p. 318) reports complete ophthalmoplegia externa and slight ptosis, with preserved reaction of the pupils, after a paresis of accommodation. Mendel (*Centr. f. Aug.*, 1885, p. 89) reports bilateral paralysis of all the recti muscles, not complete on the right side; fatal termination. Two weeks after an apparently very mild attack of diphtheria, Ewetzky (*Jahr. f. Aug.*, 1887, p. 253) observed bilateral total ophthalmoplegia externa and ptosis, with paralysis of the velum palati; recovery in three weeks.

Even in these complicated cases the prognosis is good unless death is due to other causes.

Sensory disorders are also observed in the eye. Laqueur (*Mon. f. Aug.*, XV, p. 228) reports right facial and trigeminus paralysis, terminating in neuro-paralytic keratitis and loss of the eye. The paralysis of the trigeminus recovered partially, that of the facial nerve completely.

The majority of these paralyses developed after pharyngeal diphtheria. After diphtheria of the eye, Politzer (*Jahr. f. Kinderh.*, 1870, p. 335) observed purulent cerebro-spinal meningitis and purulent inflammation of several joints, evidently a pyæmia due to secondary or mixed infection. After diphtheria of the conjunctiva, with implication of the nose, Dubois (*Prog. Méd.*, May 7th, 1887) saw paralysis of the velum palati and lower limbs.

The paralyses are rarely central or nuclear, but usually peripheral. In most cases we have to deal with hemorrhages or neuritis or with hemorrhagic neuritis. This affects mainly the nerve roots and peripheral nerves, but sometimes the nuclear region, and, in rare cases, more central portions (Krause, *Neurol. Centr.*, 1888, No. 17). Decided evidences of inflammation have also been found in the muscles (Hochhaus, *Virch. Arch.*, 124, 2). It is doubtful, however, whether all post-diphtheritic paralyses develop in this way. The very frequent paralysis of the velum palati is regarded by many as a direct local effect of the virus. Typical paresis of accommodation is also explained with difficulty by hemorrhagic or inflammatory processes in any locality. It is very evident that during a certain stage of diphtheria, or after the attack, a definite ptomaine is produced and that this has a paralytic action upon accommodation, while it has no influence on the movements of the pupil. Otherwise it is impossible to understand why accommodation alone is paralyzed, and why the paralysis is not complete but is bilateral in almost every case.

Disturbances of vision after diphtheria may also be due to diplopia, which is not usually recognized at once by the patient. But disorders of sight do occur in very rare cases. Concentric narrowing of the field of vision is mentioned by Herschel (*Berl. kl. Woch*, 1883, p. 456) and Jessop (*Jahr. f. Aug.*, 1886, p. 248). Neuritis is described by Bouchut (*Gaz. des Hôp.*, 1873, p. 302), Galezowski (*Jahr. f. Aug.*, 1881, p. 401) and Nagel (*ib.*, 1884, p. 328); the latter has noticed color disturbance and concentric narrowing of the visual field.

Among the rare sequelæ may be mentioned abscesses of the orbit and cheek (Romiée, *Jahr. f. Aug.*, 1879, p. 425, and Heyl, *ib.*, 1880,

p. 419), also inflammation of the left lachrymal gland, together with swelling of the parotid, cervical and submaxillary glands and acute catarrh of the middle ear, in a boy of nine years, after pharyngeal diphtheria (Lindner, *Wien. med. Woch.*, 1891, p. 683).

Renal disease is a not infrequent sequel of diphtheria, even apart from so-called scarlatinous diphtheria. The renal affection may later involve the eye. [Extension of inflammation to the middle ears is not to be omitted from an inventory of the rather frequent sequelæ of diphtheria.—ED.]

Influenza.

Rampoldi, *Annal. di Off.*, XIX, I; Eversbusch, *Muench. med. Woch.*, 1890, No. 6; Pflueger, *ib.*, No. 27; Greeff, *ib.;* Gutmann, *ib.*, No. 48; Adler, *Wien. kl. Woch.*, 1890, No. 4; Galezowski, *Rec. d'Ophth.*, 1890, No. 2; Badal and Fage, *Arch. d'Ophth.*, 1890, p. 136; Hillmanns, Diss. Bonn, 1890; Ehrlich, Diss. Breslau, 1892.

All the complications and sequelæ which are found in acute infectious diseases in general are also observed occasionally in influenza. Very few occur, however, with any considerable degree of frequency, if we bear in mind that among the many millions of cases within the last few years everything of a striking and phenomenal character has probably been reported.

Hyperæmia of the conjunctiva is extremely frequent and may be regarded as one of the symptoms of the disease. Conjunctivitis with mucous or muco-purulent secretion is much rarer, and severe cases are extremely rare. Croupous (Pflueger) and even diphtheritic (Coppez) forms have been observed occasionally, and Rampoldi reports the occurrence of subconjunctival abscesses. When follicles were already present in the conjunctiva, the conjunctivitis appeared as so-called follicular catarrh.

There is sometimes more or less œdema of the conjunctiva and eyelids, which may last a variable period. Hordeola are not rare. Unusually large ones are sometimes called abscesses, but recover without difficulty after perforation or incision.

Mild and severe phlyctenular diseases of the conjunctiva and cornea have been reported, but, according to Greeff, they were not more frequent during the epidemics than at other times.

Conjunctival hemorrhages, which I have seen in a number of instances, develop in a purely mechanical manner as a result of the coughing spells, especially if the vessels were already in a brittle condition.

Dacryo-cystitis will probably develop only in those cases in which stenosis or chronic inflammation of the lachrymal sac has already existed.

Influenza is one of those infectious diseases in which eruptions of herpes are not infrequent; in Hamburg, herpes labialis was observed in twenty-five per cent of the cases. Herpes febrilis was also comparatively frequent on the lids and cornea. In the latter locality it was described under various names, sometimes even as neuro-paralytic keratitis (Novelli). Finzi (*Centr. f. kl. Med.*, 1890, p. 931) mentions herpes zoster of the eyelids.

Neuralgias of the eye and surrounding parts were quite frequent. Ciliary neuralgia, tenderness of the eye on pressure, pain on movement, and a feeling of pressure behind the globe may begin during the brief prodromal stage and may continue long into the period of convalescence. They are often among the most characteristic symptoms. The pain and tenderness on moving the eye were so frequent and pronounced that Eversbusch is inclined to assume a change in the muscles themselves. In my opinion this was mainly a symptom of disease of the mucous membrane in the auxiliary cavities of the orbit, whence the inflammatory irritation extended to the periosteum of the orbit and the origins of the muscles. A similar affection of the frontal sinuses gave rise to the not infrequent supraorbital neuralgias.

If we take into account the enormous number of cases of influenza, all other ocular complications and sequelæ are extremely rare. They are, in great part, embolic processes due to secondary or mixed infection. The inflammation-producers found in such conditions have been the ubiquitous pyogenic staphylococci and streptococci, very often the diplococcus pneumoniæ. The affections in question exhibit no symptoms which are characteristic of influenza. Bennet

(*Lancet*, Feb. 8th, 1890) calls special attention to the frequency of pyæmic processes after influenza.

Among uveal affections, hyperæmia of the iris is said to have been frequent. Plastic iritis was seen by Rampoldi, Delacroix (*Arch. f. Aug.*, 1891, p. 127) and Gutmann, recent vitreous opacities by Eversbusch and Gillet de Grandmont (*Rec. d'Ophth.*, 1890, No. 2). A number of plastic, purulent uveal inflammations, similar to those occurring after meningitis, were observed by Hosch (*Corresp. f. Schweiz. Aerzte*, March 1st, 1890), Nathanson (*Petersb. med. Woch.*, 1890, p. 213) and myself. Similar cases are reported by Laqueur (*Berl. kl. Woch.*, 1890, No. 36; double embolic iridocyclitis), Rampoldi, Badal and Fage. Eversbusch reports a panophthalmia due to staphylococcus pyogenes aureus, Rampoldi a similar case starting from necrosis of an old leucoma. Inflammations of Tenon's capsule, with or without suppuration, are reported by Fuchs, Greeff and Schapringer (New York *Med. Rec.*, June 14th, 1890).

Non-septic embolism of the central artery of the retina is reported by Hosch (one case) and Coppez (three cases). Hemorrhages into the eye are nowhere mentioned and are certainly extremely rare. Magnus (*Forts. d. Med.*, 1891, p. 118) observed a so-called retinitis proliferans after influenza, and Gutmann reports a hemorrhage into the vitreous.

There have also been orbital suppurations of an undoubted embolic character (Wicherkiewicz, *Internat. kl. Rundschau*, 1890, No. 8, and Borthen, *Mon. f. Aug.*, March, 1891).

Influenza has often excited an outbreak of acute attacks of glaucoma (Eversbusch, Adler, Gradenigo, Rampoldi, Stuffler, Badal and Fage). I have also seen a bilateral attack of acute glaucoma in an individual who had exhibited symptoms of glaucoma for some time in one eye, while not the slightest abnormality had been noticed in the other. I also observed an acute relapse of a serous iritis. There have been no reports of the recovery of an inflammation of the eye under the influence of influenza.

Among diseases of the optic nerve, Denti (*Arch. f. Aug.*, 1891, p. 67) observed three cases of pronounced neuritis. Novelli (*Bollet. di Ocul.*, XIII, p. 5) and Koenigstein (*Wien. med. Blät.*, XIII, 9), each one case; Lebeau (*Ophth. Rec.*, Oct., 1891) double neuritis which recovered entirely; Vignes (*Annal. d'Ocul.*, 115, p. 244) a neuritis of the left side with blindness, one week after influenza, and terminating in recovery with $V = \frac{4}{10}$. Partial or complete atrophy of the optic nerve as the result of so-called retrobulbar neuritis is reported by Bergmeister (*Wien. kl. Woch.*, 1890, p. 204), Hansen (*Centr. f. Aug.*, 1891, p. 120), Stoewer (*Mon. f. Aug.*, 1890, p. 418), etc.

Central disorders of vision hardly ever occur, apart from the visual hallucinations and illusions in influenza psychoses. Hillmann observed yellow vision, possibly of central origin. Scintillating scotoma is mentioned quite often, but hardly appears to have been more frequent than in those who did not suffer from influenza. Another remarkable sympathetic affection is reported by Bock (*Mon. f. Aug.*, 1890, Dec.), viz., that the eyelashes became white immediately after influenza. Colley (*Deutsch. med. Woch.*, 1890, No. 35) observed Basedow's disease after influenza.

Apart from blepharospasm, affections of the muscles are rare. The most frequent was unilateral and bilateral paresis of accommodation, usually beginning suddenly, of a variable course, and sometimes associated with disturbance of speech and deglutition, as in diphtheria. A few other cases of muscular paralysis have also been reported, viz., paralysis of the superior rectus (van der Bergh, *Annal. d'Ocul.*, 1890, p. 79; Badal and Fage), abducens paralysis (van der Bergh, *l.c.*, Coppez, Badal and Fage, Valude, *Annal. d'Ocul.*, 1890, Jan., Rampoldi), coincident paralysis of the levator palpebræ and superior rectus (Stoewer, *l.c.*), bilateral paralysis of accommodation with ophthalmoplegia externa (Uhthoff, *Deutsch. med. Woch.*, 1890, No. 10), double total ophthalmoplegia externa without implication of the levator palpebræ superioris (Gayet, *Jahr. f. Aug.*, 1876, p. 351), complicated unilateral ophthalmoplegia (Gutmann, *l.c.*), polioencephalitis superior and inferior, and anterior descending poliomyelitis with fatal termination (Goldflam, *Neurol. Centr.*, 1891, p. 162), etc.

Renal affections and even acute hemorrhagic nephritis were quite frequent, but no mention is made of their effects upon the eye.

Among the rare phenomena are acute inflammation of the left lymphatic gland (Lindner, *Wien. med. Woch.*, 1891, No. 16) and temporary blindness after epistaxis in a child of seven years (Sedan, *Recueil d'Ophth.*, March, 1890).

Nervous disorders of vision, weakness of accommodation and the like, are occasionally very obstinate, corresponding with the often very protracted convalescence.

Whooping-Cough.

Conjunctivitis and epiphora are frequent in the prodromal stage. Photophobia and mydriasis often indicate the transition from the catarrhal to the convulsive stage (*Deutsch. med. Woch.*, 1891, p. 768).

The violent coughing spells may produce hemorrhages into the eye. They are found most frequently beneath the bulbar conjunctiva, and are absorbed in from one to three weeks.

Hemorrhages into the lids, with or without conjunctival hemorrhages, are not uncommon. They are also absorbed without leaving a trace.

Orbital hemorrhages, unless extensive, are usually not noticeable for several days, when the extravasated blood appears beneath the conjunctiva, in the eyelids and adjacent parts. Small hemorrhages are harmless; large ones, which are rare, produce exophthalmus.

Cerebral hemorrhages give rise to corresponding general and local symptoms, for example, hemianopsia, as in Silex's case (*Berl. kl. Woch.*, 1888, p. 841). But this is evidently not a pure case, because the disorder of vision occurred in a girl of seven years, three months after the whooping-cough and immediately after the administration of two doses of morphine. On the other hand, whooping-cough may exhibit violent cerebral symptoms, such as disorders of vision, aphasia, paralysis, etc., which cannot be attributed to hemorrhages (Troitzky, *Jahr. f. Kinderh.*, XXXI, 3, p. 291). This category includes cases of sudden double blindness with intact reaction of the pupil, normal ophthalmoscopic appearances, and without albumin in the urine (Alexander, *Deutsch. med. Woch.*, 1888, p. 204; Jacoby, *New York med. Woch.*, 1891). As in uræmic amaurosis

(which may also occur in pertussis), the blindness may be preceded
by headache and vomiting, and is probably due to slight meningitis
of the convexity. Meningitis is not a very rare complication of
whooping-cough and occasionally leads to optic neuritis with or without termination in total or partial atrophy and corresponding impairment of vision. The disease of the optic nerve may appear under the
symptoms of an ischæmia, retrobulbar neuritis, or retrobulbar hemorrhage.

Knapp (*Arch. f. Aug. u. Ohr.*, V, 1, p. 203) observed bilateral
blindness from retinal ischæmia, whitish optic nerve and narrow
vessels, as in quinine poisoning. Vision gradually improved, but
the case terminated fatally at the end of six weeks.

According to Loomis, blindness in pertussis occurs almost exclusively in those who die later of pneumonia. Jacoby (*l.c.*, Case I)
observed sudden maximum mydriasis and immobility of the pupil,
together with double optic neuritis and merely quantitative perception of light; recovery in a very short time. In Alexander's (*l.c.*)
Case 2, there was double neuritis with termination in partial atrophy; a similar case is reported by Callan (*Jahr. f. Aug.*, 1884, p.
388). Landesberg (*ib.*, 1880, p. 283) observed hemorrhage into the
optic nerve with termination in atrophy; in one case he saw obliteration of the two upper branches of the retinal artery of one eye, in
another case subluxation downward of the right lens, as the result of
violent coughing.

Muscular paralyses also occur, either as the result of hemorrhages
into the nuclear region, the nerve roots or the nerve trunks, or as
the result of meningitis. Central (conjugate) disorders of the muscles may also occur in both ways. Rosenblatt (*Jahr. f. Aug.*, 1883,
p. 314) reports dilatation of one pupil, paralysis of the motor oculi
with and without facial and acoustic paralysis, unilateral ptosis and
paralysis of the opposite upper extremity, etc.

Phlyctenular affections of the conjunctiva and cornea also occur
after whooping-cough. Like other infectious diseases which appear between the ages of two and four years, it is not infrequently
the exciting cause of internal squint, especially in children who are
predisposed to it by refractive errors.

Mumps.

This is accompanied not infrequently by conjunctivitis, occasion-ally by epiphora, œdema of the lids, chemosis of the conjunctiva (from compression of the vessels of the neck, according to Hatry). The enlargement of the parotid is sometimes accompanied or followed by enlargement of the lachrymal glands. There is more or less acute enlargement of these glands, terminating in resolution. Cases of this kind are reported by Rider (*Jahr. f. Aug.*, 1873, p. 471), Schroe-der (*Mon. f. Aug.*, 1891, Dec.), Seeligsohn (*ib.*, 1891, Jan)., Gordon Norries. A more chronic case of swelling of the lachrymal and parotid glands is reported by Fuchs (" Beitr. z. Aug.," 1891, 3); an excised piece of the former exhibited the structure of lymphoma. In parotitis of the left side, Burnett (*Jahr. f. Aug.*, 1886, p. 313) no-ticed on the right side considerable swelling of the orbital tissue with exophthalmus, mydriasis, paralysis of accommodation, swelling of the lids and double vision. The ophthalmoscope showed distended retinal veins. During an epidemic of mumps in the garrison at Lyons, Hatry (*ib.*, 1876, p. 374) noticed, in ten cases, varying grades of congestion and inflammation of the retina and optic nerve, with corresponding impairment of vision.

Hirschberg (*Centr. f. Aug.*, 1890, p, 77) and Scheffel (*ib.*, p. 136) report cases in which the lachrymal gland, but not the parotid, was enlarged. It is doubtful whether these cases may be called mumps of the lachrymal gland, in the sense that they have the same cause as epidemic parotitis, although in Scheffel's case the submax-illary and sublingual glands were also enlarged.

The rare sequelæ of mumps correspond to those of other infectious diseases. Tolon (*Jahr. f. Aug.*, 1883, p. 466) reports unilateral optic neuritis, terminating in atrophy and blindness; there were also cere-bral symptoms. The primary cause was evidently a complicating meningitis. Schiess (18. Jahresber., p. 3) observed metastatic irido-cyclitis which necessitated enucleation; "the iris was remarkably thickened, had a peculiar medullated structure, and its vessels were very large." Boas (*Mon. f. Aug.*, 1886, p. 273) reports a muscular paralysis, a paresis of accommodation after mumps. The causal connection of Adler's (*Centr. f. Aug.*, 1889, Nov.) three cases of subepithelial keratitis appeared to be very doubtful.

Pest and Yellow Fever

often lead to affections of the eye, but they need not detain us here, because they exhibit no characteristic features.

Beri-Beri.

In beri-beri or kakke, the eye muscles are often attacked in a characteristic fashion. The facial, motor oculi, abducens and trochlearis nerves may take part in the spasms and paralyses. The face and lids also take part, in many cases, in the initial œdema of the disease.

Laurençao (*Jahr. f. Aug.*, 1872, p. 219) saw a number of cases of atrophy of the optic nerve, and also an attack of glaucoma excited by beri-beri. The amblyopia mentioned by Da Costa Alvarenga (*ib.*, 1881, p. 325) was probably diplopia due to paralysis of ocular muscles.

Kessler (*ib.*, 1889, p. 506) repeatedly observed contraction and insufficient filling of the retinal arteries, white sheaths around the arteries and veins, and a whitish and bleached papilla. The same writer (*Centr. f. d. med. Wiss.*, 1891, p. 760) found anatomically, in two cases, œdematous changes in the retina and optic nerve, with dilatation of the peri-choroidal and scleral lymph spaces, and to a less extent of the optic nerve sheaths. There was moderate interstitial optic neuritis.

Vertige Paralysant.

This affection (Gerlier's disease) is a peculiar infectious malady which occurs in cowherds or in individuals who sleep in cow-stables (Epéron, *Rev. méd. de la Suisse Romande*, 1887, No. 1; 1889, No. 1; Haltenhoff, *Prog. Méd.*, 1887, No. 26). The chief symptoms are dulness, exhaustion, amblyopia and occasionally diplopia with normal ophthalmoscopic findings, ptosis, paresis of the flexors of the fingers (which are used in milking), attacks of vertigo, etc. From 1874 to 1886 Haltenhoff (*l.c.*) treated nine cases of this disease,

which usually runs a favorable course. Although the ophthalmoscopic findings are generally normal, Epéron (*l.c.*) observed two
cases of marked congestion of the papilla, associated, in one case,
with peri-papillary hemorrhages. According to Mauthner the symptoms point to the region of the muscle nuclei. It appears to me to
be at least equally probable that there is a basal cause, a very mild
form of basilar meningitis or meningeal œdema. Perhaps the whole
process is merely an abnormal innervation of the vessels. In the
latter event the ptosis would be sympathetic in origin. Anatomical
lesions have not yet been discovered.

Pellagra.

This peculiar disease begins with intestinal catarrh; it leads to
erythematous swelling of those parts of the skin which are exposed
to the air, and then to dark pigmentation of the integument. In
about half the cases, cerebral symptoms and psychoses finally develop.
The characteristic sign on the part of the eye is night-blindness, evidently as the result of the great impairment of nutrition.

According to Tebaldi (*Jahr. f. Aug.*, 1870, p. 374), ophthalmoscopic findings are not uncommon. Among fifty cases he found congestion of the papilla eighteen times and occasionally sinuous and
varicose veins; the latter were seen particularly in obstinate relapses.
In eleven cases the papilla was anæmic and pale and the vessels narrow. Stroppa (*ib.*, 1872, p. 363) saw two cases of atrophy of the
optic nerve. Neusser (*ib.*, 1887, p. 303) reports diplopia and amblyopia. According to Rampoldi (*ib.*, 1885, p. 318), the most frequent
eye symptoms in pellagra, apart from torpor retinæ, are retinitis
pigmentosa, atrophy of the optic nerve, disappearance of the choroidal
pigment, then marantic ulcers and necroses of the cornea, opacities
of the lens and vitreous. These symptoms do not warrant any conclusion with regard to the nature of the disease.

The disease is generally attributed to the ingestion of spoiled
maize, and Lombroso claims to have produced the symptoms by administration of the tincture of such maize to poorly nourished individuals. In this event pellagra would not be an infectious disease,
but a chronic ptomaine poisoning. Several writers have found a

definite bacillus which is said to be the cause of the spoiling of the grain, and this would then constitute the indirect cause of the disease.

Malaria.

This is probably due to an animal parasite, the plasmodium malariæ.

In malarial attacks certain eye symptoms may form part of the attack or may even be a substitute for it. A characteristic feature is their periodical occurrence (this is also true of other diseases in malarial regions) and the rapid curative action of quinine. Neuralgias, especially the supraorbital form, occasionally the ciliary form, are very frequent. Muscular paralyses (ptosis, Adelsheim, *Jahr. f. Aug.*, 1888, p. 385) and spasms (spasm of accommodation, Stilling, *ib.*, 1875) have also been noted. Amblyopia (central) and even complete blindness, without ophthalmoscopic findings or loss of the pupillary reaction to light, have been observed as part of severe attacks and also in latent malaria (Dutzmann, *Wien. med. Presse*, 1870, p. 514; Vachi, *Neurol. Centr.*, 1888, p. 634). Blindness has also been observed after the attack (Koslowsky, *Jahr. f. Aug.*, 1878, p. 281); it was associated with congestion and œdema of the papilla, and probably peripheral in character.

Peunoff (*Centr. f. Aug.*, 1878, p. 881) states that the pupil is always dilated during the paroxysms, more markedly in the hot than in the cold stage. When the dilatation is pronounced, it is still recognizable fifteen to eighteen hours after the attack. Otherwise it disappears in five to six hours. There was also congestion of the retina and optic nerve (as always happens in mydriasis).

Hilbert (*Centr. f. Aug.*, 1881, May) describes conjunctival catarrh as a rare form of latent intermittent fever; it occurred four times in the tertian type, was attended with splenic enlargement, and was cured by quinine. Meisburger (*Jahr. f. Aug.*, 1883, p. 299) also mentions conjunctivitis, Adams (*ib.*, 1881, p. 318) iritis, Selueck (*ib.*, 1889, p. 504) five cases of plastic iritis, Dubelir (*ib.*, 1883, p. 299) blindness with exophthalmus; intermittent strabismus has also been described (*Jahr. f. Aug.*, 1870, p. 462). Baas (*Mon.*

f. Aug., 1885, p. 240) reports a case of blue vision which appeared every other day between 10 and 12 A.M., and was cured by quinine.

Retinal hemorrhages may occur in severe cases at the beginning of the disease, but they are more frequent in later stages. According to Mackenzie (*Jahr. f. Aug.*, 1877, p. 215) these hemorrhages are more frequent in the quotidian than in the tertian type. According to Peunoff (*l.c.*) many attacks are complicated with conjunctivitis and ciliary injection, and even iritis may develop. Severe attacks are attended by coma and cerebral symptoms, among which double cortical blindness is occasionally found.

According to Sulzer (*Arch. d'Ophth.*, 1890, p. 193), hyperæmia of the disc, slight obscuration of the fundus, photophobia and "seeing sparks" are observed in twenty per cent of the cases during and between the attacks. These cases exhibited a marked predisposition to macular affections, similar to those observed in looking at an eclipse of the sun, but the prognosis is favorable.

As in other infectious diseases, a series of complications is found after the malaria has lasted a long time. According to Sulzer (*l.c.*) these include, 1, neuritis, usually bilateral, in severe cases with pigmentation of the papilla, in eight per cent with termination in atrophy of the optic nerve; 2, diffuse vitreous opacities; 3, multiple retinal hemorrhages; 4, sudden and permanent blindness. Other complications which have been observed are: iritis (Peunoff), suppurative choroiditis (Peunoff, *l.c.*, Landesberg, *Jahr. f. Aug.*, 1880, p. 324), chorio-retinitis (Poncet, *Annal. d'Ocul.*, p. 201), unilateral retrobulbar neuritis with large central color scotoma (Uhthoff, *Deutsch. med. Woch.*, 1880, p. 303), two cases of unilateral atrophy of the optic nerve without preceding neuritis (Bull, *Jahr. f. Aug.*, 1877, p. 213). Macnamara (*Brit. Med. Journ.*, March 8th, 1890) reports a paresis of the external and inferior recti, Uhthoff (*l.c.*) an abducens paralysis. Bagot (*Ann. d'Oc.*, 1891, Nov.) attributes two cases of double soft cataract to malaria.

A deposit of pigment is found not infrequently in the fundus after malaria of long standing. Such deposits are undoubtedly due to pigment emboli, as the result of gradually developing melanæmia or formation of pigment flakes in the blood. In the retina these emboli are comparatively harmless, but in the cerebral cortex they may give rise to serious symptoms.

Poncet (*l.c.*) found on autopsy changes in the retina, and particularly in the choroid, much more often than the ophthalmoscopic findings indicated. These changes included œdema, hemorrhages, lesions of the walls of the vessels and thromboses, small inflammatory foci with and without pigment, etc. He is inclined to attribute the larger proportion of the cases of malarial amblyopia and amaurosis without findings to this " chorio-retinitis palustris," but this is probably true of very few cases. The majority are undoubtedly of central origin, due to similar changes in the occipital cortex or to pigment emboli in that locality. Hemianopsia is also observed occasionally (De Schweinitz, *Med. News*, 1890, No. 27; Peunoff, *Jahr. f. Aug.*, 1883, p. 301). The latter observed in his own person an attack (lasting twenty-four hours) of complete blindness, aphasia, left hemiplegia and anæsthesia. In two comatose patients, he observed double complete blindness with aphasia and paralyses which disappeared in a few days. The central disorders of vision, which occur at the very onset, and the retinal hemorrhages of the same period, must be regarded as septic, the result of ptomaine poisoning.

Cortical blindness may also be uræmic in character, inasmuch as renal diseases and their sequelæ are found quite often in chronic malarial poisoning. A " scorbutic terminal stage" sometimes develops, and may be manifested by hemorrhages in all parts of the visual apparatus. Quinine may also give rise to disorders of vision, but these are peripheral in character (*vide* p. 351). [Keratitis as the result of malaria is not infrequent, and presents features which are more or less typical. It attacks by preference the epithelium and superficial layers, is non-suppurative; ulcerations are superficial. It is chronic in duration. There is often anæsthesia of the surface. The opacity is apt to run in streaks, yet may present itself in patches. One will find marked tenderness of the supra-orbital nerves as they pass out of the orbit, which is the most valuable pathognomonic sign, and when concurring with the conditions described indicates the absolute necessity of quinine in effective doses as an adjuvant to local treatment.—ED.]

Eruptions of herpes are frequent in intermittent fever. It also appears upon the cornea and has been described under the most various names (p. 253). Night blindness, torpor retinæ, is not infrequently one of the most striking symptoms of malarial cachexia.

Trichinosis.

This disease is also due to an animal parasite. The œdema of the lids, which often appears at the very beginning, is occasionally important in diagnosis. It is only in severe cases that the trichina emigrates into the external eye muscles, and the smooth internal muscles always escape (*vide* p. 269).

The chronic infectious diseases which require consideration are syphilis, tuberculosis, scrofula and leprosy. Prior to the discovery of their specific bacilli, the two latter diseases were regarded as "constitutional affections," and thus form a suitable transition to the latter.

Syphilis.

Vide Alexander, "Syphilis u. Auge," Wiesbaden, 1888, 1889; Schubert, "Ueb. syphilit. Augenleiden," Berlin, 1880; Manz, *Jahr. f. Aug.*, 1872, p. 220, etc.

The acquired form of syphilis will first be discussed. The eye may be the point of entrance of the infection, and the hard chancre may be found upon the lid or its free margin and upon the conjunctiva. The most frequent localization is the inner angle of the eye, upon the caruncle, where a small loss of substance is apt to occur from rubbing; next in frequency stands the lower lid and its free border, and not very rarely the palpebral conjunctiva. Hard chancre of the bulbar conjunctiva is extremely rare (Róna, *Mon. f. prak. Dermat.*, 1891, p. 462). A case of chancre of the cornea has also been reported (Jullien, "Mal. vénér.," p. 585). Like all diseases of the eye which are due to rubbing the eye, hard chancre is much more frequent on the right side.

The induration is usually very hard and extensive, and lasts a long time. A hard spot may be distinctly felt even at the end of a year and a half. Recovery is usually quite complete, and the indu-

ration is associated merely with swelling of the lids, epiphora, chemosis, conjunctivitis, etc.   Partial entropium and trichiasis are not
infrequent terminations (Krelling, *Viertelj. f. Dermat. u. Syph.*,
XV, p. 1).

The swelling of adjacent lymphatic glands may also be very pronounced; they may extend from the lobe of the ear to the acromion
and supraclavicular fossa.   From the glandular enlargements in
syphilitic primary lesions, Lavergne and Perrin concluded that the
lymphatics of the inner part of the eyelids pass to the submaxillary
glands, those of the outer half to the glands of the parotid region and
the pre-auricular glands.

Hard chancre of the lids might possibly be mistaken for epithelial
cancer.   The age of the patient will usually, though not always, decide the diagnosis.   In two cases I have seen chancroid at the inner
angle of the eye in young people of twenty years.   The enlargement
of the glands is found in both diseases, so that sometimes the appearance or absence of secondary syphilitic manifestations must decide
the diagnosis, unless a microscopical examination can be made.

The integument of the lids is also involved in the eruptions of
the early period of the disease.   According to Michel, this is especially
true of roseola or acnelike eruptions.   When they affect the border
of the lids, they may give rise to loss of the eyelashes.   Ulcerative
eruptions, even rupia, are not uncommon upon the lids (Alexander,
l.c., p. 12).   Eruptions upon the conjunctiva are very rare (Gutmann, *Deutsch. med. Woch.*, Feb. 16th, 1888; Sichel, *Jahr. f.
Aug.*, 1880, p. 292).

Syphilitic iritis generally appears at this early stage, commonly
associated with eruptions, but sometimes at an early period.   At the
beginning it is unilateral and does not differ from plastic iritis due
to other causes.   At a later period the other eye is often affected and
relapses are also frequent.   Syphilis is by far the most frequent
cause of iritis.   According to some statistics three-fourths of all iritides are syphilitic, according to others (Arlt) only one-fourth.

It is only in fifteen to twenty per cent of the cases that the iritis
appears in a manner which is characteristic of syphilis (iritis gummosa, condylomatosa, papulosa).   Yellow or dirty orange-colored

nodules appear in the tissues of the inflamed iris; they are surrounded by a narrow red zone and rarely exceed 2–3 mm. in diameter. Their favorite location is the vicinity of the border of the pupil, especially below, but they are found occasionally in other parts. A small hypopyon is not very uncommon; it is sometimes attended with a striking subsidence of the previously violent pains. The little nodules may disappear entirely under suitable local and general treatment; they usually leave a broad synechia and often a discolored atrophic patch. The term iritis papulosa is probably more correct than iritis gummosa, in view of the stage of syphilis in which iritis is generally found, but there is really only a quantitative difference between a syphilitic papule and a gumma. Larger, confluent gummy tumors of the iris occasionally appear in the later stages of severe syphilis. They may fill the entire chamber and their growth may lead to loss of the eye. These are usually associated with severe inflammatory changes in the other parts of the uvea.

Sometimes there is merely a nodule-like thickening of the tissue of the iris, but even in the apparently diffuse, non-papular iritis the microscope shows little nodules due to arteritic changes, proliferation of epitheloidal cells and accumulations of round cells.

Iritis gummosa, papulosa, or condylomatosa will suffice for a positive diagnosis of syphilis. At the most it might be mistaken for an abscess of the iris, but this is prevented by the previous history (foreign body in the iris). Certain granulomata are easily distinguished from the yellowish-red syphiloma by their gray or grayish-red color.

I have recently seen typical iritis papulosa as the first and sole secondary symptom, three weeks after a hard chancre; rapid recovery followed antisyphilitic treatment. Two months later syphilitic eruptions of a severe ulcerating form appeared upon the skin.

No part of the eye and adjacent structures is spared during the further course of syphilis, apart from the lens, which is only involved secondarily.

Syphilis is manifested either by inflammations, mainly interstitial, of any part of the visual organ, but chiefly of the uvea and retina, the nerves and meninges, or by specific new formations. The latter are composed at first of a granulation-like, later by a necrobiotic tissue (syphiloma, gumma) which may develop in any region. They may be circumscribed like a tumor, single or multiple, or diffused over a large area, especially at the base of the brain.

A common feature of all syphilitic diseases is a peculiar affection of the small arteries (arteritis syphilitica), which leads to narrowing and finally occlusion of the vessels. This lesion may also occur separately, without inflammatory phenomena or specific new growths. In the latter event the symptoms are insignificant, although they result occasionally in a cerebral hemorrhage; in young people this is always suggestive of syphilis. While the inflammations and gummy neoplasms may subside, the arteritic changes only resolve in part, and the consequent impairment of nutrition may give rise, particularly in the nervous system, to certain diseases (tabes, etc.), usually of an atrophic character, which are much less frequent in non-syphilitics.

So long as the inflammation or neoplasm remains interstitial, complete restoration is possible; destroyed nerve fibres or ganglion cells cannot be destroyed.

As a general thing, the diffuse superficial inflammations belong to the early, the gummous lesions to the later stages of syphilis, but both forms are often observed at the same time. The coalescence of the small foci (which are also present in the former variety) into larger, tumor-like or more diffuse new formations, merely shows the diminished power of resistance of the tissues.

Aside from the ordinary plastic form, iritis may also appear in the serous form, as an asthenic or cachectic inflammation, or, on the other hand, it may occur, when the inflammation is very acute, as iritis gelatinosa or fibrinosa. As in all syphilitic inflammations, hemorrhagic forms are comparatively rare. The remarks made concerning the iris are also true of the ciliary body and choroid, which are more or less implicated in iritis. Cyclitis and choroiditis may develop in every possible degree of violence and duration. The inflammation may be diffuse or focal (disseminated choroiditis); the two forms can only be differentiated with the ophthalmoscope, on account of the implication of the pigment epithelium in the disseminated form. The microscope shows that the inflammation is mainly focal in both forms, but specific changes in the arteries are also widely diffused.

In the mild forms, choroiditis occurs in the ordinary disseminated variety, with very slight disturbance of vision, and may come to a standstill at any stage. It does not differ from the variety due to other causes. Certain varieties, such as that associated with numerous small atrophic spots surrounded by a pigmented zone (choroiditis areolaris) are perhaps more frequent in syphilitics than in others. According to Graefe and Foerster, syphilitic choroiditis disseminata is more frequent at the posterior pole of the eye; according to Galezowski, it occurs chiefly at the equator. Flocculi in the vitreous, associated with disseminated choroiditis, arouse the suspicion of syphilis, but there are undoubted syphilitic affections without these flocculi.

More frequently the choroidal attacks are severe and are complicated by numerous and dense vitreous opacities. The ophthalmoscope may show disseminated foci, but the process is usually diffuse, *i.e.*, the pigment epithelium is not notably affected at the start. The attacks do not always yield promptly to anti-syphilitic treatment, and may lead to loss of sight, either by implication of the iris (occlusion of the pupil and increased tension) or by detachment of the retina. As a general thing, the severe forms belong to the late stages of syphilis and may be very painful, especially when the ciliary body is profoundly involved. In part, however, these are not true syphilitic affections, but must be regarded as sequelæ in an individual weakened by a previous infectious disease; in such cases antisyphilitic treatment is useless.

As a matter of course the retina is very often, if not always, affected in these choroidal inflammations, so that it would be more correct to speak of a chorio-retinitis. Hitherto it has been the custom to call the disease choroiditis when the ophthalmoscope showed changes in the pigment epithelium, and retinitis when these changes are absent.

At the start, the clinical symptoms are usually insignificant and of a general character, such as "seeing sparks," floating spots, occasionally subjective colored vision. These symptoms are evidently due to the irritation of the outer layers of the retina by the diseased choroid (Schenke). Even when the disturbance of vision is more pronounced,—cloudy vision to complete, usually positive scotoma (generally central, in rare cases even annular)—it is explained in the main by the impaired function of the outer layers of the retina. Torpor and anæsthesia retinæ, micropsia, etc., are sometimes the most striking symptoms.

The ophthalmoscopic appearances depend upon the implication of the pigment epithelium. We find choroiditis disseminata in every possible degree, with pigmentation of the retina similar to, but much more irregular than typical retinitis pigmentosa; or there is merely more or less dense and extensive opacity of the retina.

In the former case, anatomical examination shows numerous points of adhesion between the choroid and retina, at which the pigment emigrates into the latter. The retina exhibits irregular patches of cellular infiltration and also the appearances of typical specific arteritis. In a case examined by me, which presented the clinical symptoms of annular scotoma, the changes in the pigment epithelium were visible with the ophthalmoscope over the entire fundus.

The second form, in which the opacity of the retina predominates and the changes in the pigment epithelium are insignificant, is observed chiefly in the region of the macula lutea. It has been called central relapsing retinitis (von Graefe), because it is only in this locality that it produces notable disorder of sight. Diffuse opaque patches are not uncommon in other parts of the fundus, but a peripheral amblyopic spot in the monocular field of vision must be very large in order to produce any striking loss of sight. In the macula lutea and fovea centralis every impairment of vision at once becomes noticeable.

The initial clinical symptoms are vague and point to irritation or loss of function of the outer layers of the retina. They include distorted vision, photopsiæ, photophobia, hemeralopia, micropsia, " seeing sieves or gratings" (this indicates destruction of individual groups of rods and cones), etc. These symptoms progress gradually or quite suddenly into a central, often very dense, and usually positive scotoma, which is seen as smoke or mist. Diffuse implication of the entire retina may produce complete blindness.

The ophthalmoscope shows a diffuse gray opacity of the fundus over a greater or less area, occasionally more marked along the vessels. This opacity must be attributed, at least in part, to dust-like opacity of the vitreous. The optic nerve is usually reddened and there is some venous congestion of the fundus; hemorrhages into the retina are extremely rare.

If the disease attacks mainly the macular region, the opacity is here most dense, but is not infrequently overlooked at the start, when the examination of this region without atropine is difficult. The retinal vessels in the macula are noticeably dilated and often apparently increased in number, because small, otherwise invisible vessels can be seen on account of the congestion. The specific arteritis may be seen as whitish-yellow sheaths around the arteries, and we often find in the affected parts of the retina small, yellowish or brownish, roundish spots grouped like bunches of grapes (Ostwalt, *Berl. kl. Woch.*, Nov. 5th, 1888, and Hirschberg, *ib.*, Nov. 12th, 1888).

Apart from the specific arteritis, which may also exist independently of retinitis, the microscope shows very little in the retina. More or less extensive choroidal changes are always demonstrable.

The disease may begin in the first six months, but more frequently in the second, third, or fourth years after infection; it may be unilateral or bilateral. It is generally very obstinate and often relapses. The final termination is quite favorable in the majority of cases, although permanent impairment of vision, even yellow atrophy of the optic nerve and complete blindness, may be left. There may also be a new formation of connective tissue upon the inner surface of the retina and extending into the vitreous (retinitis proliferans). This hardly ever happens except when hemorrhages have occurred, and these are rare in syphilitic affections.

As a matter of course there is every possible transition between the two forms of disease just described. The more severe the disease, the more dense are the vitreous opacities, and the greater the likelihood that shrinking of the vitreous, detachment of the retina and phthisis bulbi will set in. I have found that typical central retinitis is met with chiefly in young people, and is rare beyond the age of forty. Disseminated choroiditis, with more or less abundant vitreous opacities, is more frequent in older people. Foerster (*l.c.*, p. 191) appears to have had the same experience.

Larger gummous neoplasms also grow from the choroid, iris and ciliary body, although rarely from the latter (Mauthner, Woinow, Alt). It is usually impossible to secure their resolution without causing destruction of the eye.

Inflammations of the sclera are frequent as a part of extensive syphilitic uveitis, and may recover or may leave scleral staphylomata. They are rare as an independent disease in syphilis and do not always yield promptly to specific treatment, so that their syphilitic nature may remain doubtful. Gummata may also start from the sclera; more frequently they start from the uvea and involve the sclera secondarily.

Syphilitic disease of Tenon's capsule is very rare; it produces the same symptoms as tenonitis from other causes and yields to anti-syphilitic treatment. In acquired syphilis, specific disease of the cornea is disproportionately rare. Diffuse interstitial keratitis, which is so frequent and characteristic in congenital syphilis, is observed only in rare cases and at a late period in acquired syphilis. In the majority of cases it should be regarded as a sequel, because anti-syphilitic treatment usually exerts no influence upon it. It runs essentially the same course as the hereditary syphilitic variety, but may merely affect a part of the cornea or give rise to opacities at the edges (Hock, *Wien. Klinik*, 1876).

Mauthner (Zeissl's "Lehrb. d. Syph.") describes a true keratitis punctata in syphilis. Opacities as large as the head of a pin come and go, with or without ciliary injection, in the various layers of the cornea; they never grow larger and never lead to suppuration. The disease is extremely rare. Alexander (*l.c.*, p. 45) has seen a similar condition associated with true iritis.

Under the term gumma of the cornea, Denairé describes roundish gray diffuse opacities, associated with iritis (probably the form just mentioned), and Magni describes small opaque patches at the extreme periphery of the cornea. I have observed a grayish-yellow infiltration, about 3 mm. in diameter, situated beneath transparent epithelium, about 2 mm. from the rim of the cornea, to the outside and above. This had developed in an elderly individual three years after syphilitic infection. Antisyphilitic treatment had no very striking effect, but the infiltration slowly disappeared, leaving a grayish, distinctly depressed patch without any development of vessels. This may very readily have been a gumma; and the imperfect effect of mercurial treatment may have been due to the fact that it was situated in a non-vascular tissue.

Conjunctival catarrh, congestion and the like are often found in syphilis without any real relation to this disease. They are sometimes produced, however, by undoubted syphilitic affections of the lids, lachrymal organs, etc. The observations of Goldzieher (*Centr. f. Aug.*, 1888, p. 103) and Sattler (*Prag. med. Woch.*, 1888, No. 12) are the only ones reported concerning a specific, trachoma-like disease of the conjunctiva, which yields only to anti-syphilitic treatment. The direct connection with syphilis of the "recurrent conjunctival hyperæmia," of which four cases were described by Alt (*Centr. f. Aug.*, 1890, p. 373), is somewhat doubtful.

True gummata also occur upon and beneath the conjunctiva (Trousseau, *Ann. de Dermat. et Syph.*, IX, No. 7, two cases in the ninth and twelfth months in severe syphilis; Estlander, *Mon. f. Aug.*, VIII., p. 259, etc.). They may also develop in the lachrymal caruncle (Taylor, *Jahr. f. Aug.*, 1875, p. 452). As a general thing, however, the gummata of the conjunctiva start from the tissue of the upper lid, particularly from the cartilage. They may develop even during the first year of the disease, grow to the size of a pea or a bean, and generally ulcerate rapidly and cicatrize. If the inferior maxillary and pre-auricular glands are greatly swollen, the gumma may present a close resemblance to the initial lesion. A non-gummous tarsitis, an acute enlargement of the cartilage without ulceration, also occurs in the early stages of syphilis and disappears after proper treatment.

Even the lachrymal glands, which were long regarded as exempt, have been found to be gummatous in rare cases, or to be attacked by an interstitial inflammation (Streatfield, Albini, Adler and Alexander, *l.c.*, p. 36). The tumor produced by their enlargement rapidly subsides under specific treatment, while they previously resisted all therapeutic efforts.

Syphilitic disease may also appear in the orbit and its bony walls. Gummy periostitis leads to caries and necrosis, especially at the rim of the orbit. Periostitis of the deeper parts necessarily produces symptoms of tumor: protrusion of the globe away from the tumor, impairment of mobility toward the side of the tumor, more or less impairment of vision according to its position (especially at the optic foramen and superior orbital fissure), anæsthesia of the conjunctiva and cornea with its sequelæ (neuro-paralytic keratitis), neuralgias and paralyses. In addition there are signs of inflammation: swelling of the lids, chemosis, more or less conjunctivitis and, at a later period, signs of orbital suppuration and the formation of fistulæ. The symptoms are not characteristic of the syphilitic origin of the lesion. The diagnosis depends upon the previous history and the results of specific treatment.

The tear passages are often attacked in the form of dacryocystitis, dacryo-cysto-blennorrhœa, and stenosis. The disease extends in most cases from the nasal mucous membrane. The bony walls of the lachrymo-nasal canal are not infrequently involved. This results frequently in cicatricial, even bony occlusion, and the prognosis is thus materially aggravated.

Lesions of the cranium and its contents are probably the most serious manifestations of the disease. Arteritis obliterans may cause sudden elimination of entire vascular tracts in the brain and lead to hemorrhage and softening. This is very significant of syphilis when the patient is young and does not suffer from cardiac disease. Gummy growths may also develop. They appear as superficial infiltrations, particularly at the base of the brain, inclosing the optic nerve, chiasm, tractus, motor and sensory nerves; but also upon the convexity, where they give rise to cortical symptoms, such as hemianopsia, aphasia, Jacksonian epilepsy, etc. They may also occur as tumors, either single or multiple, in any part of the brain or its meninges. The symptoms may advance rapidly or slowly (*vide* page 138) and with more or less irritation and inflammation of surrounding parts.

Syphilis is often manifested by a multiple gummy basilar neuritis, or every possible disease of the brain and cord may be simulated. A characteristic feature of brain syphilis is the striking variability of the symptoms, so that, for example, Oppenheim (*Berl. kl. Woch.*, 1887, p. 666) regards "oscillating" bitemporal hemianopsia as a criterion of syphilis of the base of the brain. Under suitable treatment, apparently impossible recoveries are sometimes effected; on the other hand, apparently mild symptoms may prove very obstinate.

Eye symptoms are very common in cerebral syphilis, but we will discuss only two, viz., diseases of the optic nerve and paralyses.

Central and peripheral affections of the optic nerve are very frequent tokens of syphilis. They are usually secondary, as, for example, the neuritis of uveal and retinal diseases, which generally recovers, but occasionally terminates in yellow atrophy; we find also neuritis due to meningitis or gummy growths, and terminating in recovery, partial or total atrophy; choked disc from cerebral gumma, simple pressure atrophy or retrobulbar neuritis in diseases of the skull and orbit, etc. Tabetic atrophy of the optic nerve should no longer be included among syphilitic diseases, even though it develops in an individual who was formerly syphilitic.

The optic nerve may also be affected in a more independent manner and, as a rule, at a relatively early period, viz., between the eighth and twentieth months. In this optic neuritis, which may be single or double, isolated or associated with neuritis of other basilar nerves, the microscope discloses the specific character of the affection by the arteritis and interstitial gummy processes. These are found not only in the nerve, but also in the chiasm and tractus, so that there may be various peripheral disorders of vision. The prognosis is comparatively favorable, the more so, the earlier the disease begins after infection. Later the secondary diseases of the optic nerve predominate, and the nervous elements, which cannot be replaced, are more apt to be destroyed.

Badal (*Arch. d'Ophth.*, VI, p. 301) found 139 cases of disease of the optic nerve and 144 paralyses among 631 cases of ocular syphilis.

Syphilitic paralyses of eye muscles may also be due to various causes. The muscle itself may be diseased, specific neoplasms in the orbit (especially at the superior orbital fissure) may affect the nerves and muscles, neoplasms at the base of the skull may compress the basilar nerves; growths within the brain may injure the motor nerve roots and nuclei, or, if situated in the corona radiata and cortex, may give rise to conjugate deviations and paralyses. In all these cases the paralyses are usually combined with other evidences of syphilis. Isolated paralyses of the eye muscles are not infrequently the first symptoms of syphilis. They are due either to neuritis and perineuritis of the nerve roots and at the base of the brain, or they are nuclear in origin; other causes are exceptional.

Although every muscle is occasionally paralyzed in syphilis, certain combinations are especially frequent. In about three-fourths of the cases the motor oculi is affected, in about one-fourth the abducens, the trochlearis in only one to two per cent, and the facial with equal rarity. Paralysis of the two latter nerves is usually combined with paralysis of the motor oculi or abducens. Perhaps a half of all paralyses of the ocular muscles are syphilitic in origin. Indeed, every spontaneous ocular paralysis or ophthalmoplegia necessitates a differential diagnosis with regard to syphilis.

An especially frequent form in syphilis is unilateral ophthalmoplegia interna, or paralysis of the sphincter of the pupil and of accommodation. According to Alexander, three-fourths of such cases are syphilitic, according to Uhthoff only one-fourth. We can easily understand the fact that syphilis attacks these two nuclei separately when it is remembered that the condition is due to arteritis, that the nutritive territory of the third ventricle is different from that of the aqueduct of Sylvius, and that the vessels, like those of the retina, are end-arteries (Cohnheim).

On the whole, syphilitic paralyses of the eye muscles are found in the later stages of the disease, rarely in the first six months. They may develop rapidly or slowly, are usually obstinate, and recover very slowly, in many cases not at all. Those which recover have no tendency to relapse. Naunyn observed seventy per cent of recoveries. According to this writer, there is no hope of recovery if evidences of improvement do not appear after vigorous treatment for two weeks.

Paralyses may also occur at a very late period, even after syphilis has long been cured. In this event they often exhibit a different character. They appear and disappear rapidly, often relapse, and are not really syphilitic, but are the forerunners and symptoms of cerebral and spinal affections (tabes, multiple sclerosis, general paresis, psychoses, etc.).

These diseases of the nervous system may also be observed years after the paralyses. Basilar, and particularly nuclear, paralysis indicates syphilitic disease of the arteries in the brain. The disease of the vessels leads to such a predisposition of the central organs that, even after the syphilis is cured, comparatively insignificant influences will give rise to the diseases in question.

In syphilitic coma, which occurs with or without warning symptoms (headache, vertigo, convulsions, sudden blindness, etc.), the symptoms of cortical paralysis are associated, according to Althaus (*Deutsch. med. Woch.*, Feb. 3d, 1887), with those of irritation of the pons and medulla oblongata, *i.e.*, the pupils are narrow and do not react, and the eyes are deeply sunken. This would differentiate the condition from alcoholic and uræmic coma, in which the pupils are large. In coma after opium poisoning and pontine hemorrhage, the pupils would be still narrower. It is doubtful, however, whether the combination described by Althaus is really constant in syphilitic coma. Moreover, the other forms of coma may also occur in syphilitics.

The diabetes which occurs occasionally in syphilis—usually as the result of new-formations or inflammation at the floor of the fourth ventricle—will hardly ever give rise to an affection of the eye, on account of its usually temporary character. Such an effect may be produced, however, by syphilitic renal diseases. The latter develop in early syphilis in an acute, subacute, or more chronic form, as in scarlatina, and may also give rise to uræmic amaurosis. At a later period we find only chronic interstitial changes (contracted kidneys, sometimes associated with gummy deposits and waxy changes), which are the most frequent causes of retinitis and neuro-retinitis albuminurica.

Congenital syphilis differs from the acquired form in many respects. The cases are always comparatively mild, because the severe ones die either in utero or soon after birth. Intra-uterine infection runs its course much more rapidly than extra-uterine infection, and the fœtus often dies at an early period from tertiary syphilis. Even in the new-born, acquired syphilis runs its course much more slowly than the intra-uterine variety. This, together with the development of primary and secondary symptoms, distinguishes the former from the latter. But with regard to the eye diseases, especially diffuse interstitial keratitis, which appear at a later period, I have been unable to detect any difference between congenital syphilis and that acquired immediately after birth.

Diseases of the eyes (even those which have run their course) are not infrequently congenital. They include affections of the uvea, in its broadest sense, from simple choroiditis disseminata (chiefly with numerous small, round spots with deeply pigmented areolæ) and hardly noticeable impairment of vision, to the most severe plastic forms of uveitis and irido-choroiditis with occlusion of the pupils and its results. The secondary symptoms include cataract, staphylomas of various kinds, even passing into general pathological enlargement of the eye, slowly progressive inflammations with temporary or permanent increase of tension, excavation of the optic nerve, detachment of the retina, etc. The cornea may also undergo parenchymatous disease in utero, and result in more or less extensive, congenital opacities, with or without vascular development. There may even be ulcerations of the cornea with their results, such as corneal staphyloma, anterior synechia, anterior polar cataract. Such corneal findings are extremely rare (in only one case have I seen congenital anterior polar cataract with the signs of former perforation of the cornea, as a symptom of congenital syphilis), while they are seen much more often as the results of extra-uterine disease (blennorrhœa neonatorum).

Another series of congenital diseases of the eye are the results of intra-uterine meningitis. To this category belong the majority of cases of congenital neuritis and post-neuritic atrophy of the optic nerve.

Extra-uterine uveal diseases are also observed in congenital syphi-

27

lis. They include choroiditis from the mildest to the most severe
forms, with retinitis and pigmentation of the retina, usually with,
more rarely without, opacities of the vitreous; at a later period they
give rise not infrequently to cataract, especially of the narrow striped
or punctate variety. The coincidence of visible vitreous opacities
with disseminated choroiditis is, in a measure, characteristic of its
specific origin, but exceptions to this rule are quite frequent. Moreover, choroidal affections of an asthenic type and iritis serosa in a
perfectly non-characteristic form occur much more frequently in
hereditary syphilis than in other conditions, although this is denied
by various writers.

Plastic iritis is rare in children, especially in the new-born; in
the majority of the cases it is due to hereditary syphilis. Hutchinson (Ophth. Hosp. Rep., VIII, p. 217) reports five cases of severe
plastic iritis in children from one and one-half to eight years old, in
whom every form of syphilis could be excluded. Gummata or pap-
ules of the iris are very rare. Cases are reported by Alexander (*l.c.*,
p. 196), Trousseau (*Ann. de Dermat. et Syph.*, VI, p. 415), Watson
(Ophth. Hosp. Rep., XI, 1, p. 65) and Liebrecht (*Mon. f. Aug.*, 1891,
p. 184) ; the last case is very doubtful.

Diffuse interstitial keratitis (keratitis parenchymatosa, keratitis
scrophulosa of Arlt) which is very characteristic of congenital syphilis, begins between the age of six years (rarely earlier) and the
period of puberty. [I have seen cases at the age of eighteen and of
thirty-two years.—ED.] This disease does not occur exclusively in
congenital syphilis, but this is true of such a large percentage of cases
(according to Horner two-thirds, according to Mauthner four-fifths)
that it always rouses a lively suspicion of such a cause. Arlt formerly denied any connection with congenital syphilis and called the
disease keratitis scrophulosa, but later he receded somewhat from
this position. In my own experience, an extremely large percentage
have been due to syphilis. There is no doubt, however, that a series
of cases occur in the non-syphilitic, but I have rarely observed it in
pronounced scrofula.

Diffuse interstitial keratitis may be the sole symptom of congenital syphilis. More frequently, however, there are other evidences of

syphilis, such as deformities of the skull, deafness, or other signs of meningitis, depression of the nose, arthritides and their sequelæ, especially in the knee-joint, enlargement of the glands, etc. According to Hutchinson, a peculiar formation of the teeth, which is most marked in the upper middle, permanent incisors (Fig. 20), is extremely characteristic of congenital syphilis; and in fact this is true of a large percentage of cases. The same condition is observed not so very rarely, however, in children

FIG. 20.

who do not exhibit any other evidences of congenital syphilis. This fact limits to a certain extent the diagnostic value of the malformation of the teeth. In many cases the previous history of the parents furnishes the most certain proofs of congenital syphilis. It is not alone in feeble and anæmic children that syphilitic disease of the cornea is observed, but, in rare cases, in those who are well-developed physically and mentally. Diffuse interstitial keratitis, when it occurs after the period of puberty, is usually a relapse and generally runs a more severe course than the original disease.

Interstitial keratitis is an emigration keratitis which starts from one spot in the rim of the cornea (more rarely from several spots) and gradually traverses the entire membrane. In reality, therefore, the disease is uveal in character. Hence, severe cases are complicated so often with lesions of the choroid, especially its anterior part, and of the iris and ciliary body. In other cases the interstitial keratitis is secondary to extensive choroidal disease.

In congenital syphilis the walls of the orbit are found diseased not infrequently, particularly its free border and the walls of the lachrymal sac and lachrymo-nasal canal. Usually we have to deal with periostitis or caries, which exhibit no characteristic features, much more rarely with true gummatous disease. Gummy tumors develop occasionally in the orbit, upon the periosteum, in the lids, etc., but they are very rare.

Diseases of the bony walls of the tear passages often lead to acute and chronic inflammations of the lachrymal sac and lachrymo-nasal canal with their sequelæ, especially to bony stenoses and adhesions.

Other symptoms on the part of the eyes which are due to hereditary syphilis are extremely rare. A few cases of muscular paralysis

have been reported (Graefe, *Arch. f. Ophth.*, I, 1, p. 433, paralysis
of the left motor oculi; Mackenzie, quoted by Alexander, *l.c.*, p.
207, double abducens paralysis and ptosis; Lawford, *Ophth. Rev.*,
April, 1890, two cases of incomplete paralysis of the motor oculi). To
these I can add a case of isolated paralysis of the right pupil, which
lasted two years, in a girl of nine years. The father was syphilitic
and, strange to say, also suffered from unilateral mydriasis; a sister,
aged seven years, suffered from diffuse interstitial keratitis, and the
two remaining sisters had died, with specific symptoms, soon after
birth. Barlow (*Lancet*, 1877, No. 8) reports alopecia of both eyebrows, and Scheffelt (*Arch. f. Aug.*, XXII, 4) observed disease of
the retinal veins which were converted, in part, into white bands,
as the first symptom in a man of eighteen years.

With few exceptions all these hereditary syphilitic affections of
the eye are to be regarded as occurring in syphilitic infection which
has run its course, *i.e.*, as analogous to the post-syphilitic diseases
of the nervous system. Hence, specific treatment is useful only in
the first years of childhood. In later years it may even act injuriously by producing weakness of the general system. [I have seen a
case of hereditary syphilitic keratitis which occurred in a young man
aged twenty-eight, for the first time, was successfully treated by hot
fomentations and specific remedies, and who also passed successfully
through another attack two years later with the help of the same
treatment.—ED.]

Tuberculosis and Scrofula.

In tuberculosis the eye may serve as the point of entrance of infection, usually from the conjunctiva (tuberculosis or lupus of the
conjunctiva). The intact conjunctiva is never infected by tubercle
bacilli (Valude). There must always be a loss of substance, whether
due to injury or to ulceration. Ulcers then develop with a more or
less extensive, infiltrated base. The base is often covered with abundant, bleeding granulations. The microscope reveals tubercular
nodules in the latter and in the edges of the ulcer, and miliary nodules are not infrequently visible to the naked eye. The process extends to the adjacent tissues and may attain considerable dimensions,

especially in the lids.   After a while the glands in front of the ear or below the lower jaw become swollen.   A trachoma-like tubercular infection of the conjunctiva may also develop.

In many cases conjunctival tuberculosis long remains the sole localization.   Early destruction of the neoplasm may then lead to recovery.   In other cases the conjunctiva is attacked secondarily, either by propagation from the vicinity (skin, lachrymal sac, nose, intraocular tubercular neoplasms), or by means of emboli.   The latter method is probably the least frequent.

When miliary nodules are not visible to the naked eye, the diagnosis is assured by the demonstration of tubercle bacilli in bits of the tissues.   Bacilli have also been found in the tears (Amiet, Diss. Zuerich, 1887; Burnett, *Arch. f. Aug.*, XXIII, p. 336).

A distinction between tuberculosis and lupus of the conjunctiva can no longer be made.   Although the latter is a local affection, the presence of tubercle bacilli decides the diagnosis.   Accroding to Neumann, the tubercle undergoes cheesy degeneration, the lupus nodule does not.   Inoculation of lupus material into the anterior chamber produces tuberculosis (Trousseau, *Arch. d' Opth.*, 1889, Nov.–Dec. ; Pagenstecher and Pfeiffer, *Berl. kl. Woch.*, 1883), but according to Parinaud (*Gaz. hebd.*, 1884), this remains local and does not become general as in inoculation with tubercular material.   The chief difference in such cases appears to me to reside in the nutrient medium, *i.e.*, in the individual attacked.   In tuberculosis, the products of disassimilation offer very little or no resistance to the proliferation of the bacilli, in lupus they diminish their vital activity to a notable extent.   As a matter of course, a sharp boundary cannot be drawn and every possible transition is met with.   Nor is it impossible that lupus may be converted at a later period into typical tuberculosis.

Tangl (*Centr. f. Aug.*, 1891, p. 14) found tubercle bacilli in a chalazion and therefore regards it as a tubercular new-formation. This opinion has met with decided opposition (Weiss, *Mon. f. Aug.*, 1891, p. 206; Deutschmann, "Beitr. z. Aug.," II, p. 109).   Without denying the possibility that rare cases of tubercular disease of the lids may for a time present the appearances of chalazion, typical chalazion has not the slightest clinical or anatomical connection with tuberculosis.

At the start, acute tuberculosis of the lid may look like a large hordeolum and may not exhibit the characteristic changes until a late period.   Yet no one will regard hordeolum in general as a tubercular new-formation.

In the further course of tuberculosis every part of the organ of sight may be invaded. There may be tumor-like new-formations in the eye and in the cerebral organs, or superficial, diffuse inflammations of a tubercular character. In the latter event the uvea is very often, the retina much less frequently, attacked; usually the tubercle nodules are only visible with the microscope. The latter may or may not be associated with tubercular meningitis.

In much rarer cases the tubercles are visible with the ophthalmoscope. Single or multiple, whitish-yellow, round patches, over which the retinal vessels pass, are found in the fundus. They are very rarely surrounded by a striking pigmented zone, and this may be important in differential diagnosis. The individual nodule may become smaller and even disappear, but it usually grows, coalesces with adjacent ones, and projects more and more distinctly from the fundus. Plastic, purulent inflammatory processes in the retina and vitreous, detachment of the retina, etc., develop at a later period. These make further observation with the ophthalmoscope impossible and terminate in loss of the eye, generally after perforating externally.

On the whole, choroidal tubercles which are visible with the ophthalmoscope belong to the terminal stage of tuberculosis and often develop shortly before death. This diminishes their diagnostic value, because the diagnosis is no longer doubtful at that period.

Tubercles also form in the anterior parts of the choroid, in the ciliary body and iris, but only the latter are visible from the start. Terson (*Arch. d'Ophth.*, X, p. 7) claims to have successfully removed tubercle of the iris. According to Michel, hemorrhages from the ciliary body may be the first symptom of tuberculosis in this region.

Tubercles in the choroid, which are occasionally bilateral (Haugg, Diss. Strasburg, 1890), may apparently constitute for a long time the only manifestation of tuberculosis. In such cases large nodules may form and perforate externally. They can only be distinguished with the microscope from gummy nodules in the same locality.

Tubercles rarely develop primarily in the sclera, and very rarely in the cornea (Panas and Vassaux, *Jahr. f. Aug.*, 1885, p. 341; Roy and Alvarez, *ib.;* Rachet, Thèse de Paris, 1887). In the latter event there is evidently infection of corneal ulcers from tuberculosis of the conjunctiva.

The optic nerve, tractus, chiasm and central parts are also affected very often in tuberculosis, either directly, or by inflammation and the formation of tumors in the vicinity, or by "remote action." Visual disorders with and without ophthalmoscopic findings, spasms, paralyses, sensory disturbances, etc., are frequent symptoms of tubercular meningitis or of tumor-like new-formations; the latter are especially apt to be situated in the anterior fossa and the region of the chiasm.

Tubercular processes within the orbit are rare, but its bony walls, and occasionally those of the lachrymal passages, are often diseased in the form of periostitis and caries. In the large majority of cases these must be attributed to syphilis or tuberculosis.

Superficial, diffuse diseases of the membranes of the eye, especially of the uvea, also occur in tuberculosis. The bacilli are not uniformly diffused, but give rise to very small, scattered nodules which are apt at a later period to coalesce. According to Wagner (*Muench. med. Woch.*, 1891, No. 15), tubercular iritis constitutes fifty per cent of all iritides, but this figure is much too high. Wagner has apparently included all cases of so-called iritis serosa, but the miliary foci of the latter affection are merely accumulations of more or less degenerated round cells.

Iritis attended by the formation of nodules often appears to be of a tubercular nature. These nodules are grayish-red in color (gumma nodules are yellowish-red), of the size of a pin's head or larger, are multiple, and often come and go for a long time. A similar condition is observed occasionally in the absence of all signs of inflammation of the iris (lymphoma?).

It is probable that, under the influence of the products of the bacilli circulating in the blood, inflammations may be produced if some other factor is added, as, for example, a trauma. In this event the products of the inflammations will naturally be free from bacilli, and yet, with a certain degree of propriety, may be called tubercular, *i.e.*, they have developed under the influence of tuberculosis. We must assume this in a series of inflammations, especially of the different sections of the choroid (also in many other infectious diseases).

So-called scrofula has been recognized as latent tuberculosis since the demonstration of tubercle bacilli in the cheesy glands. It may be converted into typical tuberculosis by means of any influence which causes the bacilli to enter the general circulation.

The term scrofula is applied to a condition in which, during the course of protracted inflammations of the mucous membranes (particularly of the respiratory organs, but very often of the digestive canal), the corresponding lymphatic glands suffer tubercular infection from losses of substance and undergo cheesy degeneration. The patients are sometimes vigorous individuals, and in such cases recovery may be effected by calcification of the degenerated glands (or by operative removal), after recovery of the disease of the mucous membrane. But the majority of the patients are in a more or less debilitated condition, and usually the term scrofulous is applied only to them.

Such individuals exhibit a decided tendency to eczematous (phlyctenular) affections of the integument of the lids, the conjunctiva and cornea. Arlt applies the term scrofulous conjunctivitis to eczematous disease of the conjunctiva, but, strange to say, he does not apply this term to the analogous disease of the cornea but to diffuse interstitial keratitis.

A decided tendency to phlyctenular disease of the conjunctiva and cornea is also observed in other asthenic conditions, especially those left over after protracted and severe diseases, for example, during convalescence from infectious diseases.

On the other hand, one or more outbreaks of eczema may appear upon the conjunctiva and cornea in otherwise healthy individuals to whom the term scrofulous would not be proper. It is therefore preferable to drop the term conjunctivitis scrofulosa, etc., and to employ the term which is applied to the corresponding disease of the skin, such as eczema, or, if this is not considered desirable, phlyctenulae (vesicles).

The term "scrofulous" was formerly much abused. It meant cod-liver oil, etc., and often entire neglect of the disease of the skin or eyes, although suitable local treatment is all important. This often causes surprisingly rapid recovery of the glandular enlargements when they are due not to bacillary, but merely to inflammatory swelling.

In "scrofulous" diseases of the eye the local treatment of the respiratory mucous membrane, especially of the nose, is the most important for the purpose of preventing the endless relapses.

The term "scrofulous" was formerly applied to those normal but feeble individuals who, as the result of improper nourishment, have lost to a certain degree their power of resistance to external influences. In children this is manifested chiefly in the integument and certain mucous membranes, in the form of chronic catarrhs attended by ulceration and eczemas. The term scrofula was employed especially when numerous lymphatic glands became enlarged on account of infection from the ulcers, either from the reception of infectious germs or of the products of decomposition. At the present time the term scrofula, from an anatomo-pathological standpoint, is confined to those cases in which the glands undergo cheesy degeneration as the result of the absorption of tubercle bacilli, while from a clinical standpoint the term is still used in many cases in which the enlargement of the glands is due to less harmful influences.

Leprosy.

Vide Bull and Hansen, *Jahr. f. Aug.*, 1873, p. 218; Pedraglia, *Mon. f. Aug.*, X, p. 65.

For a long time this disease appeared to be confined, in Europe, to a few small localities, but in recent times it has evidently spread, for example, in the Baltic provinces. It is due to a micro-organism which resembles the tubercle bacillus very closely. Prior to its discovery leprosy was regarded as a constitutional disease.

Unlike the infectious diseases hitherto considered, this has a very long period of incubation, on the average four to five years, sometimes apparently ten to fifteen years; not infrequently the period of incubation is shorter. The primary lesion is comparatively insignificant and is usually overlooked. According to Eklund infection often takes place from inoculation into the conjunctival sac by means of towels (*Jahr. f. Aug.*, 1879, p. 256).

Lepra anæsthetica and tuberosa differ merely in the amount of development of the specific new-formation, which is a granulation-like tissue containing typical bacilli (with marked implication of the interstitial tissue of the nerves), in a nodular or superficial form. Both forms occur in the tissues of the organ of sight. According to Lopez (*Arch. f. Aug.*, XXII, 2 and 3), the eye is affected in half the cases, the eye with its appendages in all cases.

Anæsthetic patches and nodules often develop in the eyebrows and lids; the hairs fall out, the nodules grow to the size of a hazel-nut and disappear mainly through ulceration, rarely from central softening. Cicatricial ectropium of the lids may develop under either process.

The nodules are situated in or beneath the skin; they are accompanied by cellular infiltrations, especially along the vessels. The secondary effects are: paralysis of the orbicularis palpebrarum, paralytic ectropium, traumatic conjunctivitis and keratitis, pannus of the lower half of the cornea, iritis, etc. (Parinaud, *Ann. d'Ocul.*, 115, p. 140). According to Meyer numerous very small nodules are found in such cases of pannus.

Anæsthesia of the cornea may result in neuro-paralytic keratitis. Bull and Hansen did not observe its development, despite anæsthesia of the trigeminus, even when flies wandered about the cornea and conjunctiva.

We often find growths upon the conjunctiva, which spread to the cornea and also invade the interior of the eye. Eversion of the lids from the eye, interference with their closure, epiphora, etc., may also be caused by the leprous connective-tissue tumors. The latter form round, hard, whitish, pale yellow or red, shining growths, which are not tender on pressure, and are situated at the rim of the cornea and beyond it. They are gradually lost toward the fornix, but terminate abruptly toward the cornea. The secretion is slight, and the nodules usually do not ulcerate.

The adjacent cornea is cloudy, later it is entirely opaque or looks as if strewn with flour; it may become very ectatic.

The conjunctival nodules may grow more and more over the cornea. They may shrivel and soften, sometimes after the lapse of years. The usual termination is in shrivelling of the globe.

Nodules may also develop in the deeper layers of the cornea; there may also be simple punctate keratitis with or without iritis.

Bacilli are often found in the tears, nasal secretion and fluid of the mouth (Besnier, "Sur la Lèpre," Paris, 1887).

According to Bull and Hansen, the sclera, unlike the uvea, is never attacked primarily and independently.

In the nodular form, iritis may develop during the first year; in the anæsthetic form, it is usually secondary, due to corneal ulcers accompanying paralysis of the orbicularis. Acute iritis is almost always associated with vitreous opacities. More rarely there are nodules both in the iris and in the cornea. They are grayish in color, start from the periphery and are generally situated in the lower half of the iris. They may gradually fill the entire anterior chamber and cornea and produce staphyloma of the adjacent sclera. Small nodules in the iris may disappear spontaneously and then undergo several relapses.

In many cases the iritis terminates in complete occlusion of the pupil, while cyclitis and cyclo-choroiditis may develop secondarily. Secondary opacity of the lens is also observed not infrequently.

According to Bull and Hansen, iritis and irido-choroiditis are more amenable to treatment than corneal affections. These writers successfully removed iris nodules by operation.

It is not to be wondered at that hemeralopia is often present in the later stages of such a chronic disease. A scorbutic terminal stage is observed not infrequently, and may lead occasionally to hemorrhages into the eye.

CHAPTER IX.

CONSTITUTIONAL DISEASES.

THE so-called constitutional diseases make up a category which is constantly growing smaller with the increase of our knowledge. Apart from the few congenital conditions (hæmophilia) or developmental anomalies (chlorosis), they may be regarded as chronic diseases of uncertain origin.

Some of the diseases formerly included in this category, such as scrofula, leprosy, and probably rheumatism, have been found to be chronic infectious diseases. Many are probably to be regarded as chronic diseases of various organs, with which they often exhibit great similarity, for example, with the condition in contracted kidneys. To this class belong many forms of diabetes, particularly the severe forms (pancreas), Addison's disease (suprarenal capsules), Basedow's disease and myxœdema (thyroid gland), leukæmia (spleen, lymphatic glands, medulla of the bones), etc. Other constitutional anomalies are, in the main, diseases of a special system, such as rachitis and osteomalacia of the osseous system, scurvy, amyloid and general atheromatosis of the vessels, gout of the joints, serous membranes and cavities. In all cases there are disturbances of respiration, digestion, absorption, or secretion.

Certain of the constitutional anomalies also exhibit a great resemblance to chronic infectious diseases, in so far as "specific" neoplasms may develop, for example, in the tumor cachexia of malignant tumors and leukæmia. As no specific micro-organism has been found in these cases, we must assume that the tumor cells and the leukæmic leucocytes themselves play the part of the specific inflammation-producer.

In a measure we may distinguish between quantitative and qualitative constitutional anomalies. In the former there is an absence of definite organic disease, while the latter receive from such disease their peculiar characteristics. The former exhibit in the eye, as in the rest of the body, mere symptoms of general weakness without visible anatomical findings: weakness of accommodation and convergence, retinal asthenopia, neuralgic pains on using the eyes, night blindness, often associated with xerosis of the conjunctiva (*vide* p. 34), concentric narrowing of the field of vision, with paroxysmal loss of sight, especially as the forerunner of attacks of syncope, etc.

Anœmia.

The symptoms just mentioned are the only eye symptoms observed in pure cases of anæmia, cachexia and marasmus. It is only in very advanced cases that the papilla is notably paler and may even be chalky-white; the blood in the vessels is evidently lighter, but the color of the fundus is not changed appreciably. Pulsation of the retinal vessels is occasionally visible (Becker).

A rather striking fact is the great frequency of congestion of the conjunctiva, also called dry catarrh, in anæmias of all kinds. One of its main causes is probably insufficient sleep or insomnia.

The eyeball and lids may be more or less sunken on account of wasting of the orbital fat, and this furnishes one of the characteristic peculiarities of the "marantic" facial expression.

The arcus senilis is merely evidence of a local fatty degeneration of the corresponding parts of the cornea. It is only when it appears at an unusually early period that it possesses a certain significance in the diagnosis of "senium præcox."

Other eye symptoms in pure anæmia are more or less accidental complications. They are very numerous because anæmic conditions are observed so frequently.

Spontaneous œdema of the lids, icteric color of the conjunctiva (slight hæmatogenous jaundice) and spontaneous hemorrhages (especially into the retina [1]), which occur in very advanced cases, consti-

---

[1] The retinal hemorrhages of pernicious anæmia are often converted, in the centre, into white patches. At first this was looked upon as a peculiar phenomenon, but it is simply a process of absorption which often occurs in other non-septic hemorrhages. Vessel changes may be present or absent in pernicious anæmia.

tute transitions to qualitative tissue changes and toxic phenomena. These give to the condition the character of pernicious anæmia, which is not necessarily fatal in all cases. Fraenkel (*Deutsch. Arch. f. kl. Med.*, XX) has also described parenchymatous changes in the external eye muscles in such cases. They were pale and clay-colored, the transverse striation was absent in great part, the fibres were filled with yellow or brown pigment, or were finely granular. Some fibres were narrow and waxy. · Neuritis, retrobulbar neuritis, atrophy of the optic nerves, etc., also occur under such conditions, especially when there has been considerable loss of blood. The ophthalmoscopic appearances sometimes resemble those of retinitis albuminurica, although the urine does not contain albumin.

If the general disorder of nutrition is acute in its onset, a hemorrhagic diathesis develops not infrequently. This is shown by spontaneous hemorrhages without visible tissue changes in all parts of the body, including the eye and adjacent parts. These hemorrhages exhibit no unusual symptoms, and this is also true of the hemorrhages of scurvy. A hemorrhagic nuclear paralysis (Cavalie, *Jahr. f. Aug.*, 1879, p. 224) or neuritis (Lawford, *Brit. Med. Journ.*, 1882, II, p. 119) is observed occasionally in scurvy. It goes without saying that hemeralopia is frequent in scurvy.

So-called marantic keratitis is a traumatic or desiccation keratitis which develops as the result of narcotic toxines and toxalbumins which are produced in the final stages.

Plethora, Corpulence, Obesity.

When these conditions produce eye symptoms they are similar to those found in anæmia. It is to be noted that in general obesity the eyelids (and the scrotum) are unaffected, and this gives the individual a peculiar appearance.

In the higher grades of plethora and corpulence, abnormal products of disassimilation or excessive amounts of the normal products can be demonstrated, for example, increased amount of uric acid in the urine. This furnishes a transition to true gout. On the other hand, there is not infrequently a transition into qualitative constitutional anomalies, particularly diabetes, so that it is sometimes diffi-cult to draw a line between these conditions.

Schoeler (*Berl. kl. Woch.*, 1887, No. 52) observed three cases of corneal and conjunctival xerosis after a course of treatment for obe-sity. Excessive weakness, or even actual collapse, is not infre-quent and compels the observance of caution.

The anæmic and marantic conditions may be regarded as equiva-lent to insufficient nutrition or absorption, the plethoric conditions as equivalent to insufficient respiratory activity with regard to the in-gested food. Hence the frequent shortness of breath in plethora, even in the absence of respiratory diseases, and the accumulation of imperfectly oxidized substances in the body (fats, uric acid). This is true to a certain degree, and then pathological and toxic products of disassimilation appear. These give rise to anatomical changes in the tissues (necrobioses and inflammations), particularly in the walls of the vessels. The latter are characteristic of the qualitative con-stitutional anomalies, although they may not occur for quite a long time.

As a general thing, signs of quantitative disturbance of nutrition are alone present for a certain length of time, and even these may be very slight or almost entirely absent, despite the presence of abnor-mal products of disassimilation or the increase or diminution of nor-mal products. At a later period, toxic and inflammation-producing substances are created (auto-intoxication). These may give rise to tissue changes, even passing into hemorrhagic inflammation, or a narcotic action which may terminate in profound coma. Both effects are sometimes produced paroxysmally, with comparatively free inter-vals. Some of these constitutional conditions like diabetes and leukæ-mia may present very active outbreaks.

In the course of the constitutional anomalies there are very often secondary infections with all their sequelæ, partly through the agency of inflammation and suppuration producers (pyæmia, erysipelas, etc.), which are everywhere present, partly through other bacilli, such as the tubercle bacillus. These secondary infections may ex-hibit peculiarities in their course on account of the modified nutrient medium offered by the diseased individual. The course of the sec-

ondary affection is usually more rapid and severe than in healthy individuals. In other cases, however, there may be a certain immunity against different micro-organisms. This may even be confined
to special organs.

Chlorosis.

Apart from the typical symptoms of exhaustion, slight grades of
chlorosis produce no striking changes in the eye. At the beginning
they correspond to the degree of anæmia and nutritive disturbance,
which are always present at the same time.

Despite marked anæmia, there are very often signs of congestion
of the conjunctiva, such as a tired feeling, a sensation of sand in the
eyes, of weight or roughness of the lids, etc. These are the signs of
conjunctival asthenopia or so-called dry catarrh, and are relieved
most effectually by diluted laudanum (5 : 10–15).

In very severe cases of chlorosis there may be visible pulsation
of the retinal arteries (Becker), but this is confined, as a rule, to the
optic papilla. This symptom is to be regarded as an evidence of
considerable diminution of the arterial pressure. In other respects
the ophthalmoscopic appearances are normal.

The œdema of the skin, which is not uncommon in severe cases,
is also observed upon the lids, particularly the lower lid. The optic
nerve is pale, the retinal vessels narrow and filled with light-colored
blood, a condition transitional to that of pernicious anæmia; hemorrhages into the fundus are never observed. In such an event
we would hardly be justified in regarding the case as one of pure
chlorosis.

A number of cases have been reported in which bilateral neuritis
and neuro-retinitis, with or without hemorrhages and with or without whitish exudation in the retina, have been associated with chlorosis (Gowers, *Jahr. f. Aug.*, 1880, p. 236; Herschel, *ib.*, 1882, p.
329; Eddison and Teale, *ib.*, 1888, p. 530; Bitsch, *Mon. f. Aug.*,
1879, p. 144). In view of the rarity of these cases and the frequency
of chlorosis, this association may be a mere coincidence.

In several cases of severe chlorosis without any other demonstrable disease, I have observed whitish shining patches in the retina;

they were mostly grouped around the fovea centralis in the well-known stellate shape of retinitis albuminurica. The remainder of the fundus was normal, and there was very little or no impairment of vision. The affection was always unilateral, lasted more than six months, and, with improvement of health, finally disappeared without leaving a trace. The appearances were probably due to small foci of fatty degeneration in the nerve-fibre layer of those parts of the retina which are destitute of capillaries. Such cases are probably at the extreme boundary of pure chlorosis, the characteristic feature of which is the absence of anatomical changes in the tissues.

Hœmophilia.

In rare cases this condition gives rise to hemorrhages into the organ of sight. Bramwell (*Jahr. f. Aug.*, 1886, p. 307) reports a cerebral hemorrhage associated with eye symptoms, Priestley Smith (*ib.*, 1888, p. 530) an orbital hemorrhage following an injury. Other eye symptoms are, in the main, those of "weakness of the eyes" as a part of the general condition of weakness. After profuse hemorrhages of hæmophilia, the disturbance of vision may appear as amblyopia or amaurosis after loss of blood. (These patients are known as "bleeders.") In Grossmann's case (*Arch. d'Ophth.*, II, p. 122), a boy of fifteen years suffered, on the fifth day after violent epistaxis, from impairment of vision passing into complete blindness; complete atrophy of the optic nerves finally developed.

Spontaneous hemorrhages into the lids and conjunctiva are not very frequent, and nothing has been reported concerning extravasations into the interior of the eye.

Addison's Disease.

This affection is not infrequently associated with degenerative and inflammatory diseases of the nervous system, particularly of the sympathetic (Flenier, Wiesbadener Congress, 1891).

The eyelids and face are often deeply discolored, the conjunctiva hardly ever, but the latter is often more or less jaundiced. Huber (*Deutsch. med. Woch.*, 1885, No. 38) saw a patch upon the conjunctiva; Schroetter (*Wien. med. Blaett.*, 1886, No. 21) saw patches on
28

the gums, buccal mucous membrane and sclera; Feuerstein (*ib.*, 1888, No. 35) saw several dark-brown patches, as large as millet-seeds, upon the conjunctiva.

The organ of sight also takes part in the general symptoms of weakness and is occasionally attacked as the result of complications, for example, renal diseases. It is well known that bronzing of the skin is not always connected with disease of the supra-renal capsules —and also that disease of these organs does not always produce bronzing.

Diabetes.

Vide Leber (*Arch. f. Ophth.*, **XXI**, 1, 3, and **XXXI**, 4), Foerster (*l.c.*, p. 217), Hirschberg (*Centr. f. Aug.*, 1891).

There are several varieties of diabetes mellitus, and the excessive amount of sugar in the urine is perhaps the only feature which is common to all of them.

Apart from toxic forms (curare, phloridzin, chloral, sulphuric acid, arsenic, alcohol, carbonic oxide, etc.) and the symptomatic mellituria of injuries to the head and of various cerebral diseases which affect the floor of the fourth ventricle, the remaining forms of diabetes mellitus appear to be due mainly to disease or abnormal function of the pancreas. Sooner or later this gives rise to profound changes of nutrition.

The abnormal increase of sugar in the urine may exist for a long time without causing any eye symptoms, and these do not appear in the mild forms, including the temporary toxic and traumatic glycosurias.

In the severe forms, however, the organ of sight is often affected. On the one hand, it exhibits symptoms of weakness (of accommodation and convergence), on the other hand, symptoms of auto-intoxication by pathological products of disassimilation. The latter include diseases of the vessels with their sequelæ, fatty degeneration and inflammatory proceses in the tissues themselves.

The frequent development of cataract is due to disease of the vessels, particularly in the ciliary processes, and to the consequent disturbance of the nutrition of the lens. Apart from the causation, this cannot be distinguished from cataract due to other causes at the same age. As it occurs generally in young people, we have to deal chiefly with rapidly progressive, broad-striped, soft cataracts, which are often associated at the start with visible swelling of the lens. The operation is as satisfactory as in other individuals of the same age and nutritive condition. It is only when the nutrition is very much impaired that it offers a serious objection to operation.

A spontaneous clearing up of the opacity of the lens is seen occasionally, particularly when the diabetes improves or recovers. As a matter of course, this is only possible so long as the cortical substance of the lens has not been destroyed.

The cataract is characteristic only when it appears in both eyes, in young people, and without visible cause. Frey (*Lond. Med. Rec.*, May, 1887) reports double diabetic cataract in a girl of nine years, a year after the beginning of the disease. At a later age the development of cataract is much less characteristic. It is to be remembered that an individual suffering from cataract of old age may be attacked by diabetes, and that, on the other hand, an old diabetic may suffer from cataract, although in the latter event the diabetes is not necessarily the connecting link.

The development of cataract in diabetes mellitus has been explained by various theories. It has been attributed to: 1, the general marasmus, although very feeble diabetics are often not attacked by cataract; 2, the removal of fluid from the lens through the medium of the sugar dissolved in the tissue juices. In fact, sugar has been detected in the lens in two-thirds of the cases, and still more often in the aqueous humor and vitreous. The presence of a certain amount of sugar for a certain period would, according to this theory, produce cataract in all diabetics, and this is controverted by experience. Moreover, the opaque as well as the clear lens may contain sugar in diabetes (Leber), and the cataract of one eye may contain sugar while the other does not (Becker). 3. Conversion of the sugar in the aqueous humor into lactic acid is also said to be the cause. This is a pure hypothesis; the fluid in the aqueous is distinctly alkaline in diabetic cataract, and the opacity does not start in the anterior cortical substance.

We must assume, therefore, that the cataract develops under the same conditions as spontaneous cataract, as the result of processes in the choroid and particularly in the ciliary processes, which furnish the nutritive supply to the lens. The swelling of the lens at the onset and the proliferating processes in its elements warrant the inference that irritating substances are present, which stimulate the living cells to proliferation and then cause their destruction.

Changes have been found in a number of cases in the pigment epithelium of the uvea, especially of the iris and ciliary processes (dropsical swelling, detachment during iridectomy), *i.e.*, in those parts which are most important to the nutrition of the lens. The iris has also been found more or less changed, partly to a condition of simple atrophy, partly to that of slight inflammation. Other uveal inflammations may also occur. The development of cataract appears to be due, therefore, to toxic substances circulating in the blood and not to the harmless sugar; it is a symptom of auto-intoxication. The latter condition includes an entire series of tissue changes in the vessels, interstitial tissues and parenchyma, which lead to obliteration and dilatation of the vessels, œdema and hemorrhage, fatty degeneration and other necrobioses, even to inflammatory affections. This condition is analogous to that found in long-continued albuminuria, to which the severe forms of diabetes exhibit great similarity (although not in regard to the development of cataract). The similarity of the symptoms of severe diabetes to certain chronic poisonings is also evident.

The following diseases of the eye may also occur in diabetes:

Iritis and irido-cyclitis, the former often associated, according to Leber, with hypopyon or fibrinous exudation. The simple forms are also frequent, and may vary from the chronic to the acute varieties; they may even be attended by hemorrhages into the chamber.

Spontaneous short-sightedness, at the age of forty to sixty years, without opacity of the lens, must be attributed to swelling or hardening of the crystalline or to a diffuse affection of the choroid, in which the pigment epithelium is affected very slightly or not at all, but the sclera is softened and yielding.

Vitreous opacities, without other visible signs of uveitis, are in great part the result of hemorrhages. The latter may be unilateral or bilateral (Mackenzie, Ophth. Hosp. Rep., IX, 2, p. 134).

Relapsing scleritis has been brought into connection with diabetes by various writers, but the relationship has not been established positively.

The changes in the retina are inflammatory to a very slight extent. The ophthalmoscope shows changes in the vessels and hemorrhages and whitish patches. According to Hirschberg, they are always found in diabetes which has lasted more than ten to twelve years, and are terminal symptoms of the disease.

Some of these symptoms are the result of coincident albuminuria (owing to epithelial necroses, fatty degeneration and interstitial changes in the kidneys), and differ in no respect from retinitis albuminurica. Others are found irrespective of changes in the kidneys.

Hirschberg (*l.c.*) distinguishes two principal forms : 1, a variety with small, light, shining patches, usually with punctate hemorrhages (retinitis centralis punctata diabetica). The patches are always found in both eyes, the impairment of vision develops gradually, and the optic nerve is not affected. Implication of the optic nerve, opacity of the adjacent retina and dilatation of the vessels distinguish the albuminuric affection from the diabetic form, although sometimes the last-mentioned symptoms are present in diabetes and absent in albuminuria. 2, a hemorrhagic form, either in the shape of small punctate hemorrhages, larger hemorrhages associated occasionally with vitreous opacities, hemorrhagic infarctions and venous thromboses (Michel, *Deutsch. Arch. f. kl. Med.*, XXII, p. 439), or even a hemorrhagic glaucoma (Galezowski, *Rec. d'Ophth.*, 1873, p. 90; Knapp, *Arch. f. Aug.*, X, 1, p. 99; Hirschberg, three cases). This form is less characteristic than the other. The disturbance of sight depends upon the location and size of the hemorrhages.

Neuritis, neuro-retinitis, choked neuritis, so-called retrobulbar neuritis with secondary atrophy of the optic nerve, etc., are also observed occasionally in causal connection with diabetes, but the ophthalmoscope shows nothing characteristic.

On the whole, hemorrhages are the most frequent among diabetic diseases of the retina. As in albuminuria, the retinal affections at-

tended with hemorrhages are more unfavorable, as regards prognosis, than those which consist solely of degenerative foci and exudations.   Similar hemorrhagic lesions are found even more frequently
within the cranial cavity, where they not alone threaten the organ
of vision but also life.   Hence every diabetic disease of the retina
has an ominous prognostic significance.

A hemorrhage behind the retina occasionally leads to its detachment.

In long-continued, severe diabetes vascular changes (particularly ·
sclerosis) are found not only in the retina, but in almost all the
tissues of the eye and in many other tissues of the body, especially
the central nervous system.   These also give rise to spontaneous
hemorrhages into the conjunctiva.

Various central and peripheral disorders of vision may develop
during the course of diabetes, such as amblyopia and amaurosis without findings (almost always bilateral, according to Cohn, and almost
always unilateral, according to Galezowski), homonymous defects in
the field of vision (Leber, *l.c.*, p. 288), homonymous hemianopsia
(Schiess, *Jahres.*, 1886, p. 50).   On the whole, diabetic hemianopsias
are rare, but they are usually obstinate (Leber, *l.c.*, p. 295).   Hemiplegia, aphasia, ataxia, etc., also occur occasionally.   Central or
paracentral scotomata in an almost normal visual field, as in toxic
amblyopia and probably from a similar cause (auto-intoxication),
have been reported by Bresgen (*Centr. f. Aug.*, 1881, Feb.), Samuel
(*ib.*, 1882, July), Lawford (*Jahr. f. Aug.*, 1882, p. 293), Stanford
Morton (*ibid.*), Edmunds and Nettleship (*ibid.*).   As the last-mentioned cases occurred in smokers, it is not improbable that the tobacco
was especially injurious on account of the influence of the diabetic
auto-infection.   [A central scotoma for red has been noted by many
observers in cases of diabetes.   See Leber in Graefe and Saemisch's
"Handbuch," etc.   I have noted it several times.—ED.]   According
to Hirschberg such cases present a bad prognosis.

Many of the last-mentioned affections are due to hemorrhages or
circumscribed spots of softening; others are co-ordinate symptoms
of a lesion situated at the floor of the fourth ventricle.

Of paralyses of the ocular muscles, which are usually incomplete and temporary, paralysis of accommodation is the most frequent. Like paralyses in general, this may be one of the first symptoms. Pure paralysis of accommodation in middle life is especially suggestive. Paralyses which develop during advanced stages of diabetes are cured with great difficulty.

Abducens paralysis was observed by Gutmann (*Centr. f. Aug.*, 1883, Oct.) and Landesberg (*Jahr. f. Aug.*, 1884, p. 323). Kwiatowski (*ib.*, 1879, p. 224) reports trochlear paralysis; Seegen, double ptosis; Rolland (*Arch. f. Aug.*, 1888, p. 257) and Galezowski (*Rec. d'Ophth.*, 1878, p. 3) report paralysis of the motor oculi; Legrange and Ogle report facial paralysis.

These paralyses are due to nuclear and peripheral hemorrhages or to peripheral neuritis. The latter lesion is also the cause of the frequent neuralgias and of the rarer anæsthesiæ and sensory disturbances. Anæsthesia of the first branch of the trigeminus may give rise to neuro-paralytic keratitis; neuritis may cause herpes zoster ophthalmicus.

In diabetic coma, which may be associated with cortical visual disorders, the odor of acetone is often very strong in the expired air, but this sign may also be absent. During the coma the urine is usually free from sugar, even if it has been very abundant immediately prior to the attack. Hence the cause seems to be an abnormal product of decomposition of the glucose. Diabetes is often combined with renal disease, and the latter may produce its characteristic symptoms in the visual organs.

In advanced diabetes there is a great tendency to the formation of furuncles, which may also appear upon the eyelids. Obstinate ezcema of the border of the lids is very frequent, and ulcerative processes in the cornea are not uncommon.

Among 52 diabetics, Lagrange (*Arch. d'Ophth.*, 1887, Jan.) found 13 cases of cataract and 13 of hemorrhage into the interior of the eye. Among 144 diabetics, Galezowski (*Jahr. f. Aug.*, 1883, p. 207) observed 5 cases of paresis of accommodation, 4 of parenchymatous and purulent keratitis, 7 of iritis, 4 of glaucoma, 46 of cataract, 27 of retinitis, 31 of amblyopia, 10 of paralysis, 4 of hemianopsia, 3 of detachment of the retina and 3 of atrophy of the optic nerve.

Among 169 diabetics, Mager (*Berl. kl. Woch.*, 1879, No. 21) found only 5 cases of cataract.

It is evident, therefore, that the urine should be examined for sugar in every case of cataract which develops at an unusual period and without any known cause, as well as in every case of neuritis, retinitis, neuro-retinitis and atrophy of the optic nerve, in every doubtful visual disorder, neuralgia or ocular paralysis. Hirschberg found diabetes in one per cent of all his private patients.

Diabetes Insipidus.

In this condition there is a very notable increase in the excretion of urine, without the presence of sugar. It is called polyuria or polydipsia, according as the increase of the amount of urine or the increased thirst is regarded as the primary event.

It is well known that diabetes insipidus is produced by destruction of a certain portion of the floor of the fourth ventricle. In almost all cases in which visual symptoms and polyuria are associated, the latter is a symptom of the same cerebral disease which has given rise to the former. For example, van der Heyden (*Jahr. f. Aug.*, 1875, p. 191) reports optic neuritis and symptoms of brain tumor; Handfield Jones (*ibid.*) reports pain in the eyes, inequality of the pupils, epileptoid attacks; David (*ibid.*, 1889, p. 492) reports hemianopsia; Raynaud (*ibid.*, 1874, p. 407) reports Raynaud's disease with bulimia, polydipsia and visual disorders which occurred paroxysmally. Some doubt attaches to Gayet's case (*ibid.*, 1875, p. 191) of right abducens paralysis with polydipsia and polyuria.

It is said that opacity of the lens may also develop in pure diabetes insipidus, but this is undoubtedly extremely rare.

In phosphaturia, Courssérant observed hemianopsia; Dor (*ibid.*, 1877, p. 208) found phosphaturia seven times among eight cases of soft cataract in young people. It would seem, therefore, as if diabetes mellitus, diabetes insipidus and phosphaturia were somewhat similar disturbances of nutrition—in the broadest sense of the word —and also resembled the excessive production of uric acid (from imperfect oxidation, particularly in obese individuals) which not infrequently precedes true diabetes. It is only when the abnormal products of disassimilation exert a toxic action that the symptoms appear which are so frequent in the terminal stage of severe diabetes. The excessive excretion of sugar, phosphates, or uric acid in the urine is merely the most striking symptom of the disorder of disassimilation. The by-products of the latter, for example, ptomaines which give rise to necrobioses, necroses and inflammations of the tissues, or act as narcotic poisons, are evidently a much greater source of danger to the body and the visual organ than the sugar, uric acid, or phosphates.

Basedow's Disease.

The main symptoms of Basedow's disease are thyroid enlargement, palpitation of the heart, nervous disorders and exophthalmus. The latter symptom, like all the others, may be absent or, in rare cases, it may be almost the only symptom. It occurs in every possible degree, and there may even be paroxysmal or permanent luxation of the eyeball in front of the lids. In such cases, suture of the palpebral fissure is the only means of preventing the loss of the protruded eye. As a matter of course, traumatic inflammations of the conjunctiva and cornea are often present.

The development of the goitre is usually preceded by increased heart's action, but the opposite condition may also obtain. Like all the other symptoms of the disease, exophthalmus may occur paroxysmally, may increase or diminish, or may be permanent and uniform.

In addition to exophthalmus, there are also other signs of irritation of the sympathetic, such as spasm of Mueller's muscle (Graefe's symptom). As a result of this spasm, the palpebral fissure is dilated and the upper lid lags behind when the patient looks downward. In this way the white sclera often becomes visible above the cornea in looking downward. Sharkey (*Brit. Med. Journ.*, Oct. 25th, 1890) found Graefe's symptom absent only twelve times among six hundred and thirteen cases. For a long time it may be the sole symptom of the disease.

Moderate dilatation of the pupils with correct reaction to light and inequality of the pupils are also observed in many cases, and are evidences of spasm of the sympathetic. Even the exophthalmus may be due, in part, to spasm of the smooth muscular fibres (protrusor bulbi) which close the inferior orbital fissure. In great part, however, it is the result of œdema, increased congestion and direct increase of the orbital tissues. Hence the exophthalmus often diminishes after death.

A distinct vascular murmur can sometimes be heard in the orbit with the stethoscope. It resembles the placental bruit and is probably explained in a similar way (Donders, *Arch. f. Ophth.*, XVII, 1, p. 102). Sinuosity and atheroma of the ophthalmic artery have sometimes been found post-mortem.

Unilateral exophthalmus is not very rare (Voelkel, Diss. Berlin, 1890). Among 20 cases, Emmert (*Arch. f. Ophth.*, XVII, 1, p. 203) found 1 unilateral case; among 32 cases, Griffith saw 25 of double, 4 of right-sided, and 3 of left-sided exophthalmus. This symptom often differs in degree on the two sides, being more marked upon the side on which the thyroid enlargement is greater.

Hack (*Deutsch. med. Woch.*, 1885, No. 25) observed disappearance of the exophthalmus on one side after cauterization of the nasal mucous membrane on the same side, and Bobone (*Ann. d'Ocul.*, 96, p. 260) claims to have obtained improvement in the same way. A few other cases of this kind have been published, but they are very exceptional.

The excessive sweating which is not uncommon in Basedow's disease is also a sympathetic symptom. Occasionally there is very marked epiphora.

The tremor may extend to the eyelids (Liebrecht, *Mon. f. Aug.*, 1890, Dec.) and may even be confined to the eyes (Freund, *Deutsch. med. Woch.*, 1891, No. 8). This symptom and likewise the nystagmus, which is observed occasionally, must be attributed to diminution of cortical motor innervation.

Slight impairment of corneal sensibility is quite common and is to be regarded as the cause of the diminution in the frequency of winking movements (Stellwag's symptom). In combination with the increasing exophthalmus it facilitates the development of traumatic keratitis, which may assume the xerotic or neuro-paralytic form. In some cases the corneal sensibility is very considerably impaired, but complete anæsthesia is rare.

In advanced cases the ophthalmoscope often shows pulsation of the retinal arteries, occasionally extending far into the periphery of the fundus. Becker (*Mon. f. Aug.*, 1880, p. 1) found arterial pulsation six times in seven cases. It is probably due to the fact that, as a result of the arterial paralysis, the blood pressure falls notably during diastole and becomes so slight, in comparison with the intraocular pressure, that it causes an entrance of blood into the retinal arteries only during the intensified systole of the heart. At the same time the intraocular pressure may be normal or even diminished. Sinuosity or dilatation of the retinal veins is often noticed.

Various muscular disturbances have been reported, such as external ophthalmoplegia (Schoch and Koeben, Diss. Berlin, 1854; Fischer, *Arch. gén.*, II, p. 521; Stellwag, *Wien. med. Jahrb.*, XVII, p. 25; Chvostek, *Wien. med. Presse*, 1872, p. 497; Roth, *ibid.*, 1875, p. 680) and complicated ophthalmoplegia (Warner, *Brit. Med. Journ.*, Oct. 28th, 1882; Bristowe, *ibid.*, May 6th, 1886; Jendrassik, *Arch. f. Psych*, 1886, p. 301; Ballet, *Rec. d'Ophth.*, VIII., p. 451). Implication of the trigeminus, facial, glossopharyngeal, spinal accessory and hypoglossal nerves, hemianæsthesias, hemiplegias, color blindness, paralyses of the extremities, epilepsy, loss of smell and taste, etc., have also been reported. Isolated paralyses of the eye muscles (trochlearis, abducens, motor oculi) are rare (Féréol, *Un. méd.*, 1874, p. 153; Finlayson, *Brain*, 1890, March). Liebrecht (*Mon. f. Aug.*, 1891, p. 182) also mentions paralysis of convergence, but this was possibly due to notable insufficiency of the interni. Complete paralysis of all the external eye muscles and the orbicularis has also been reported (*Neurol. Centr.*, 1888, p. 554). The paralyses appear to be chiefly nuclear, basilar, or root paralyses.

Basedow's disease is often combined with other affections of the nervous system, such as syringomyelia, hemichorea, hemiplegia, polio-encephalitis, multiple sclerosis, bulbar paralysis (the autopsy gave negative results), hysteria, neurasthenia, spinal and cerebral symptoms of all kinds (conjugate paralyses and spasms, central disorders of vision), psychoses, trophic disorders, etc.

Other constitutional anomalies are also observed, such as diabetes mellitus and insipidus, Addison's disease (in the face the eyelids are chiefly pigmented; Drummond, *Brit. Med. Journ.*, May 14th, 1887, six cases; Mackenzie, *Lancet*, 1890, II, five cases; even the conjunctiva may be pigmented, Oppenheim, *Muench. med. Woch.*, 1887, No. 52).

Peripheral visual disorders with ophthalmoscopic findings are rare (Story, *Ophth. Review*, 1883, p. 161, double optic neuritis; Emmert, *Arch. f. Ophth.*, XXVII, 1, p. 203, atrophy of the optic nerve). These disorders without ophthalmoscopic findings are quite frequent, especially concentric narrowing of the field of vision, with or without impairment of central vision or the color sense. Such symptoms are commonly bilateral, as in neurasthenia.

As a matter of course, conditions of exhaustion of the eye (weakness of accommodation and convergence, asthenopia retinæ) are noticeable. Headache, vertigo, insomnia, impairment of memory and the like are probably due to the same cause. Extreme anæmia may lead to œdema of the lids and finally to a sort of hemorrhagic diathesis in which the eye also takes part. The frequent menstrual disorders (the majority of the patients are females) are to be regarded as results, not as causes of the disease. [Diarrhœa is not infrequent. —Ed.]

According to Ballet (*Rev. de Méd.*, 1888, May and July), the disease points toward the nuclear region of the facial and ocular muscles; he thinks it is located in the medulla oblongata, perhaps as far down as the cilio-spinal centre. According to this writer the disease is a functional bulbar neurosis; the scanty anatomical changes which are found there appear to be secondary. Moebius' theory is much more plausible. He believes that we have to deal with the toxic effects of products of disassimilation, which are furnished by the diseased thyroid gland. These appear to act chiefly on the vasomotor cells and the cortical motor cells. As in all other diseases of the kind, we find the signs of increasing weakness and diminishing nutritive energy, sometimes passing into spontaneous necrosis of peripheral parts. It is only in late stages that we find local and diffuse necrobioses, vessel changes, interstitial inflammations, hemorrhages, multiple neuritis, etc. These occur in chronic toxæmias in general, and always exhibit a certain predilection for the region of the cerebral muscle nuclei and nerve roots. The principal symptoms are toxic phenomena, but these are preceded for a long time by general nutritive disturbances. I desire to call special attention to one point. The poisonous action of the noxious substances produced in the thyroid gland are noticeable, in the main, in the adjacent cervical sympathetic. The oculo-pupillary fibres are chiefly affected; the upper fibres which pass to the vessels of the head are affected to a much less extent. This is easily understood in view of the fact that the current in the veins and lymphatics flows toward the heart. The general symptoms are those of a chronic toxæmia, the local toxic symptoms are those of a constant, sometimes a remittent or intermittent irritation of the sympathetic at the site of closest proximity to the diseased thyroid gland.

It is also easy to understand that these local symptoms may be absent or vary in intensity, if only a certain portion of the gland is affected, and that the symptoms are more pronounced, as a rule, on the side on which the gland is more involved.

Hence, the earliest possible enucleation of the larger part of the diseased gland is indicated so long as the principal symptoms are still local (goitre, palpitation, exophthalmus). But as the symptoms are rarely dangerous at this period, the operation is not performed, as a rule, until the chronic general toxæmia is too far advanced. Hence, the operative treatment has remained very unsatisfactory despite a few good results (Rehn, *Berl. kl. Woch.*, 1884, No. 11). [See also Mannheim: "Der Morbus Gravesii," Berlin, 1894. Most of the authorities advise against operation; see p. 124 *et seq.*—ED.]

Myxœdema.

In many respects myxœdema (Ord, Lond. Clin. Soc., May 25th, 1888) is the direct antithesis of Basedow's disease. A characteristic feature is the absence of the thyroid gland (cachexia strumipriva) or loss of its function. In the latter event the gland may even be enlarged.

Myxœdema is characterized by a non-œdematous general swelling of the skin (no pitting on pressure), with pallor, coldness and cyanosis, resulting in a stupid, dull expression of the face, thickening of the lids, lips, tongue, hands and feet.   The pulse is slow, the voice hoarse and deep; speech, thought and motion are slow and difficult, memory and intelligence are impaired.   The patients also suffer from mental irritability and depression, pain in the head and back, weakness of the limbs, loss of the hair and teeth, atrophy or absence of the thyroid gland, general anæmia and cachexia and, not infrequently at a later stage, from albuminuria.   The patients are often cretins or idiots, or the affection is the result of extirpation of goitre.

Ord has demonstrated an increase of mucin, particularly in the subcutaneous connective tissue (Horsley found it even in the blood of operated monkeys), so that the disease has been called mucinæmia. Virchow found irritative processes in the subcutaneous tissues. Whether changes in the infundibulum and pineal gland, which are not uncommon, are a cause, a symptom, or a result of the disease, cannot be decided.

In addition to implication of the lids in the cutaneous affection,— the latter may even begin in this locality (*Jahr. f. Aug.*, 1889, p. 558 and 561)—amblyopia has been reported in several cases.   Wadsorth (*ib.*, 1884, p. 390) saw double atrophy of the optic nerve.   The case of cataract after extirpation of a goitre (Landsberg, *ib.*, 1888, p. 308) seems to be very doubtful.

Sollier (*Neurol. Centr.*, 1892, p. 25) reports two cases of Basedow's disease combined with myxœdema, *i.e.*, with general swelling of the skin.   The symptoms of Basedow's disease first appeared and, after diminution in the size of the thyroid, the symptoms of myxœdema were superadded.

It is evident that myxœdema is a profound disturbance of disassimilation and nutrition.   Its most striking result appears to be the increase of mucin, like the increase of sugar in diabetes mellitus. The symptoms of auto-intoxication are due, however, to other unknown products of abnormal disassimilation.

Rachitis.

This is often the cause of so-called laminated cataract, owing to a temporary, intense disorder of nutrition in the lens.   If it develops at a very early period (even intra-uterine) a central cataract forms. When this adheres to the capsule of the lens at the anterior and posterior poles, it contracts in the course of time into a thread-shaped opacity in the axis of the lens, the so-called spindle-shaped cataract. Both are merely varieties of laminated cataract. In laminated cataract proper, the lens substance which has developed prior to the nutritive disturbance remains transparent, as well as that formed at a later period. If the nutritive disturbance returns, a multiple laminated cataract may develop.

At the beginning, the laminated cataract appears to be a cortical or total cataract. It is only at a later period that the opacity moves away from the capsule of the lens on account of the new formation of normal cortical substance. The opaque mass of the lens is absorbed in part, sometimes with the exception merely of a layer of punctate opacities which may be overlooked on superficial examination.

Laminated cataract is usually present in a similar form in both eyes, but it is not infrequently unilateral. It is sometimes congenital. Similar forms may also develop after injury to the capsule of the lens, which subsequently closes up. The large majority of cases develop in the first years of life in rachitic children.

The violent and long-continued convulsions are the cause of the disease, not the rachitis as such. But the real cause is not the general convulsions and the consequent concussion of the body (Arlt), but the nutritive disturbance of the lens due to violent spasm of the ciliary muscle (associated probably with spasm of the muscular coat of the vessels). I have been able to demonstrate the latter symptom with the ophthalmoscope during general convulsions (epilepsy). It is necessary, however, that these spasms occur at a period when the growth of the lens is still active, i.e., before the age of six years. At a later period this result follows spasm of the ciliary muscles only under special circumstances, when the muscular coat of the vessels is affected (ergotin poisoning), or when the individual is very anæmic (Saemisch). In the latter event the connecting link is supplied not infrequently by a uveal disease (serous iritis).

During childhood, rickets is by far the most frequent cause of general convulsions, with the exception of meningitis. The latter ter-

minates much more frequently in death and hence plays no great part
in the etiology of laminated cataract.

The opacity does not appear immediately after the convulsions,
but usually a few days, sometimes even two weeks, later. In rare
cases it disappears spontaneously. If the cataract is congenital or
has developed at an early period and is very dense, the symptoms of
psychical blindness may develop after a subsequent cataract operation, as the result of imperfect development of the tracts between the
brain and eye.

It is a very interesting fact that an analogous condition is ob-
srved, under similar circumstances, in another organ which is genetically co-ordinate with the lens, viz., the enamel of the permanent
teeth (Horner). The teeth, particularly the upper incisors, often exhibit horizontal grooves and ridges (Fig. 21), *i.e.*, places in which
the enamel is alternately present and absent. This produces
a characteristic appearance which is entirely different from
that of hereditary syphilis. The change is only seen distinctly during the first few years after the appearance of the
teeth. The latter soon crumble at the parts which are destitute of
enamel, so that later only stumps of teeth remain.

Fig. 21.

It is not surprising that laminated cataract may be present without rachitic teeth, and *vice versa*, because both conditions are the results of local spasms in the ciliary body and the vessels of the enamel
respectively. These local spasms may be entirely absent, or they are
not always present at the same time in both organs. When present,
they warrant a probable diagnosis of rachitis in infancy, unless some
other cause can be demonstrated.

In osteomalacia no special eye symptoms have been reported,
apart from evidences of weakness.

In conclusion, we will discuss a few constitutional anomalies
which are closely related to the chronic infectious diseases.

Leukœmia.

Like the chronic infectious diseases, leukæmia also possesses a
specific new-formation. This consists of lymphoid tissue like that
found in the spleen, lymphatic glands and medulla of the bones.

At the onset of the disease the symptoms of progressive weakness predominate, as in every notable disturbance of nutrition. At a later period the signs of auto-intoxication make their appearance. Hence leukæmia should really be considered among the diseases of the circulatory apparatus, if we were certain that the changes in the spleen, medulla of the bones and lymphatic glands were primary. They may be due, however, to some noxious agent which is already in the organism, especially as true leukæmia is often preceded for a long time by disorders of digestion and nutrition.

The disease usually runs a very chronic course, and there may be no eye symptoms, apart from the general symptoms of weakness; this often lasts until death. On the other hand, eye symptoms may lead to the recognition of the disease.

In a series of cases, retinitis occurs (almost always bilateral) as in albuminuria. There are two principal varieties, one accompanied by hemorrhages, the other by whitish patches; the former is more frequent. The hemorrhages often exhibit a white centre (symptoms of absorption), but this is not pathognomonic, as Poncet believed. The retina itself may be normal or more or less cloudy, occasionally milky and strewn with hemorrhages (Perrin, *Gaz. des Hôp.*, 1870, p. 191). According to Roth's findings (*Virch. Arch.*, 49, p. 441), this opacity appears to be due to finely granular cloudiness of the nerve-fibre layer.

Post-mortem examination shows that the vessels are, in the main, unchanged (Poncet, *Gaz. des Hôp.*, 1874, p. 360; Gallasch, *Jahr. f. Kind.*, VI, 1); hemorrhages, œdema and cellular infiltrations are found in the papilla, retina and choroid. The latter may be so extensive as to form neoplasms which are even visible with the ophthalmoscope (Becker). Oeller found the choroid thickened sixfold by cellular infiltration. In addition, the retina exhibits hypertrophy of the radial fibres and sclerotic nerve fibres. The peripheral œdema of the retina, which is occasionally mentioned, has no significance; it occurs normally at a certain age.

In other cases, however, vessel changes have been demonstrated: dilatation and sinuosity, thickening and round-cell infiltration of the adventitia in spots, visible occasionally with the ophthalmoscope

29

as a whitish sheath around the vessels (Tillaux, *Rec. d'Ophth.*, 1878, p. 461); also advanced fatty degeneration of the walls of the vessels, especially of the peripheral branches (Roth), and sclerosis of the vessel walls (Kramsztyk, *Jahr. f. Aug.*, 1878, p. 226). Venous thromboses also occur (Michel, *Deutsch. Arch. f. kl. Med.*, XXII, p. 439). The choroid may be unaffected, or it may also exhibit hemorrhages, infiltrations and vessel changes.

Other lesions which have been observed are vitreous hemorrhages, occasionally very profuse (Saemisch, *Mon. f. Aug.*, 1869, p. 305), simple and retrobulbar neuritis (usually hemorrhage into the optic nerve), occasionally an enlargement of the papilla like choked disc (Oeller), or a fungus-shaped enlargement due to cellular infiltration (Poncet).

Hemorrhages into the conjunctiva and face (eyelids) belong to the terminal stage and are not very frequent. In acute cases they may be the first symptom which calls attention to the disease.

In severe forms of leukæmia, a light or orange yellow color of the fundus may be visible, but this is very rare. Liebreich ("Atlas"), Litten (*Jahr. f. Aug.*, 1889, p. 491), Becker (*Arch. f. Aug. u. Ohr.*, I, 1; p. 94), Kramsztyk (*l.c.*) report cases of this kind. I have seen this condition in a single case. Quincke (*Jahr. f. Aug.*, 1880, p. 236) mentions a very pale red fundus. It is usually stated that the color of the fundus presents no striking change even in advanced leukæmia. The contents of the vessels may or may not be very pale.

Leukæmic neoplasms occur in the lids and orbit, and also within the eye and brain. Friedlaender (*Virch. Arch.*, 1878, 2, p. 362) reports nodular lymphomata in the medullary substance and cortex of the brain (and the internal granular layer of the retina), with ampullæ of the optic nerve (*vide* p. 143). These produce the clinical symptoms of brain tumor. Leber (*Arch. f. Ophth.*, XXIV, 1, p. 295) observed double hemorrhagic retinitis, symmetrical swelling of all the eyelids and double exophthalmus, from leukæmic proliferations in the orbits. Guaita (*Jahr. f. Aug.*, 1890, p. 239) reports thickening of all the lids in pseudo-leukæmia. Osterwald (*ib.*, 1881, p. 451) observed leukæmic tumors in the orbit. Birk (*ib.*, 188, p. 298) found the posterior part of both orbits filled with a "lymphatic" new-formation.

May (*ib.*, 1884, p. 351) describes a circumscribed leukæmic infiltration of the facial nerve in the Fallopian canal; Becker (*l.c.*) a leukæmic tumor of the retina. Perhaps Oeller's case should be regarded as a leukæmic tumor of the choroid, but it may also have been an extensive hemorrhage. Gallasch found the lachrymal glands converted by enormous lymphoid infiltration into tumors as large as a pigeon's egg. Delens (*Jahr. f. Aug.*, 1886, p. 475) reports the cure of leukæmic tumors of the orbit as the result of cholera.

Under the term iritis leukæmica, Michel (*Arch. f. Ophth.*, XXVII, 2, p. 25) describes a case of bilateral chronic iritis with occlusion of the pupils in a woman of thirty-six years, who also suffered from slight enlargement of the spleen and lymphatic glands and temporary increase in the number of white blood globules. The excised iris contained nodules of lymphoid and epitheloidal cells, which were partly calcified; these were not visible to the naked eye. It is doubtful, however, whether this case should be regarded as leukæmia.

The enormous increase in the number of white blood globules is the most striking and characteristic symptom of leukæmia, but it does not constitute the source of danger to life. This is due to pathological products of disassimilation, which may not only exert a "formative" irritation in various parts (leukæmic tumors), but may also have a direct toxic action. This is either acute and resembles scurvy, or it is less violent and consists of the production of necrobiotic and inflammatory processes in the vessels and tissues.

To a certain extent the cells of leukæmic tumors may be regarded as infected and infecting tumor cells. They thus constitute a transition to malignant tumors (malignant multiple tumors of the lymphatic glands are actually called pseudo-leukæmia) whose clinical symptoms, in the later stages, are very similar to those of leukæmia.

Tumor Cachexia.

This may be regarded as an infectious disease whose micro-organism is the malignant specific tumor cell (sarcoma, cancer). It is possible that the tumor cell itself is infected by a bacterium and thus becomes malignant. The tumor cell is the element of the specific neoplasm, as opposed to the "granulation tumors" (syphiloma, leproma, tubercle, etc.), which develop solely from the action of the specific bacterium upon the infected tissues.

Malignant tumors may long remain local and give rise to merely local destructions. Even then, secondary infection from without may give rise not only to local inflammation and suppuration, but also to general infection, to septicæmia and pyæmia.

It is also evident that malignant tumors produce substances which are chemically irritating and productive of inflammation. At the periphery of the neoplasm these substances cause emigration of cells and proliferation in the adjacent tissues (zone of reactive inflammation); on the other hand, they exercise a disturbing effect on the general nutrition, may even have a direct toxic action, and thus give rise to the so-called tumor-cachexia. In such cachectic conditions hemorrhages may develop in all the organs, including the eye,—for example, retinal hemorrhages (Mackenzie, *Jahr. f. Aug.*, 1883, p. 298, in cancer of the stomach).

It is also possible that purely toxic actions may produce conditions similar to those in other chronic and subacute poisonings, viz., renal diseases and their sequelæ, fatty degenerations and hemorrhages, especially in the retina and brain, multiple neuritis, peripheral and nuclear paralyses of the ocular muscles, etc. The partial paralysis of the left motor oculi nerve in cancer of the pylorus, which was observed by Bettelheim (*Jahr. f. Aug.*, 1888, p. 571), probably belongs to this category, although such purely toxic effects are rare. The chief symptoms are due to " specific neoplasms," to emigrating and proliferating tumor cells in the lymphatic and blood channels.

As a matter of course, the tumor may also originate in and about the eye (retinal glioma, choroidal sarcoma, cancroid of the conjunctiva and lids, sarcoma of the conjunctiva, etc.) and then give rise to metastases which may, in turn, produce ocular symptoms,—for example, when the metastases are situated in the nervous system. Metastases in one eye are rarely secondary to a growth in the other eye (choroidal sarcoma); a more frequent event is the spread from one eye to the other (glioma of the optic nerve and retina) or the co-incident affection of both eyes (sarcoma of the choroid and glioma of the retina). Tumors in other parts of the body rarely produce metastases in the eyes, but a number of cases have been reported, three-quarters of which started from tumors of the breast. The eye is secondarily affected in a much larger number of cases by metastases in the orbits, but particularly within the skull, where they may give rise to all the symptoms of a primary tumor.

In the eye itself, the choroid is almost always the structure which is affected by metastasis, but Elsching (*Arch. f. Aug.*, XXII, 2 and 3) has also described a metastatic tumor of the optic nerve.

The eye possesses diagnostic importance not alone in the living but also in the dead, inasmuch as it presents certain important, although not infallible, signs of death.

The so-called sclerotic patch, a desiccated portion of the sclera to the inner or outer side, or below the cornea, in the palpebral fissure, opacity and insensibility of the cornea, abolition of the pupillary reaction to light, xerosis of the cornea and conjunctiva, usually develop shortly after death. They are also observed occasionally during life (cholera, cholera infantum, etc.).

According to Poncet (*Arch. gén. de Méd.*, 1870, p. 498), conversion of the red color of the fundus into yellowish-white, complete disappearance of the retinal arteries and narrowing of the veins, are positive and suddenly developing signs of death. As a matter of course, this is really only a sign of the cessation of the circulation of the blood in the retina and choroid. Nevertheless, this change in the color of the fundus is one of the most positive early signs of death.

To make a brief *résumé* of the previous considerations, we find that in the secondary ocular affections of the most varied diseases—apart from those which have a direct or indirect mechanical action upon some part of the body which is optically efficient—there is a certain uniformity which points to a similar mode of development. We have to deal, in fact, with toxic symptoms in the broadest sense of the term, from the most acute to the most chronic, as we have described on p. 360.

The differences are due mainly to the fact that in one case the virus is introduced into the body, in another it is produced in the body, either with or without the aid of living elements.

Improper articles of diet give rise to changes in the stomach and intestine, an unsuitable composition of the atmosphere entails changes in the respiratory organ, and, secondarily, changed chemical conditions of nutrition in the entire organism. If the latter are sufficiently intense and prolonged, true organic disease will be produced, very often combined with vessel changes and their sequelæ. This takes place particularly in the small arteries and capillaries, where the noxious substances which have been absorbed in other parts pass the walls of the vessels. These substances act first upon the walls of the vessels and the interstitial connective tissue, and they affect the parenchyma (cells, nerve and muscular fibres) only when they have been very abundant from the start.

In these auto-intoxications we also find that a certain virus attacks certain organs more than others. These organs are either the site of the greatest accumulation of the virus, or their constituents are especially sensitive to the virus, evidently on account of the presence or absence of certain chemical substances. The predominant affection of certain organs then impresses a peculiar character upon the entire clinical history. In this regard, a spontaneous contracted kidney or cirrhotic liver exactly resembles diabetes or Basedow's disease.

The clinical history is often complicated by secondary infections with living inflammation-producers of all kinds, and it is a general rule that an infection during an already existing constitutional anomaly is more serious and obstinate than infection in a healthy individual.

In infections, the purely toxic action is supplemented by that of the infectious emboli and thrombi. After destruction or elimination of the germs, the purely toxic "constitutional anomaly" remains. This either returns to the normal condition—complete recovery—after more or less prolonged convalescence, or a permanent condition of ill health remains which exhibits a different character, according to the quantity and quality of the tissue changes and destructions. There may then be a tendency to definite diseases of certain tissues and organs, which are often characteristic of the previous disease. In such cases the changes in the visual organ often possess the highest diagnostic importance.

It is well known that the organized producers of inflammation themselves rarely constitute the dangerous element, but that this consists, as a rule, of certain of their metabolic products, *i.e.*, of chemical substances. In like manner, the proliferation and destruction of the bacteria depend upon the presence of certain chemical substances which are present or are produced in the infected body. The proliferation of the microbes is not only counteracted by a certain degree of concentration of their own metabolic products, but also by certain chemical elements and metabolic products of the diseased body. Hence, infection with a certain microbe may fail (immunity) at one time, produce a mild affection at another time, and a severe disease a third time. The same micro-organism may produce very different processes in different individuals, and, on the other hand, different organisms may produce very similar affections in different individuals. This may often be very clearly demonstrated during implication of the visual organ in the morbid process.

Hitherto the study of the diseases in question has been mainly morphological. Our knowledge of the chemical changes is still very imperfect, and usually refers merely to certain striking substances (sugar, albumin in the urine, etc.) which, with few exceptions (tetanotoxin in traumatic tetanus, and a few others), are not the real toxic agents. The chemical metabolic products of the microbes and other chemical constituents of the infected body are, however, at least as important as the specific micro-organism in the production of a definite clinical complex, and sometimes even more important.

Here a large field, which has hardly been touched, is opened to physiological and pathological chemistry.

INDEX.

* 9 7 8 3 3 3 7 7 7 9 1 1 5 *